FOOD SECURITY
New Solutions for the
Twenty-first Century

Other related Iowa State University Press titles—
for additional information on these or any Press title, call
ISU Press at 1-800-862-6657 or visit our secure interactive
web site www.isupress.edu.

The Development and Growth of Brazil's Soybean Industry
by Philip F. Warnken. 0-8138-2196-7

*Economic and Business Principles in Farm Planning
and Production*
by Sydney C. James and Phillip R. Eberle. 0-8138-2880-5

Evaluating Natural Resource Use in Agriculture
edited by Thyrele Robertson, Burton C. English, and
Robert R. Alexander. 0-8138-2958-5

How to Farm for Profit: Practical Enterprise Analysis
by Donald M. Fedie. 0-8138-2560-1

The Making of the 1996 Farm Act
by Lyle P. Schertz and Otto C. Doering III. 0-8138-2608-X

Marketing Grain and Livestock
by Gary F. Stasko. 0-8138-2832-5

Natural Resource and Environmental Economics
by Tony Prato. 0-8138-2938-0

*Public-Private Collaboration in Agricultural Research: New
Institutional Arrangements and Economic Implications*
edited by Keith O. Fuglie and David E. Schimmelpfennig.
0-8138-2789-2

FOOD SECURITY

New Solutions for the Twenty-first Century

Proceedings from the
Symposium Honoring the
Tenth Anniversary of
the World Food Prize

EDITED BY
Amani E. El Obeid
Stanley R. Johnson
Helen H. Jensen
Lisa C. Smith

IOWA STATE UNIVERSITY PRESS / AMES

Amani E. El Obeid, Pre-doctoral Research Associate, Center for Agricultural and Rural Development, Iowa State University, Ames.

Stanley R. Johnson, Vice Provost for Extension and C.F. Curtiss Distinguished Professor of Agriculture, Iowa State University, Ames.

Helen H. Jensen, Professor, Department of Economics; and Head, Food and Nutrition Policy Division, Center for Agricultural and Rural Development, Iowa State University, Ames.

Lisa C. Smith, Research Fellow, International Food Policy Research Institute, Washington, D.C., and Visiting Professor, Department of International Health, Emory University, Atlanta, Georgia.

Iowa State University Press
2121 South State Avenue, Ames, Iowa 50014

Orders:	1-800-862-6657
Office:	1-515-292-0140
Fax:	1-515-292-3348
Web site:	www.isupress.edu

This text was produced from camera-ready copy, prepared and edited by the Center for Agricultural and Rural Development (CARD).

Authorization to photocopy items for internal or personal use, or the internal or personal use of specific clients, is granted by Iowa State University Press, provided that the base fee of $.10 per copy is paid directly to the Copyright Clearance Center, 222 Rosewood Drive, Danvers, MA 01923. For those organizations that have been granted a photocopy license by CCC, a separate system of payments has been arranged. The fee code for users of the Transactional Reporting Service is 0-8138-2910-0/99 $.10.

♾ Printed on acid-free paper in the United States of America
First edition, 1999

International Standard Book Number: 0-8138-2910-0

Cataloging-in-Publication information is available.

The last digit is the print number: 9 8 7 6 5 4 3 2 1

cov

Contents

FOOD SECURITY: NEW SOLUTIONS FOR THE TWENTY-FIRST CENTURY

◆ Foreword

John Ruan
Chair, The World Food Prize Foundation

This symposium, "Food Security: New Solutions for the 21st Century," was organized to commemorate the tenth anniversary of The World Food Prize and to draw attention to the priority and opportunities for more successfully addressing global food security issues in the twenty-first century. The emphasis was on opportunity. We have made great progress on improving food security during the last decades of the twentieth century. We have in hand the technologies and development know-how to more fully solve food security problems in the twenty-first century. Still, there are obstacles. We hope that this symposium will help encourage the moral resolve of the global society for giving the food security problem a high priority and investing the resources to solve it.

The 1996 World Food Prize laureates were also honored at this symposium. Dr. Henry Beachell and Dr. Gurdev Singh Khush exemplify by their deeds and lifetime commitments the resources we can bring to reducing food insecurity. The work of these two laureates and their coworkers has gone far to improve the production of rice, the staple food of a large share of the global population. Moreover, their work has combined genetics with production systems to raise the incomes of many of the rural poor in Asia and other rice-growing areas. These two men are among the real heroes of the twentieth century, and The World Food Prize Foundation was pleased to honor them.

The symposium participants included more than 100 agricultural leaders, scientists, policymakers, and representatives from major national and multinational organizations concerned with development assistance and food security. We were especially pleased that nine of the previous recipients of The World Food Prize could attend and participate in the symposium. And, not the least, we were fortunate to have at the symposium a number of students—representatives of the new generation that will carry the work of food security, poverty reduction, and agriculture into the next century.

We organized the symposium to provide the best information on the current food security situation and to assemble available projections based on alternative policy, technology, and other scenarios. We also included sessions in which the participants could use this background information and their experience and science to make observations and recommendations for better addressing food security in the twenty-first century. These recommendations are summarized in chapters 15, 16, and 17 of this book, and I invite their careful study by those who will lead the initiatives to reduce food insecurity in the twenty-first century.

There is a great job ahead, if we are to make real progress in reducing food insecurity. More than 800 million of the global population are food insecure and suffer from insufficient energy intake and malnutrition—nearly 20 percent of the population. Two billion of the global population live on less than $2 per day. Many of these may be at times food insecure and suffer from malnutrition. We are only coming to fully understand the consequences of poor nutrition for the capacities of individuals, households, and nations to reach their fuller human potential, but it is great.

Regions that should receive priority in the quest to reduce food insecurity were identified during the symposium. Sub-Saharan Africa has a high prevalence of food insecurity and developmental problems that have resulted in widespread poverty. It is also a region with a rapidly growing population. If the young in this region are to have a bright future, we must reduce poverty and improve food security in this area. South Asia also has a very large number of the food insecure. Here the issue is more one of poverty than of the nations' capacity to make food available. But nonetheless, the large numbers speak to the priority for broad-based development and poverty reduction as a means of reducing food insecurity.

The growth of the urban population in the developing nations suggested yet another food security issue: How will these populations be served by the changing food systems? Here our symposium addressed the importance of postharvest technology as a factor in increasing the availability of and access to food. Growing urban populations that are in poverty suggest added dimensions to the food security problems for the twenty-first century.

Finally, the symposium reinforced the importance of research and development in agriculture. It called attention to trends in funding and organization of research and development for agriculture that have implications for the targeting of these benefits to the populations that are food insecure. How will the changing mix of private and public funding for agricultural research and development impact the nations in which the most food insecure reside. Broader development perspectives and added attention to environment were other issues that surfaced in the rich dialogue of the symposium.

This proceedings summarizes and synthesizes the symposium. The World Food Prize Foundation has been pleased to support its publication and dissemination. The new millennium will provide many new challenges and opportunities. One of those opportunities will be to significantly decrease the numbers of the global population who are food insecure. We feel that The World Food Prize, by recognizing the leaders among those who have contributed to food security, can help to draw attention to the problem of food security and the opportunity that is before us all for solving it.

◆ Acknowledgments

The symposium "Food Security: New Solutions for the 21st Century" was sponsored by The World Food Prize Foundation, Des Moines, Iowa, in celebration of the tenth anniversary of The World Food Prize. The symposium was planned by Herman Kilpper, Executive Director, Al Clausi, Robert Havener, and Stanley R. Johnson. Special assistance with the planning and organization of the symposium was provided by Amani El Obeid, Judith Pim, and Nancy Beltramo.

Technical and staff support for the publication of the book was provided by the Center for Agricultural and Rural Development (CARD) at Iowa State University, Ames, Iowa. This book benefited greatly from the editorial expertise of Judith Pim and the technical assistance of Betty Hempe at CARD, and was copyedited by Linda Chimenti, formatted by Ruth Bourgeois, and proofread by Bonnie Harmon. We express special appreciation to Judith, Betty, Linda, Ruth, and Bonnie for their careful and timely assistance.

◆ Contributors

IRMA ADELMAN. Forsyth Hunt Chair holder and Professor in the Graduate School, University of California, Berkeley.

HENRY M. BEACHELL. 1996 World Food Prize laureate (corecipient); former research scientist, International Rice Research Institute, Manila, Philippines; and former rice breeder, RiceTec, Alvin, Texas. Currently resides in Pearland, Texas.

GEORGE H. BEATON. Consultant, GHB Consulting, Ontario, Canada; and professor emeritus, nutritional sciences, University of Toronto, Canada.

ROBERT F. CHANDLER, JR. 1988 World Food Prize laureate; and retired, original founder and first director, International Rice Research Institute, Manila, Philippines. (Deceased 1999)

A. S. CLAUSI. Consultant on food research and development; chairman, International Food Technologists Foundation Board; chairman, Research Directors' Roundtable; former chairman and current member of The World Food Prize Council of Advisors; and retired senior vice president, General Foods Corporation. Currently resides in Greenwich, Connecticut.

COLIN DENNIS. Director-general, Campden & Chorleywood Food Research Association, Gloucestershire, United Kingdom; visiting professor in food science, Queen's University of Belfast; and honorary professor of biological sciences, University of Birmingham, United Kingdom.

JACQUES DIOUF. Director-general, Food and Agriculture Organization of the United Nations, Rome, Italy.

JEANNE DOWNEN. Director, Partnership and Household Livelihood Security Unit, CARE, Atlanta, Georgia.

AMANI E. EL OBEID. Pre-doctoral research associate, Center for Agricultural and Rural Development, Iowa State University, Ames, Iowa.

KENNETH J. FREY. Retired, distinguished professor, Department of Agronomy, Iowa State University, Ames, Iowa.

HELEN A. GUTHRIE. Retired, professor emerita of nutrition, Pennsylvania State University, State College, Pennsylvania.

RICHARD L. HALL. Retired, vice president, science and technology, McCormick and Company; and previously served on The World Food Prize Council of Advisors. Currently resides in Baltimore, Maryland.

ROBERT W. HERDT. Acting vice president, The Rockefeller Foundation, New York, New York.

HANS R. HERREN. 1995 World Food Prize laureate; director-general, International Centre of Insect Physiology and Ecology, Nairobi, Kenya; and member of the Food and Agriculture Organization panel of experts.

CHARLES E. HESS. Director, International Programs, College of Agricultural and Environmental Sciences; and professor emeritus, Department of Environmental Horticulture, University of California, Davis.

JIKUN HUANG. Director, Center for Chinese Agricultural Policy, Chinese Academy of Agricultural Sciences, Beijing, People's Republic of China.

JOSEPH H. HULSE. President, Siemens-Hulse International Development Association, Ontario, Canada.

HELEN H. JENSEN. Professor, Department of Economics; and head, Food and Nutrition Policy, Center for Agricultural and Rural Development, Iowa State University, Ames, Iowa.

MARTIN C. JISCHKE. President, Iowa State University, Ames, Iowa.

STANLEY R. JOHNSON. Vice provost for Extension; director, Center for Agricultural and Rural Development; executive director, Food and Agricultural Policy Research Institute; and C. F. Curtiss Distinguished Professor of Agriculture, Iowa State University, Ames, Iowa.

EILEEN KENNEDY. Executive director, Center for Nutrition Policy and Promotion, United States Department of Agriculture, Washington, D.C.

GURDEV SINGH KHUSH. 1996 World Food Prize laureate (corecipient); and principal plant breeder and head, Division of Plant Breeding, Genetics and Biochemistry, International Rice Research Institute, Manila, Philippines.

EDWARD F. KNIPLING. 1992 World Food Prize laureate (corecipient); retired, former director of Research for the United States Department of Agriculture; former director, Entomology Research Division of the Agricultural Research Service; and former science advisor and administrator, Agricultural Research Service. Currently resides in Arlington, Virginia.

VERGHESE KURIEN. 1989 World Food Prize laureate; and chairman, National Dairy Development Board and the National Cooperative Dairy Federation of India, Ltd., Anand, India.

UMA LELE. Advisor, Agricultural Research Department, The World Bank's Vice Presidency on Environmentally Sustainable Development, Washington, D.C.

LUC M. MAENE. Director-general, Fertilizer Industry Association, Paris, France.

ROBERT S. MCNAMARA. Member of The World Food Prize Advisory Committee; former president, Ford Motor Company; former secretary of defense of the United States; and retired president, The World Bank, Washington, D.C.

M. PETER MCPHERSON. President, Michigan State University, East Lansing, Michigan.

NORMAN MYERS. Consultant in environment and development; and Fellow, Green College, Oxford University, Oxford, United Kingdom.

ROBERT E. SMITH. President, The Institute of Food Technologists; retired senior vice president of research, Nabisco Inc.; and president, R. E. Smith Consultants Inc., Newport, Vermont.

SAMUEL E. STUMPF. Retired, professor of philosophy of law, School of Law; and research professor of medical philosophy, School of Medicine, Vanderbilt University, Nashville, Tennessee. (Deceased 1998)

M. S. SWAMINATHAN. 1987 World Food Prize laureate; and president, M. S. Swaminathan Research Foundation, Chennai, India.

ROBERT L. THOMPSON. President, Winrock International Institute for Agricultural Development, Morrilton, Arkansas.

DONALD L. WINKELMANN. Chairman, Technical Advisory Committee, Consultative Group on International Agricultural Research, Santa Fe, New Mexico.

YANG YANSHENG. Vice president, the Chinese Academy of Agricultural Sciences; vice chairman, Board of the China Crop Science Society; and head, Sugar Crop Commission, China National Committee for Varietal Evaluation of Agricultural Crops, Beijing, People's Republic of China.

MUHAMMAD YUNUS. 1994 World Food Prize laureate; and managing director, Grameen Bank, Dhaka, Bangladesh.

◆ Welcome to the Symposium

Martin C. Jischke

Let me add my welcome to everyone in attendance today. It is an honor for me to be here this morning for the tenth anniversary of The World Food Prize, and to share the platform with some of the world's most distinguished scholars, business leaders, and humanitarians.

On behalf of all Iowans, it is a pleasure to welcome you to our state. I especially want to acknowledge the work of one of these individuals, John Ruan. John is one of Iowa's leading citizens, business executives, and philanthropists. He has made enormous contributions to this city, state, and nation, including the institution I serve—Iowa State University. Now, through his generous support of the prestigious World Food Prize, his influence truly touches all corners of the globe.

It has been said that nations are happy according to the amount of food that is divided by their citizens or according to the quantity of food which a day's labor will purchase. More simply stated, a hungry person is not a free person.

Radical improvements in food production in this century have increased the food supply to record levels, keeping pace with the appetite of a growing global population. But as you know, many forces still prevent certain nations and regions from getting the food they need and deserve. The problems are complex. They range from a lack of transportation to move the food to where it is most needed to people simply not having the income to purchase food when it is available.

These challenges will not be solved easily or inexpensively, for they require a vision and the ability to formulate new ideas and approaches for addressing food security issues. However, efforts by individuals, such as The World Food Prize laureates, give our world hope that widespread hunger and malnutrition will shift from a political issue to a subject of historical interest.

The World Food Prize recognizes individuals whose outstanding accomplishments have contributed to improving the quality, quantity, and availability of food. Because the prize emphasizes the importance of a healthy, safe, and sustainable food supply worldwide, it calls attention to what has been already achieved and what can and needs to be accomplished in the future. The World Food Prize laureates are indeed role models who will inspire others to make significant and measurable contributions to improve the food supply.

The prize is a constant reminder, according to John Ruan, that "one person really can make a difference toward alleviating hunger throughout the world." I would like to thank John and everyone associated with this recognition, and everyone who has come to participate in this important symposium, for the significant roles you are playing in this very crucial world issue.

1996 WORLD FOOD PRIZE LAUREATE

Acceptance Speech

Henry Beachell

First, I would like to thank my parents, the late William and Alice Beachell, for their love and support and for teaching me the difference between right and wrong. I send best wishes to my sisters, Alice Rock and Kathryn Brown, and to Albert Brown who for health reasons were unable to be here today. I wish to thank my wife Mary; her sister Ruth White; Mary's daughter Ann Singleton and children Alby, Caroline, and Lizzie; my brother William and his wife Wanda; and my other relatives, friends, and colleagues who are in attendance.

It is a great honor to be selected to receive the 1996 World Food Prize and to have the pleasure to share it with my respected colleague Gurdev Khush. I wish to express my sincere thanks to those who placed my name in nomination for the award and to those members of the committee who selected me. I also extend my gratitude to Mr. John Ruan and Dr. Norman Borlaug, founders of The World Food Prize Foundation.

I accept this prestigious award not only for myself, but also for my esteemed colleagues in science, extension, and administration as well as the organizations, both private and public, who supported me. It is only through the collective efforts of so many people around the globe that I receive The World Food Prize today.

The World Food Prize recognizes the great need in the world today for agricultural sustainability for all nations, cultures, and ecosystems. Agricultural sustainability advocates high and stable production of nutritional food, sound environmental management, economic profitability and just distribution of its product, taking into account those consumers most in need of adequate staple food.

The World Food Prize also honors me by selecting breeding as its focus. It gives special recognition to plant breeding and, in particular, my field, rice breeding. Breeding is central to crop improvement, but it cannot exist alone. The development and evaluation of crop varieties represent the combined effort of multidisciplinary teams. I feel honored to have had the opportunity to be a team player in rice research and development based on this premise for more than 65 years.

Early Experiences: Rice Agriculture in the United States

My career spans nine decades if you include my farm and high school vocational agriculture training, so I will try to be concise. It is almost as long as the history of modern plant breeding. Although Gregor Mendel discovered the laws of genetic inheritance in plants and animals in 1853, we did not recognize their potential impact until the late nineteenth century.

I was born into agriculture on a farm, not in the great state of Iowa but on the other side of the wild Missouri, down in old Nebraska (Waverly). My farm experience began on my family's wheat farm near Grant, Nebraska, in the northern Great Plains of the United States. It was strengthened by high school vocational agriculture training (the Smith-Hughes Act). Those were the days of low input, labor-intensive agriculture. The small grain binder, the steam-powered stationary grain threshing machine, the mower, and the hay rake were the principal mechanized machines of the day. All but the thresher were horse or mule powered. During this period, farm production practices were mostly static, allowing a crop variety to remain well adapted for a long period. For example, Turkey Red was the principal wheat variety planted throughout the Midwest for many years, as were the U.S. rice varieties Carolina Gold, Blue Rose, and Caloro.

During World War I, farm labor shortages triggered the switch from horse and mule power to small gasoline-powered tractors. This change had little effect on the intensity of crop production, i.e., the use of other inputs, such as fertilizers. Following the war, grain supplies in the United States were adequate so there was little need to intensify crop production further until the beginning of World War II.

I received a bachelor's degree in agronomy and genetics from the University of Nebraska in 1930. Due to the very limited job opportunities during the Depression, I immediately enrolled in graduate studies at Kansas State University. I completed a master's degree in plant breeding and genetics in 1933. At Kansas State University, I had the good fortune to have the late Dr. John H. Parker as my major professor. Dr. Parker was perhaps the greatest plant breeding instructor of the period. Dr. Parker encouraged me and all of his students to develop our abilities through working interdisciplinarily and maintaining a multi-user focus for the successful development of new varieties. He stressed that plant breeders develop varieties not only for the farmer but also for the miller, processor, and ultimately the consumer. In 1939 I took additional graduate studies at Texas A&M University in plant physiology, genetics, and statistics.

On March 2, 1931, I accepted a United States Department of Agriculture (USDA) appointment as junior agronomist in charge of rice varietal improvement at the Texas A&M Rice Experiment Station near Beaumont, Texas. This began my career as a rice breeder, which now spans nearly 66 years. My superior in the USDA–Agricultural Research Service (ARS) was the late Jenkin W. Jones, who headed the USDA rice research program. Jones was a renowned authority on rice research—truly a great scientist and friend.

In 1931 the U.S. rice industry was near bankruptcy due to its labor-intensive farm methods and very low prices, a result of low demand for rice. The combine harvester introduced to the northern U.S. small grain fields in the mid-1920s was not adapted to the muddy rice fields of the Gulf coast. Therefore, the labor-intensive binder, stationary thresher, and sack storage methods continued. Around 1940, artificial grain drying, bulk storage, and the self-propelled combine were perfected. Shortly thereafter, war surplus airplanes were used for seeding and applying fertilizers, pesticides, and insecticides. It was in this way that rice production and processing became one of our most highly mechanized food crops. Serious labor shortages during World War II hastened the shift.

About this time, artificial fertilizer production costs dropped to a price that the rice farmer could afford. Research in progress since the early 1930s showed that dramatic yield increases were possible when nitrogen and other fertilizers were applied.

These changes brought about the need for new and improved rice varieties adapted to the new farming practices. The old varieties planted from 1910 to 1940 had served the industry well. Most of them were developed by the late Sol Wright, a Crowely, Louisiana, rice farmer and plant breeder. Wright developed Blue Rose, Early Prolific, Lady Wright, and Edith in the 1910s and 1920s, and they were so successful that farmers and millers often referred to Wright as the "Burbank of the Rice Industry." Sol Wright's varieties we now know were selected from an early generation natural cross found in a rice farm field.

Varietal improvement had been in progress at the Rice Experiment Stations of the state universities of Arkansas, California, Louisiana, and Texas A&M since early in the century. But, improvement was limited to selection within introduced varieties, mostly from Asia.

The Modernization of Rice Breeding in the United States

With Wright's passing, the rice industry anticipated the need for expanded plant breeding and other research and approached the U.S. Congress for federal funding. Consequently, in 1930 Congress appropriated funds to place rice breeders at four state rice experiment stations: the universities of Arkansas, California, Louisiana, and Texas A&M. The breeding programs set up in the 1930s made available to farmers the adapted rice varieties the industry needed in the 1940s and beyond.

In 1941 the Texas rice industry recognized that additional funding would be needed to finance the modernization of the rapidly developing industry. There was little hope for any further increases in state or federal funding. The result was that Texas rice farmers initiated the organization of a *privately* operated crop (rice) improvement association that would provide private funding to strengthen the Texas A&M rice research program. In 1943 the Texas Rice Improvement Association (TRIA) was organized, and a three-way memorandum of understanding between Texas A&M University, the USDA–ARS, and the TRIA was approved by all parties. The TRIA accumulated funds through the production and sale of high-quality seed rice to farmers, with all profits going to the rice research program. In 1946 TRIA purchased 603 acres of prime rice land that formed the enlarged Texas A&M Rice Research and Extension Center. Today the center is one of the finest centers in the Texas A&M University system.

Texas A&M released several new varieties that were earlier maturing, slightly shorter in plant height, and better adapted to the new production and harvesting methods. They included Texas Patna (165 days), Bluebonnet and Bluebonnet 50 (135 days), Century Patna (130 days), and Belle Patna (105 days).

The very early Belle Patna matured early enough for Texas and Louisiana farmers to grow a ratoon crop (a second harvest produced from the stubble of the first crop) that produced an additional yield of 25 to 40 percent. In all, the Texas A&M program released nine rice varieties. These varieties made up as much as 75 percent of the U.S. long grain rice crop for over 15 years, and were also popular in Central and South America.

Parboiled, instantized (precooked), and other processed whole grain rice products started in the early 1940s. At that time, breeders were limited to cooking rice to evaluate cooked rice properties. By 1949, some newly released varieties were not suitable for some processed products. Rice breeders needed a grain quality testing laboratory to develop and conduct screening tests for evaluating specific cooking and processing properties of rice varieties and breeding lines. The rice industry recognized this need and, through the combined efforts of Texas A&M, USDA–ARS, TRIA, Uncle Ben's, General Foods, Campbell Soup, and other agencies, funds were provided to hire a cereal chemist and build a national rice quality laboratory at the Texas A&M Rice Research and Extension Center in Beaumont, Texas. Today such laboratories are in operation at the International Rice Research Institute (IRRI) in the Philippines and in many other rice-producing countries. They are in operation in both the public and the private sectors.

As farmers adopted more intensified production practices, they needed shorter, more sturdy-stemmed rice varieties. However, we could not find a suitable semidwarf rice variety similar to those of wheat and grain sorghum. In 1952 I met Norman Borlaug, who told me about semidwarf wheat varieties that the late Orrville Vogel, a former classmate and friend at the University of Nebraska, had developed at Pullman, Washington. I was so impressed that in 1955 I made a personal trip to Pullman to see the dwarf wheats. Continued research, including mutation breeding, failed to produce a suitable semidwarf rice.

To support rice farmers in their transition to mechanized and intensified crop production, semidwarf rice varieties were needed. The intensified production practices were dictated by food-rice shortages in Asia as post World War II growth gained momentum.

During a 1962 consulting visit I made to the newly dedicated IRRI in the Philippines, Peter Jennings, head of IRRI plant breeding, and his colleagues T. T. Chang and the late Sterling Wortman informed me that a semidwarf rice was being grown in Taiwan. Before returning to the United States, I visited Taiwan and observed, first hand, Taichung Native No. 1, the Taiwan semidwarf rice variety. The semidwarf gene conditioning plant height, as in other cereal crops, was found to be simply inherited and could be readily incorporated into U.S. varieties.

Rice Breeding and Farming in Asia

In October 1963, I retired from the USDA-ARS and accepted a Rockefeller Foundation appointment as a rice breeder attached to the IRRI, in Los Banos, the Philippines. For the next 19 years—the first nine in the Philippines and the next ten in Indonesia—I served as a member of the team of IRRI research scientists put together by 1988 World Food Prize laureate Bob Chandler and Sterling Wortman that triggered the "Rice Green Revolution" in Asia. Bob Chandler was by our side continuously. Thank you, Bob, for your help and friendship. My present director, Robin Andrews at RiceTec, Inc., is made of the same cloth. IR8 was the first IRRI semidwarf variety and was the prototype from which the semidwarf rice varieties planted throughout Asia and elsewhere were derived. This advance, which brought about the Green Revolution in rice, is well documented. The revolution is still in progress with continued success.

Rice yields throughout tropical Asia have more than doubled in many areas during the past 30 years. As important as semidwarf rice varieties were, the intense management production practices that were developed were of equal importance. Major ingredients of the improved management practices were commercial fertilizers and how to apply them.

While highly successful, the Green Revolution nonetheless created problems, not all of which have been solved after 30 years. For example, high rates of nitrogen fertilizer and semidwarf varieties are essential to produce high grain yields. However, fertilizers produced a soil and plant environment that was more favorable to insects and diseases. Outbreaks of epidemic proportions occurred throughout Asia. Total chemical control was not always successful; in some instances, it caused a breakdown of varietal resistance to insects and insect tolerance to chemicals. This was experienced in the rice fields of Asia. Therefore, integrated pest control practices were developed that, along with resistant varieties, are maintaining high and stable grain yields.

Strong insect and virus resistance had to be incorporated into the semidwarf varieties. Integrated pest control management practices are being improved further. We must continually improve our methods of integrated pest control management as well as fertilizer application. As these methods are improved, new varieties that are better adapted to modern farm practices must be developed. We must also keep in mind the importance of mandatory long-term environmental sustainability through agricultural technology and policy based on sound and proven scientific facts and not on political rhetoric.

An area where environmental protection and more efficient use of fertilizers might be achieved is leaf feeding (liquid) or foliar application of fertilizers. This should be given top priority in tropical Asia. Ground applications of fertilizers will continue to be important for the early applications. Leaf feeding has not been completely successful in the past because of low penetration into the plant and leaf damage. However, new chemicals and technology are forthcoming. Spray application of slow release fertilizer material has the potential to prevent waste and decrease runoff of chemicals into surface water and groundwater, particularly in later stages of the plant's development. With support from agricultural extension and research personnel, the Indonesian rice farmer with his knapsack sprayer will be able to carry out foliar fertilizer application effectively. Present day information shows that, with liquid application, the volume (weight) of fertilizer required (leaf feeding) is only a fraction of the amount required for ground feeding. For the farmer, this would reduce transportation costs and reduce pollution. In these ways leaf feeding would be more environmentally friendly.

During my assignments in the Philippines and in Indonesia, I visited many countries to get to know and understand agricultural scientists and their administrators and to understand their problems better. In fact, this activity was equally as important as the technical aspects of the breeding program. The countries with which I was intimately associated were Bangladesh, Indonesia, South Korea, Pakistan, and Sri Lanka.

As an example, during work with Korean trainees assigned to IRRI, the concept of an IR8 plant type rice variety adapted to the temperate climate of South Korea was conceived. The Tongil varieties were the result. In 1977 South Korea planted more than 50 percent of its rice crop to Tongil varieties, achieving a world record rice yield with a country average of about 6 tons per hectare (rough rice). Many IRRI trainees who have returned to their home countries now hold top research and administrative positions.

While in Indonesia I worked primarily with scientists at the Central Research Institute for Agriculture in Bogor. During this period, rice production in Indonesia increased by more than 100 percent when high-yielding, disease- and insect-resistant, early maturing semidwarf varieties were adopted.

In 1975 a severe brown plant hopper and grassy stunt epidemic placed the Indonesian rice crop in jeopardy. A rapid varietal change was essential to prevent a greatly reduced rice crop in 1976 and beyond. Only 500 tons of seed of the newly developed IR36 were available for the 1975 dry season crop. National extension and research leaders decided to distribute the 500 tons of the high-yielding, early maturing IR36 variety in lots of a few kilograms per farm to reliable seed producers scattered throughout the country. This was a well-planned program that Indonesian extension personnel supervised closely. The plan was to plant the seed in the dry season, followed immediately by a wet season planting. It was estimated that in a single season a single ton of original seed could produce 80 to 100 tons of seed. After the second season crop this would amount to 8,000 to 10,000 tons of seed from a single ton of original seed in just one year.

Exact figures are not available, but the project was highly successful and the quantity of IR36 seed for the 1976 dry season (third planting) crop was roughly 4 to 5 million tons. The result was that in one year's time there was sufficient seed to plant the most severely affected areas and the epidemic was brought under control. Annual rice production doubled when farmers adopted early maturing, pest-tolerant semidwarf varieties, high nitrogen fertilizer applications, and reliable integrated pest control management practices. Expanded irrigation projects provided additional and improved irrigation.

While in Asia I had the good fortune to be able to interact with excellent scientists and administrators in the Philippines, Indonesia, Sri Lanka, Korea, China and other countries, many of whom were former IRRI trainees from many disciplines.

Recent Advances

Upon returning to the United States in 1982, after 19 years in Asia, I wanted to continue in rice research in either the private or the public sector. I accepted a position with RiceTec, Inc., in Alvin, Texas, (then Chocolate Bayou Company) as a rice research consultant for the development of aromatic rice varieties adapted to the southern U.S. farming practices and environment. Later, following reorganization, RiceTec undertook a hybrid rice breeding and seed production program. It included the development of mechanized methods of hybrid seed production to substitute for labor-intensive methods such as those practiced in China. Recently, a RiceTec field of approximately 35 acres produced more than 9 tons per hectare. A ratoon crop yielded another 2 to 3 tons. The hybrid varieties developed are approaching the maturity, plant type, and grain milling and processing characteristics demanded by U.S. farmers and markets.

The RiceTec hybrid rice research and seed production program will have a significant impact on world rice production. Hybrid rice varieties today make up more than 50 percent of China's rice crop. As new, site-specific varieties adapted to northern China are developed, the planting of hybrids will continue to spread. India has developed several hybrids that are being planted on farm fields. RiceTec is developing hybrids specifically adapted to the United States and South America.

In China, hybrids are outyielding standard varieties by about 15 to 25 percent. In other countries similar yield increases are indicated. As the technology advances and site-specific hybrids are developed, it is predicted that hybrids will make up a significant percentage of the world's rice production early in the twenty-first century. Just when this will happen depends on the priority assigned to hybrid rice research and to the demand for more rice. Widespread rice shortages in Asia would stimulate hybrid rice adoption.

My latest collaboration at RiceTec has been with Cornell University, Ithaca, New York, and USDA–ARS scientists at Stuttgart, Arkansas, in the development of a molecular genetic seed bank of rice cultivars. Molecular geneticists urgently need a purified seed source of the varieties and breeding lines they are studying. They have been handicapped by the lack of an unadulterated seed source. After three years of purification, more than 200 varieties are being stored in the seed bank.

Closing Remarks

It is my belief that rice yields will increase dramatically in 10 to 15 years through improved site-specific varieties and through improved production practices. In general, varieties and their management will be better adapted to the social and economic conditions different farmers must face. Hybrid rice should increase yields by 15 to 25 percent. Improved varieties and management practices will reduce production costs, increase grain yields, provide more long-term productive stability and better pest and disease control. The grain characteristics of the new varieties will more nearly meet the demands of millers and processors, and the eating preferences of consumers. Varietal improvement teams in both the public and private sectors will develop the site-specific varieties and management practices required.

With the advent of molecular genetics, the interdisciplinary approach is essential. Genetic markers identified by molecular scientists will assist the breeder in the selection process as new varieties are created. The breeder, working with teams of interdisciplinary scientists, must specify varietal traits. We must learn and understand the patterns from the past in order to focus on advanced varietal improvement and management for the future.

In closing, I am grateful to have had the opportunity to be a part of the rice research programs of USDA–ARS, IRRI, the Rockefeller Foundation, RiceTec, and U.S. universities—Texas A&M, Arkansas, California, Louisiana, Texas, Mississippi, and Cornell—as well as the TRIA.

During the past 30 years, it is only by working and learning from younger scientists that I have remained at the forefront of rice breeding. Although I cannot thank each of them individually, I will never forget the support of my friends and colleagues—leaders in rice research and administration, farmers, millers, and processors in the United States, Asia, and throughout the world. It has been a pleasure and an honor to have worked with them.

I hope that I may have, in some small way, contributed to the betterment of humankind. My career over the past 66 years has been a most gratifying experience. I hope it will extend far into the future.

Note

The final product must be adaptable to the food system. In the farm setting, agronomists, pathologists, entomologists, physiologists, chemists, and economists must interact in defining the suitability of any proposed variety to its agro-ecology. At the social level, the farmer must be considered for his or her interest, need, and ability to choose one variety over another. In terms of processing and marketing, industry demands must be considered. And, it is this interaction that leads to informed decisionmaking, sound judgments, and successful selection of genetic materials. It is the interdependence of scientists, producers, processors, marketers, and consumers that guides the selection of the breeding process and affects the suitability of a particular variety to its social and natural environment.

1996 WORLD FOOD PRIZE LAUREATE

Acceptance Speech

Gurdev Singh Khush

I am greatly honored by my selection to receive the immensely prestigious World Food Prize. Thank you for giving me this honor. Through your decision you have also honored present and past staff members of the International Rice Research Institute (IRRI) and numerous rice scientists in the national programs of developing countries with whom I have had the good fortune of working during the last 29 years.

To me, this prize should be considered as a tribute to the vision of wise men in two organizations—the Rockefeller Foundation and the Ford Foundation—who first conceived the idea of an international center for agricultural research. The IRRI was established in 1960 to apply science to agriculture in order to increase the production of rice, which constitutes more than one-half of the total food consumed by one out of every three persons on the earth.

In the 1950s and 1960s, populations in the Asian countries were increasing faster than food production. Because of these trends, several authorities, such as the Paddock brothers in their now famous book *Famine 1975!*, predicted large-scale famine in the late 1970s. However, advances in wheat and rice breeding produced varieties with a yield potential two to three times higher than that of traditional varieties and ushered in the Green Revolution. These varieties respond better to modern agronomic practices. Wide-scale adoption of such varieties and proper agronomic practices have led to major increases in food production, which means that the food shortages forecast by the Paddocks and others have not occurred.

Rice varieties grown in the tropics and subtropics in the pre–Green Revolution era were tall and leafy with weak stems, were susceptible to most diseases and insects, and had a harvest index (ratio of dry grain weight to total dry matter) of 0.3. When nitrogenous fertilizer was applied at rates exceeding 40 kg per hectare, these varieties tillered profusely, grew excessively tall, fell over or lodged early, and yielded less than they would with lower fertilizer inputs. The average yields in most Asian countries were between 1 and 2 tons per hectare. To increase the yield potential of rice, it was necessary to improve the harvest index and increase lodging resistance and nitrogen responsiveness. This was accomplished through the reduction of the plant's stature by the team led by my predecessor, Dr. Henry M. Beachell, who shares the 1996 World Food Prize with me. The first improved variety of rice released in 1966, IR8, has a combination of desirable features such as heavy tillering, dark green and erect leaves, and sturdy stems. It responds to nitrogenous fertilizers much better than traditional varieties. It has a harvest index of 0.5 and double the yield potential of tropical rice.

When I arrived at IRRI in August 1967, IR8 was breaking all the yield records and was being hailed by the national programs as a miracle rice. It had its own detractors, however, who pointed out its poor grain quality. IRRI scientists were concerned about its lack of resistance to diseases and insects—and they were right. Serious outbreaks of diseases and insects were threatening rice supplies in Asia.

The first task that Dr. Robert Chandler, who was then director of IRRI, assigned to me was to develop rice varieties with genetic resistance to diseases and insects. Fortunately, IRRI scientists had assembled a large collection of varieties from different parts of the world. Plant pathologists and entomologists had evaluated this collection and identified donors, i.e., varieties with genetic resistance to individual diseases and insects. These varieties were low yielding and had poor characteristics but had genes for resistance. My job was to transfer these genes from several poor varieties into high-yielding varieties. My training in genetics at the University of California, with professors George L. Stebbins and Charles M. Rick, proved to be most helpful.

Our first breakthrough came in 1973 with the development of IR26. It was the first improved variety with resistance to brown plant hopper and was adopted rapidly by farmers in the Philippines, Indonesia, and Vietnam. It was also resistant to bacterial blight, tungro virus, and green leafhopper, which transmits the tungro disease. IR26 was not resistant to grassy stunt virus, however. Our plant pathologists, after screening more than 7,000 rice varieties, found one collection of wild species from India called *Oryza nivara* to be resistant. The gene for resistance from this wild species was bred into high-yielding varieties, which are now widely grown. Since the large-scale introduction of grassy stunt-resistant varieties, this disease is rarely observed in farmers' fields. The virus is now maintained in the laboratory. One of the varieties resistant to grassy stunt, IR36, released in 1976, became the most widely planted variety of rice, or of any other food crop, the world has ever known.

IR36 was planted on 11 million hectares of rice land in the 1980s. The popularity of this rice variety was due to its resistance to a dozen diseases and insects. Its multiple resistance properties saved farmers nearly $500 million in insecticide costs annually. Other improved features of IR36 are its excellent grain quality and shorter growth duration. It matures in 110 days as compared to 130 days for IR8 and IR26, and it produces the same amount of rice. Thus, its per-day productivity is much higher. The shorter growth duration also permitted farmers to grow two crops where only one was grown before, and three crops instead of two in some areas.

Subsequent varieties were improved in many other traits. IR64, released in 1985, has the most palatable rice, high-yield potential, short-growth duration, and multiple resistance. It has replaced IR36 in many areas and is now the most widely planted variety of rice in the world. The yield potential of rice varieties has been progressively improved at the rate of about 75 kg per year. IR72, released in 1990, outyields all the other varieties of rice.

Each year I receive hundreds of requests for seed samples of improved rice varieties and experimental lines from rice scientists all over the world. All the rice varieties and breeding lines are shared freely. IRRI's network on genetic evaluation distributes numerous breeding lines every year to all rice-growing countries. Thus, IRRI rice breeding lines are evaluated in almost every rice-growing country. Some of the lines are released as varieties and others are used in the local rice breeding programs as parents. More than 250 rice varieties from IRRI-bred materials have been selected and released worldwide. It is estimated that 50 percent of the world's rice lands are planted to IRRI-bred varieties or their progenies.

Wide-scale adoption of these materials and improved management practices have resulted in major increases in food production. World rice production doubled in a 25-year period—from 257 million tons in 1966 to 520 million tons in 1990. This increased production feeds 700 million more people annually. Indonesia used to be the world's largest importer of rice in the 1970s. Its rice production increased from 12.9 million tons in 1966 to 47 million tons in 1994.

Rice production in Vietnam increased from 9.4 million tons in 1966 to 22.5 million tons in 1994. No longer a chronic rice-deficient country, Vietnam has emerged as the fourth largest rice-exporting country, with yearly exports of 2 million tons.

Due to chronic food deficits, India was considered a "basket case" by some authorities in the 1960s. However, rice production in India increased from 46 million tons in 1966 to 122 million tons in 1995. The country has not only become self-sufficient, but was able to export 4.1 million tons of rice in 1995. Similar increases in rice production have occurred in other Asian countries. The economic miracle now under way in Asia could not have been possible without food self-sufficiency and food security.

We cannot rest on our laurels. The challenges ahead are even greater. The population in Asia is increasing at a rate of 2 percent annually. Rice consumption is going up because of rising living standards. Agricultural economists estimate that 70 percent more rice will be needed by 2025 to feed 5 billion rice consumers. In the past, increased rice production resulted from opening up more rice lands for cultivation and increased productivity. However, there are no more lands to open up for rice cultivation. In fact, some of the best rice lands are being taken over to build houses, factories, and roads. Thus, the additional rice will have to be produced on less land, with less water, less labor, and fewer chemical inputs. The rice scientists must develop rice varieties with higher yield potential and better management technologies. In 1988, IRRI scientists conceptualized a "new plant type" which will produce 20 to 25 percent higher yield. The breeding program to develop such plants was initiated in 1989, and within five years the new plant type became a reality. Further improvements are being made and, when it is widely grown, starting by the turn of the century, the new type should help feed an additional 400 million people.

One of the most rewarding aspects of being at IRRI has been the opportunity of working with numerous excellent scientists from rice-growing countries. During the last 29 years, more than 325 trainees have participated in various kinds of training programs in rice breeding. Upon return to their countries, they become active collaborators in rice research and many of them are now heading national rice-breeding programs in their countries. Most of the rice-growing countries now have a well-trained cadre of rice breeders who can develop rice varieties suitable for local conditions.

Sometimes I look back and count my blessings, which are many. IRRI has been led by exceptionally dedicated and visionary directors general who provided continuing support to my work. I am grateful to numerous colleagues at IRRI as well as those in the national programs. Their cooperation has been invaluable. The advances in rice breeding would not have been possible without the dedicated support and input of my Filipino staff.

I have also benefited greatly from interaction with students, scholars, and national program scientists. The excellent support of the Philippine government for IRRI's work has been invaluable, too.

It takes resources to mount a major effort for developing technologies to increase food production. The Rockefeller Foundation and the Ford Foundation supported IRRI for the first 10 years. Since the early 1970s, the Consultative Group on International Agricultural Research (CGIAR), jointly sponsored by the Food and Agriculture Organization of the United Nations (FAO), the United Nations Development Programme (UNDP), and the World Bank, has marshaled the resources to support the work of IRRI and its 15 sister centers. Special mention for support and confidence in IRRI's program must be made of the development agencies of the United States, Japan, Canada, Australia, Germany, the United Kingdom, Denmark, Sweden, the Netherlands, and Belgium.

There are many remaining challenges. The World Food Prize, which I am receiving today, will help attract the attention of government officials and policymakers in the donor community. We need to invest today in the future of our children and grandchildren. It is our foremost responsibility to develop the technologies which will help produce enough food for more and more people in a sustainable way, while protecting our environment. We should commend Nobel Laureate Dr. Norman Borlaug, who has been my role model, for conceiving the ideal of The World Food Prize, and also the generosity of Mr. John Ruan for making it a reality. Both of these individuals deserve our gratitude and that of generations to come.

Dr. Borlaug, Mr. Chairman, members of the Council of Advisors, ladies and gentlemen, it is with gratitude and humility that I accept the prize you have awarded me. I wish to express my deepest sense of gratitude to numerous colleagues at IRRI and in the rice-growing countries for their collaboration. Finally, I would like to thank my parents who taught me the dignity of labor; my relatives and friends for their support; my wife, Harwant, who has been my best friend, advisor, and confidant and who sacrificed her own career for the success of mine; and my children, Ranjiv, Manjeev, Sonia, and Kiran, who have all been a source of strength and inspiration for me. Thank you very much.

PART I

Food Security Problems and Solutions

CHAPTER ONE

Food Security: Problems and Solutions

Robert S. McNamara

Introduction

India, sub-Saharan Africa, and the United States have at least three things in common: all have sufficient food to feed their populations, all have large numbers of hungry people, and all are pursuing environmentally unsustainable food production practices. Globally the situation is much the same. In the face of abundant supplies—reflected in the fact that, over the past 50 years, real prices of the three major cereal crops have dropped by almost 50 percent—over 800 million human beings are malnourished, including millions in the United States, the richest nation in the world. It is morally outrageous.

What does the future hold? With one important exception, more of the same. If one looks out a quarter of a century, to the year 2020, world population will increase by nearly 2 billion to approximately 8 billion; global food production will rise proportionately; environmental degradation will increase, limiting the opportunity for future production increases; and the number of hungry will grow to well over 1 billion. But, as compared to today, when food supplies are adequate across the globe, certain regions of the world—in particular sub-Saharan Africa—will become food-deficit. This will vastly increase food insecurity in these regions.

Put very simply, food security is a function of two factors: global food availability and access to the available food, particularly by the poorest among us. Because this conference is focused primarily on agriculture, I will address the issues of food production and sustainability. But I want to stress that the heart of the problem, at least for today and for the next quarter century, is food access and, therefore, poverty. I know of no one who has the answer to this issue. It was widely debated during the meetings of the 1996 World Food Summit in Rome and it deserves the attention of all of us.

One thing is clear: there is no solution to the access problem which does not begin with acceptance of the principle that all human beings have a right to minimum levels of nutrition. Food security is a societal responsibility. That principle has not been accepted and acted upon in sub-Saharan Africa and India. It has not been accepted and acted upon even in the United States.

We have an extraordinarily rich agenda ahead of us today that will examine in detail the outlook for achieving global food security in the twenty-first century. Therefore, in these opening remarks, I will do no more than frame the issues. I will express my own view as to production prospects and I will point to action that must be taken if my production forecasts are to be realized. At the end of the day, you can judge whether the cautious optimism I will express is justified.

The Debate Over Future Global Food Production

There is general agreement among the experts on a number of fundamental points. First, to meet food requirements in 2020, when the global population will be 2 billion larger than today, food production worldwide must increase by at least 1.5 to 2 percent per year. Second, a growth rate of 1.5 to 2 percent per annum would be approximately one-third less than the rate realized in the past quarter century. Therefore, the neccessary growth in production would appear to be readily achievable. However, past growth came from three sources which cannot be expected to contribute to the same degree in the years ahead.

From 1970 to 1995, there has been an expansion of cultivated area, increased intensity of land use through additional irrigation, and very large yield increases on both rain-fed and irrigated lands. The yield increases resulted from a wider application of known technology, from technological advances derived from agricultural research, from broader use of optimal input packages (seeds, fertilizers, and pesticides), and from widespread macroeconomic policy reforms.

Third, in the years ahead, the rate of expansion of cultivated land areas and of irrigated areas will be substantially less than in the past. Therefore, increased attention must be directed toward the productivity gains which can be realized only from continuing emphasis on agricultural research, and from the broader application of "best practices" in the areas of economic policy, technology, and inputs.

Although there is broad agreement on the three points I have mentioned, there have developed two widely divergent views—the optimistic and pessimistic views—on what sustainable levels of production will amount to during the next quarter century. Alex McCalla (1994) pointed out that what he calls the "food-production pessimists" believe that the world will face a greater challenge in expanding food production in the future. According to pessimists such as Lester Brown and Hal Kane of the Worldwatch Institute, there will be a reduction in the record growth in food production experienced in recent decades as a result of the earth's limited natural resources, environmental degradation, and the shrinking backlog of yield-raising technologies. Brown and Kane argue that food production growth rates are already substantially below past levels and further production will be subject to new constraints, including an increasing food demand, limited availability of water, ineffective fertilizer use, loss of agricultural land to industrialization and urbanization, and the adverse effects of high population growth and environmental degradation (McCalla 1994).

Brown and Kane are particularly pessimistic about China, whose future production and consumption levels will have an immense impact on global food security. It is projected that by 2030 China will experience a significant reduction in grain production and an increase in imports amounting to more than current levels of total world trade in grain. They conclude that because of constraints such as reductions in grain yields and fertilizer use, declining investments in agricultural research and development, and increasing environmental problems, the world may not be able to meet future food demands. In order to avoid such a crisis, Brown and Kane propose that the solution lies with population control (McCalla 1994).

Brown and Kane's views are not unique. They were supported by a panel of experts of the American Association for the Advancement of Science. The *New York Times* of February 18, 1996, reported that the experts, meeting in Atlanta, concluded that yield increases in U.S. croplands, already near the limit of production, will not be able to meet the growing demand of the U.S. population which is expected to double within 60 years. Dr. David Pimentel of Cornell University told the conference that the United States would cease to be a food exporter by about 2020 if current trends continue.

In contrast to the pessimists are the "food-production optimists." According to McCalla (1994), the "optimists," whose views I support, do recognize that the challenge facing world agriculture is enormous. They agree that in the last 40 years the doubling of cereal output came from the three sources which I referred to earlier: area expansion, increased intensity of land use, and yield increases. They accept that the growth of land under cultivation and the expansion of irrigated areas cannot continue at past rates; they recognize, therefore, that in order to double food production in the future, productivity, i.e., yields, must increase. They accept that increasing productivity, using current agricultural practices, will put stress on the natural resource base in both developed and developing countries. But, they agree with an International Food Policy Research Institute (IFPRI) study by Rosegrant and Agcaoili (1994) which states that global food production will be able to meet global food supplies in the future provided that policies are targeted toward growth in agriculture and increased investments in agricultural research, irrigation and water development, human capital, and rural infrastructure.

The optimists assume that the action required to achieve the productivity gains will be identified, initiated, and maintained. They predict, in effect, that there will be an expansion in agricultural research—at both the national and international levels—large enough to develop new technology, full application of existing technology to farmer's fields, optimal use of key inputs, the continued reform of macroeconomic policies, the identification and reversal of environmentally unsustainable agricultural practices, and the availability of the additional financial resources required to fully implement such a program.

The remainder of today's meeting will be devoted to identifying what specific actions are required if the optimistic scenario is to prove valid, who would be charged with the responsibility for carrying out those actions, and whether it is likely they can be depended upon to do so. The alleviation of poverty is a major component of the solution to achieving global food security. Particular attention will also be paid to environmental degradation which is recognized by both optimists and pessimists as requiring immediate attention if agricultural output in the decades ahead is not to be severely constrained. I will briefly address these two issues in the following sections.

Food Security and Poverty

In 1990, approximately 780 million people in the developing world did not have sufficient food to maintain a healthy life, which means that about one in every five people in the developing world was food insecure (Garcia 1994). There are two primary causes of malnutrition: insufficient consumption of food, and poor health and disease. Insufficient consumption of food leads to protein-energy malnutrition (measured by the proportion of underweight children), iron-deficiency anemia, blindness resulting from vitamin A deficiency, and iodine deficiency disorders and goiter.

Table 1.1 provides information on the contribution of malnutrition to the global burden of disease. Among the most vulnerable groups are preschool children and pregnant and lactating women. In 1990, dietary deficiencies represented 3.4 percent of the global burden of disease. The

Table 1.1. Direct and indirect contributions of malnutrition to the global burden of disease, 1990

Type of malnutrition	Sub-Saharan Africa	India	China	Other Asia and islands	Latin America and the Caribbean	Middle Eastern crescent	World[4]
	(millions of disability-adjusted life years [DALYs][1])						
Direct effects							
Protein-energy malnutrition	2.2	5.6	1.7	0.9	1.0	1.0	12.7
Vitamin A deficiency	2.2	4.1	1.0	2.5	1.4	0.5	11.8
Iodine deficiency	1.7	1.4	1.0	1.3	0.5	1.4	7.2
Anemia	1.0	4.5	2.7	2.3	1.0	1.5	14.0
Total direct	7.0	15.5	6.3	7.0	3.9	4.5	45.7
Total DALYs per 1000 population	13.8	18.3	5.6	10.3	8.9	8.9	8.7
Indirect effects (*minimum estimate*)							
Mortality from other diseases attributed to mild or moderate underweight[2]	23.6	14.9	3.3	8.0	2.4	8.0	60.4
Mortality from other diseases attributed to vitamin A deficiency[3]	13.4	14.0	1.0	7.0	1.8	2.0	39.1

[1]DALYs measure the burden of disease. They combine healthy life years lost because of premature death with those lost as a result of disability.

[2]Based on the global burden of disease attributable to deaths from tuberculosis, measles, pertussis, malaria, and diarrhea and respiratory diseases in children under age five; in developing countries 25 percent of those deaths are attributed to mild or moderate underweight.

[3]Based on estimated deaths attributable to vitamin A deficiency in age groups 6–11 months and 1–4 years. These account for, respectively, 10 and 30 percent of all such deaths in high-risk countries and for 3 and 10 percent of all such deaths in other countries. Thirty lost DALYs are attributed to each child; death losses are redistributed to the regional classification used in this report.

[4]Includes formerly socialist economies and established market economies.

SOURCE: Binswanger and Landell-Mills 1995. (Reprinted with permission)

nutritional disease burden for young children accounted for 6 percent of their total burden of illness. Anemia accounted for 24 percent of the disease burden among women of reproductive age and 1.3 percent of the total female disease burden (Binswanger and Landell-Mills 1995).

South Asia is home to over 50 percent of the world's protein-energy malnourished population (see Figures 1.1 and 1.2). In 1990, about 54 percent (or 100 million) of the world's 184 million underweight children lived in South Asia, 30 million lived in sub-Saharan Africa, 24 million in China, and 20 million in Southeast Asia. It is expected that this situation will improve for all the regions by 2020 with the exception of sub-Saharan Africa (Garcia 1994).

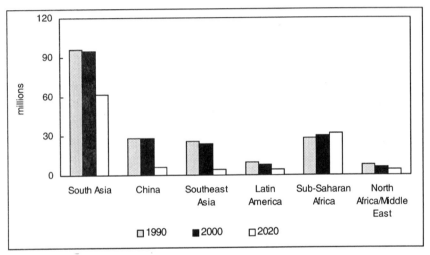

Figure 1.1. Projections of numbers of underweight preschool children, by region, optimistic scenario, 2020
SOURCE: Garcia 1994. (Reprinted with permission)

As I stated earlier, malnutrition is caused primarily by insufficient food consumption due to the unavailability of food, and to poor health and disease resulting from poverty. Let us look at these two contributing factors to malnutrition, i.e., domestic food production and poverty.

In addressing the future domestic food production in India and sub-Saharan Africa, I will examine some conservative projections to 2020 provided by Iowa State University's Center for Agricultural and Rural Development (Angel and Johnson 1992). These projections, which leave the basic structure of production, consumption, and population growth unchanged, show that sub-Saharan Africa will experience a reduction in per capita wheat production of 2.9 percent for the period 1995 to 2020. Per

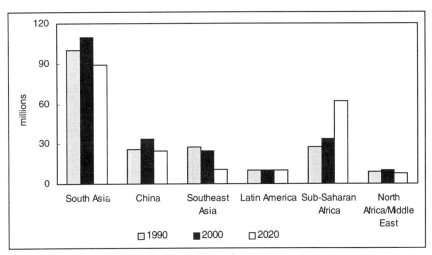

Figure 1.2.   Projections of numbers of underweight preschool children, by
region, pessimistic scenario, 2020
SOURCE: Garcia 1994. (Reprinted with permission)

capita consumption of wheat will increase by approximately 32 percent. The implication is for a very large shortfall to be covered by imports. In the case of India, the per capita wheat production figure is -2.9 percent while the per capita consumption change over the period is 0.58 percent. Clearly, India is in better shape than sub-Saharan Africa, based on this trend analysis.

Similar results are suggested for per capita coarse grains production and consumption. For sub-Saharan Africa, per capita production of coarse grains increases by 1.8 percent while per capita consumption increases by 3.6 percent. Comparable figures for India show a 7.9 percent increase in per capita coarse grain production and a 9.2 percent increase in per capita consumption. Again, the data indicate a larger food security problem for sub-Saharan Africa than for India.

Such projections are particularly important for these low-income countries where the population obtains a large share of caloric consumption from cereals. In India, caloric consumption from cereals is around 70 percent for the urban population and 82.5 percent for the rural population. In sub-Saharan Africa, cereals make up a smaller share of consumption, but this share is rising. In Togo, for example, the percentage of caloric consumption from cereals is approximately 58 percent whereas in the Ivory Coast it is 40 percent nationwide and over 50 percent in certain urban areas. It is evident that unless the current production trends are improved, many developing countries, especially those in Africa, will continue to face insufficient consumption of food and therefore, malnutrition.

Poverty is another major factor affecting food security and malnutrition. More than 1.3 billion people in the developing world exist on less than one dollar per day. Although income levels and the standard of living are increasing substantially in the upper and middle classes in developing countries, the number of people living in poverty is expected to rise, particularly in urban areas. In 1960, the gap in per capita income between the developed and the developing world was $5,700. By 1993, this number had increased to $15,400. The gap between the poor and the rich has also increased. In the past three decades, the income share of the poorest 20 percent of the world's population declined from 2.4 percent to 1.4 percent while the share of global income of the richest 20 percent increased from 70 percent to 85 percent (Leisinger 1996).

In sub-Saharan Africa, between 45 and 50 percent of the population lives below the poverty line, a proportion exceeded only in South Asia. In 1993, an estimated 40 percent of the population in sub-Saharan Africa lived on less than a dollar a day. The depth of poverty, i.e., how far incomes fall below the poverty line, is greater in sub-Saharan Africa than in any other region in the world. There are a number of causes of poverty, particularly in sub-Saharan Africa, including inadequate access to employment opportunities and to credit; inadequate physical assets such as land and capital, and low endowment of human capital; inadequate infrastructure and rural development in poor regions; inadequate access to domestic markets; environmental degradation and low productivity; inadequate access by the poor to assistance; and failure to incorporate the poor into the design of development programs (The World Bank 1996).

In addition to low income levels, another indicator of poverty is the inadequate access to social services. The availability of social services in sub-Saharan Africa is the lowest in the world. The average gross primary school enrollment rate during the early 1990s was 67 percent for sub-Saharan Africa (a decline since the 1980s), 94 percent for South Asia, and 117 percent for East Asia. Because of a lack of adequate health services, the average infant mortality rate in sub-Saharan Africa is 93 per 1,000 which is higher than South Asia's 84 per 1,000, Latin America's 46 per 1,000, and East Asia's 36 per 1,000 (The World Bank 1996). These factors translate into poor health and the prevalence of disease, thus contributing to malnutrition.

Sustainable Development

Another major factor affecting food security and sustainable agricultural development is the effect of environmental degradation. It is estimated that 21 percent of global land under pasture and 38 percent of global cropland are degraded to various degrees, primarily as a result of water and wind erosion and overuse of land. Degradation affects 65 percent of Africa's cropland, 51 percent of the cropland in Latin America, and 38 percent of Asian cropland (Leisinger 1996). This has resulted in declining yields and increased inefficient use of inputs. In addition, 17 million hectares of the world's forests are lost annually.

Poverty, high population growth rates, and urbanization are major problems in many developing countries, contributing significantly to environmental degradation and air and water pollution (Pinstrup-Andersen 1994; The World Bank 1994). Over 1.5 billion people, mostly in the developing world, live in unsanitary conditions, the majority without access to clean water. Approximately 2 to 3 million children die each year under these conditions (The World Bank 1994). In sub-Saharan Africa, in particular, rapid population growth combined with inappropriate agricultural practices have resulted in soil degradation and low productivity. The lack of well-defined ownership of resources and user rights also contributes to environmental degradation through the exploitation of fragile and scarce lands and through deforestation. The result is reduced forest land, soil erosion, changing rainfall patterns, and low land productivity.

One of the major resource problems currently facing low-income countries, particularly those in sub-Saharan Africa, is the depletion and degradation of water resources as a result of economic development and high population growth. Water is a critical resource, necessary for both agricultural production and human survival, and it is now in short supply, especially in the densely populated regions of the world.

Water is being depleted through household and industrial use as well as through withdrawals by the agricultural sector which consumes over 50 percent of the annual freshwater in most of these countries. Water resources degradation, in the form of sediment runoff, siltation, agro-chemical runoff, and industrial and household pollution has a number of causes: deforestation, cropland expansion, soil erosion, improper drainage, inadequate irrigation planning, poor enforcement of environmental regulations, and destruction of shoreline vegetation (USDA 1995). Future production gains from irrigated lands may also be adversely affected by salinization and waterlogging. It is expected that if population trends continue, and given the existing climatic conditions, per capita water availability will decrease by about 20 percent in developing countries and by 34 percent in African countries by the turn of the next century.

Given that yield increases are not expected to reach the levels required to meet the rising global food demands, expansion of agricultural lands and intensification of production are the two options available for increasing food production. However, land expansion will account for only 20 percent of production growth. Also, most of the potential cropland for expansion is in developing countries, mainly in sub-Saharan Africa and South America where about one-half of this land consists of poor soils with low fertility. Major investments will be required in order to make these lands fit for cultivation (USDA 1995). It is also crucial that the intensification of production not cause further damage to the environment.

For sustainable development and the reduction of environmental degradation, policies targeting the eradication of poverty (which is a major source of degradation) and policies that invest in human capital and that improve primary health care, education, family planning, sanitation and water supply, and rural development must be implemented. Environmental sustainability can also be achieved through investments in agricultural research and extension, yield-enhancing technologies, and new technologies that improve farmers' welfare (The World Bank 1994).

The allocation of property rights and better definition of land ownership and user rights also have a positive effect on the environment and can greatly reduce such degradation as deforestation and desertification (Pinstrup-Andersen 1994). Improving the status of women who play significant roles in agricultural production but do not have equal access to land, inputs, and information is also important. In addition, investments in transportation and infrastructure are necessary to link the producers in the remote areas of crop production to both domestic and foreign markets, currently a major problem in sub-Saharan Africa and South America (USDA 1995).

In order to achieve environmental and development sustainability, promotion of the efficient use of resources is of the utmost importance. Many governments have adversely impacted the environment by subsidizing the consumption of vital resources which has led to increased fiscal cost, inefficient use of resources, and environmental degradation (The World Bank 1994). According to Pinstrup-Andersen (1994), it is also important to incorporate environmental costs into production and consumption decisions or through government policies.

Furthermore, trade liberalization, improved management of government utilities, and macroeconomic stability all contribute positively to the environment. Pollution control, management of natural resources, and research are policies that need to be implemented by governments to positively affect the environment (The World Bank 1994). Measures should be taken to ensure the efficient use of pest controls and the development of pest-resistant plant varieties to reduce the need for chemicals.

Conclusion

We find that, throughout the world, current agricultural practices are harming the natural environment. Lands not suited to farming are being cleared and plowed; habitat is being destroyed; biodiversity is being lost; soil nutrients are being exhausted; trees are being cut and burned; rivers and water resources are silting up; irrigated land is becoming waterlogged; surface water or gradient water is becoming polluted with chemicals; and farm workers are suffering from increased exposure to toxic pesticides even as pests are becoming more resistant to them (Paarlberg 1994). Is this sustainable? The pessimists believe that the further spread of science-based, high-input farming in poor countries will only lead to faster depletion of water supplies and further reductions of biodiversity. The optimists believe that action can be taken to make modern, highly productive farm practices compatible with environmental protection. Who is right? We don't know.

Although on this issue, too, I am inclined to agree with the optimists, we lack the facts on which to judge the sustainability of current and future farming practices. It is absolutely essential, therefore, that the world develop a monitoring system to measure, by country, such factors as loss of arable land, levels of soil fertility, changes in water tables, pollution of surface waters, and destruction of vegetative cover. At the end of the day, each of you must judge the validity of the optimists' position and the likelihood that the action they hypothesize will, in fact, take place.

When arriving at that judgment, you will have to consider whether, in the future, the United States is likely to support such action. The trend is not encouraging. In recent years U.S. national policy has often moved in directions opposite to what is required to advance food security both here and abroad. I will provide three examples.

First, the number of the hungry in this country—both adults and children—will rise substantially as a result of recent legislation affecting the distribution of food stamps, school lunch programs, welfare payments to the poor, and aid to legal and illegal immigrants. There were 36.9 million poor Americans in the United States in 1992, 1.2 million more than in 1991 and 5.4 million more than in 1989. Forty percent of the poor were children and 10.8 percent were older Americans (Herron and Zabel 1995). As is the case around the world, hunger in the United States is directly linked to poverty. There has already been a rise in the number of poor persons in recent years, especially among the most vulnerable groups, i.e., children, single mothers, and minorities, reflected by the increase in the demand for food stamps and emergency food assistance (Federman, Garner, and Short 1996; Passero 1996; Herron and Zabel 1995). The recent welfare legislation is bound to make matters worse.

Second, the U.S. Agency for International Development (USAID) reduced its contribution to international agricultural research by almost 50 percent between 1992 and 1994. That action led to a cut of 10 percent in the number of international scientists and a reduction of 20 percent in local staff working in the International Agricultural Research Institutes around the globe. The destructive effects of the U.S. cuts were partially offset by emergency contributions from the World Bank.

Third, the United States joined other nations at the 1994 Cairo Population Conference in pledging to support an increase in global funding for family planning, from $5 billion in 1992 to $17 billion in 2000. The objective was to sharply reduce fertility rates as a major step toward global food security. However, the international program is now in jeopardy as a result of sharp cutbacks in U.S. support.

The last session of our conference will give each of you an opportunity to push for answers to your doubts as to whether the optimists are correct. I hope no one will leave this room at the end of the day without a clear understanding of both the magnitude of the task and the role that each of us can and must play if the world is to achieve global food security in the twenty-first century.

I want to conclude by repeating what I said earlier: the heart of the problem, at least for today and the next quarter century, is food access. I feel disgraced—and I hope all of you do as well—by our failure to address the most fundamental of needs for nearly 1 billion human beings across the globe. It is, indeed, morally outrageous.

References

Angel, Bruna, and Stanley R. Johnson. 1992. Changing International Food Markets in the 1990s: Implications for Developing Countries. In *World Food in the 1990s: Production, Trade and Aid,* edited by Lehman B. Fletcher. Boulder, CO: Westview Press, 43–129.

Binswanger, Hans P., and Pierre Landell-Mills. 1995. *The World Bank's Strategy for Reducing Poverty and Hunger.* A Report to the Development Community, Environmentally Sustainable Development Studies and Monographs Series Number 4. Washington, D.C.: The World Bank.

Federman, Maya, Thesia I. Garner, and Kathleen Short. 1996. What Does It Mean to be Poor in America? *Monthly Labor Review* 119 (May):3–17.

Garcia, Marito. 1994. *Malnutrition and Food Insecurity Projections, 2020.* 2020 Brief Number 6. Washington, D.C.: International Food Policy and Research Institute.

Herron, Nancy L., and Diane Zabel, eds. 1995. *Bridging the Gap: Examining Polarity in America.* Englewood, CO: Librarian Unlimited, Inc.

Leisinger, Klaus M. 1996. Food Security for a Growing Population 200 Years after Malthus, Still an Unsolved Problem. Saguf-Symposium, *How Will the Future World Population Feed Itself?* Zurich, October 9–10.

McCalla, Alex F. 1994. *Agriculture and Food Needs to 2025: Why We Should Be Concerned.* 1994 Sir John Crawford Memorial Lecture. Washington, D.C.: Consultative Group on International Agricultural Research.

Paarlberg, Robert L. 1994. *Countrysides at Risk: The Political Geography of Sustainable Agriculture.* Washington, D.C.: Overseas Development Council.

Passero, William D. 1996. Spending Patterns of Families Receiving Public Assistance. *Monthly Labor Review* 119 (April):21–8.

Pinstrup-Andersen, Per. 1994. *World Food Trends and Future Food Security.* Food Policy Report. Washington, D.C.: International Food Policy Research Institute.

Rosegrant, Mark W., and Mercedita Agcaoili. 1994 Global and Regional Food Demand, Supply and Trade Prospects to 2010. Paper presented at the roundtable meeting *Population and Food in the Early Twenty-First Century: Meeting Future Food Needs of an Increasing World Population* held February 14–16. Washington, D.C.: International Food Policy Research Institute.

United States Department of Agriculture (USDA). 1995. *World Food Aid Needs and Availabilities.* USDA World Agricultural Outlook Board. Washington, D.C.: USDA.

World Bank. 1994. *Making Development Sustainable: From Concepts to Action.* Environmentally Sustainable Development Occasional Paper Series Number 2, edited by Ismail Serageldin, Andrew Steer, Michael M. Cernea, John A. Dixon, Ernst Lutz, Sergio Margulis, Mohan Munasinghe, and Colin Rees. Washington, D.C.: The World Bank.

_____. 1996. Poverty in Sub-Saharan Africa: Issues and Recommendations. *Findings.* Africa Region, no. 73, October. Washington, D.C.: The World Bank.

CHAPTER TWO

Agriculture, Food Security, and the Environment: An Action Agenda for the New Millennium

Ismail Serageldin

Introduction

It was indeed a pleasure and an honor to be invited to speak at the tenth anniversary of The World Food Prize which, from the beginning, when Norman Borlaug pressed for its establishment, has been of particular interest to the Consultative Group on International Agricultural Research (CGIAR). Six of the ten World Food Prize laureates are connected with the CGIAR. That includes Henry Beachell and Gurdev Khush who are being honored this year. I join in paying tribute to their scientific achievements. They have truly placed science at the service of the poor.

This tenth anniversary event comes at a very appropriate time for the CGIAR, which is on the verge of commemorating its twenty-fifth anniversary. Indeed, World Food Prize laureate Gurdev Khush will be a featured speaker at the commemoration, and all other laureates from the CGIAR family have been invited to attend as special guests. We have set apart some time in the agenda to pay our own tributes to The World Food Prize Organization. The laureates will join us—today's CGIAR, our alumni who helped to make the CGIAR what it is today, and our partners in the emerging global agricultural system—to define parameters of research that can help fulfill the vision of a food-secure world in the twenty-first century.

Food security will require more than production: it will require attention to the issues of pervasive poverty, access by the poor to food, and the quality of the nutrition they get. It will require everything from sound policies and access to credit, especially for poor women, to a revitalized and prosperous rural world, and sustainable production systems coupled with sound environmental management. Agricultural research is a key to much of this and I will focus on this topic. I want to share with you the reasons why I consider agricultural science to be at the very heart of that vision's fulfillment. In doing so, I want to challenge you, the best minds in the business, to join us in addressing critical issues of research priority setting and evaluation.

The Basic Proposition

The challenges facing global agricultural science in the next century are arguably greater than at any previous time. And these challenges must be met in a political environment that is a dangerous mix of complacency, fiscal constraints, aid fatigue, and fundamental disagreements about the magnitude of the problem and the appropriate paths to its solution.

The global objectives of poverty reduction, sustainable natural resource management, and food security cannot be met unless rural well-being in general, and a prosperous private agriculture for small- and medium-size holders in particular, are nurtured and improved. This is my basic proposition. Central to improving the productivity and profitability of agriculture are improved technology, appropriate policies, and supportive institutions. At the core of technological improvement is agricultural research. Without steady improvement in sustainable productivity growth, the triple challenges identified cannot be met. Failure is a prospect so frightening that we must not allow it to happen.

To reduce poverty, feed the hungry, protect the environment, and improve living standards through sustainable, broad-based growth and investment in people, the world needs to confront three formidable challenges. I invite you to join me in reviewing them.

The Poverty and Hunger Challenge

The major achievements of the twentieth century include rapid increases in the living standards in the developing world, a decline in the proportion of the world's people living in poverty, rapid technological change in agriculture, and declining real prices of food.

However, rising populations and unequal participation in growth have left 1.3 billion people in the world struggling to survive on less than one dollar a day. About 800 million of them are hungry, undernourished, or malnourished. Over 500 million children under the age of five are not receiving the nutrition they need to fully develop, mentally and physically.

Ironically, nearly three-quarters of the poor and hungry live in rural areas where food is produced. These people are concentrated in poor, slow-growing countries, and often in regions with limited agricultural potential.

Lack of opportunity and poverty have been the main factors pushing the rural poor into cities. Yet, a century of migration from the countryside to cities has not been able to reduce the number of poor people living in rural areas.

The concentrated poverty and hunger of rural populations is intolerable and must be eliminated in the first decades of the twenty-first century. This requires concentrated attention to agricultural and rural development in the entire developing world, but especially in slow-growing and food-deficit countries.

The Food Security Challenge

Everyone agrees that the world's population will exceed 8 billion people by 2025, an increase of 2.5 billion in the next 30 years. Much, but not all, of the increase will occur in developing-country cities where urban populations will more than triple. Most scholars agree that given moderate income growth, food needs in developing countries would more than double, and global food demand could nearly double. The challenge to world agriculture is enormous.

Future increases in food supplies must come primarily from rising biological yields, rather than from area expansion and more irrigation. Why? Because land and water are becoming increasingly scarce. Most new lands brought under cultivation are marginal and ecologically fragile and cannot make up for the land being removed from cultivation each year due to urbanization and land degradation. The sources of water that can be developed cost-effectively for irrigation are nearly exhausted, and irrigation water will need to be reallocated increasingly for municipal and industrial use. Therefore, yields on existing land will need to more than double.

The Sustainable Natural Resource Management Challenge

Hundreds of millions of private farmers, large and small, men and women, are the stewards of the vast majority of the world's renewable resources. These farmers recognize the importance of maintaining and enhancing their productivity, and have shown that they can do so, given proper incentives. But the resources are severely undervalued by inappropriate accounting methods, policies, and institutional frameworks. Of all the world's freshwater used by people, agriculture uses more than 70 percent in irrigation. Unsustainable agricultural practices are major contributors to nonpoint source pollution. Deforestation remains a critical issue with 25 hectares of forest being lost every minute. The global challenges of desertification, climate change, and loss of biodiversity require major efforts. If yields are to double in the next 30 years, the policy and institutional failures that cause or contribute to the negative environmental impacts of agriculture must be reversed and sustainable production systems developed, encouraged, and applied.

These major challenges of overcoming poverty and hunger, increasing food production, and halting natural resource degradation require action on a broad and complex rural development front. It is not just a matter of agricultural production but of widely shared and sustainable rural growth and development.

The Role of Agricultural Science

The tasks ahead can be stated directly and relatively simply, but their accomplishment will chart new territory for agricultural science. We need to intensify the sustainability of the complex agricultural production systems at the smallholder level, while preventing damage to natural resources and biodiversity and contributing to the improved welfare of farmers. I submit to you that doubling the yields of complex farming systems in an environmentally positive manner is an enormous challenge that is not going to be easy to meet. Let me give some flavor of why I believe this to be true.

Agricultural science has done a far better job of increasing individual commodity yields in input-intensive monocultures than it has of improving the productivity of complex farming systems. The use of modern reductionist science has tended to focus on commodity yields rather than the long-run sustainability of production systems. Yet for most of the developing world, even this approach has bypassed them. Where it has been applied, e.g., semidwarf rice and wheat, it has been in favorable well-watered areas. In many intensive monoculture systems, negative externalities such as water quality, loss of biodiversity, chemical pollution, and soil degradation threaten long-run sustainability.

What is required is a systems and interdisciplinary approach to the sustainable improvement of complex systems. Although developed-country scientists have lessons to offer developing-country colleagues in productivity improvement, it is less clear that they are role models for the sustainable intensification of complex systems.

Some Critical Variables

The challenge of meeting this considerable research task is helped by some positive developments—molecular biology, information technology, and advanced systems analysis—but hindered by some additional constraints—

biodiversity preservation, environmental control, intellectual property rights, bioethics, the changing interface between the public and private sectors, and the call for greater stakeholder participation in the research process.

Radical and rapid changes in our understanding of molecular biology have spawned a potential biotechnology revolution. Yet the application of this science is occurring more slowly than its enthusiasts predicted. Nevertheless, it is enhancing our capacity to redesign plants and animals, and its full potential has not yet been realized. New information technology has revolutionized our capacity to generate and exchange information. It has basically removed the relevance of national boundaries to science and technology. Thus our capacity to learn from and share with others is creating a truly global scientific network. Computer technology and advanced systems methods have greatly enhanced our capacity to analyze complex systems and simulate biological as well as economic systems.

These are all powerful tools to help agricultural science meet its formidable challenge. But there are also greater constraints. Concerns about preserving biodiversity, about intellectual property rights, and about bioethics have the potential to alter significantly the free exchange of scientific information and genetic material. Much more attention must now be paid to the long-term environmental consequences of the application of agricultural science. There is a rapidly changing perception about the appropriate role of the public sector in agricultural research. Biotechnology and plant patenting have altered the incentives for the private sector. The current evidence suggests that in many developed countries private sector expenditure on proprietary agriculture research now exceeds public expenditure, in some cases by a wide margin.

What are the implications of these trends for complex, small farmer tropical farming systems? In many developing countries fundamental changes are occurring because of decentralization of fiscal and political power and calls for greater stakeholder participation. This challenges the model of a centralized, autocratic agricultural research establishment.

In addition to these changes are issues of global warming, dependence on petroleum-based external inputs, and cumulative contamination—the magnitude of which is not fully known. Choice making in research is clearly decisionmaking under uncertainty.

Some Difficult Trade-offs

Research and science policymakers clearly face a myriad of complex and difficult choices. Let me share a few real world choices that I, as Chair of the CGIAR, have confronted.

First, favored versus less-favored areas. This debate has gone on forever. I believe we are past treating this as a zero-one choice, but there are many critical choices of how scarce resources should be deployed. How does one weigh poverty reduction for poor subsistence farmers against low food prices for the urban poor? How does one decide how to allocate resources between traded food grains and the subsistence crops, roots and tubers?

Second, traditional versus exotic crops. The Sahel is one of the most difficult environments in the world. How can one improve the productivity and the income potential of poor farmers? Should one invest in improving the yields of indigenous crops, such as millet and sorghum, or try to improve the stress resistance of higher-yielding nontraditional maize? Should one invest in tropical wheat or potatoes versus yams, sweet potatoes, and cassava?

Third, time horizon choices. What should be the time frame of the expected impact? Often improved agronomic practices such as spacing, seeding time, weed control, and planting depth can have short-term yield impacts, whereas genetic improvement, particularly involving complex characteristics, takes much longer but has higher long-run yield potential. What is the appropriate discount rate for poor, food-deficit countries with limited research budgets?

Fourth, environmental improvement versus yield maximization. Frequently this is posed as a major trade-off in choice priorities, but it cannot persist as such. Clearly the major challenge to agricultural science is to turn this apparent win-lose situation into a win-win situation. This is a nontrivial scientific task which is long-run and interdisciplinary. How much should be invested in fundamental research as opposed to applied systems improvement?

Fifth, international versus national research investment. Here, too, there are critical choices. What should be the distribution of resources between local, national, regional, and international research? Should every country have a full-blown domestic research system? How much should donors invest in the CGIAR versus strengthening national programs? What role should multinational firms play in the global research system? What are truly public research goods, nationally and internationally?

I could go on but you have the flavor of the difficult decisions that decisionmakers face. This is why your involvement and that of many others is so important. Scientists, governments, financial managers, farmers (especially women farmers), the private sector, the civil society from both industrialized and developing countries, and international agencies must all rise to the challenge. The array of wise men and women gathered at this symposium has much to contribute to meeting the fundamental challenges for the twenty-first century. Together, you can help the world choose from among crucial and difficult alternatives: choice of scientific approach, agro-ecology, geography, species, and expenditure. These choices are made more difficult by the increasing complexity of science policy and by declining fiscal resources. But let us not be frightened by the challenge. Without challenges, science dies.

Conclusion

Let me conclude by returning to the broader context. Meeting the agricultural science challenge is only a necessary condition. Establishing food security in the fullest sense of the phrase requires much more.

To double agricultural output while preserving the national resources on which production is based will require the following:

- Rapid technological change in agriculture in the developed and the developing world. The private sector will have to undertake an increasing share of the necessary research and diffusion. Public sector financing will be needed for areas of limited interest to the private sector, such as genetic resource conservation, common property resource management, integrated pest management, and research on subsistence crops.

- Massive increases in the efficiency of irrigation water use through changes in water policies, water rights, and institutions for allocating water and by technical improvements in water conveyance and use.

- Dramatic improvements in the management of soils, watersheds, forests, and biodiversity by local and community-based institutions.

- Accelerated private investment in the rural economies of the developing world.

- Enhanced support by the international community to the agricultural and rural development programs of poor countries. Even if world food supplies grow dramatically over the next 30 years, fast-growing countries such as China and India could become major food importers. If they, and other countries, are to pursue optimal food strategies, they must have stable, long-term access to world markets. Countries may then focus on producing the goods in which they have a comparative advantage and exporting them in exchange for goods they cannot produce efficiently, including food. Only with stable, long-term access to world markets can countries comfortably refrain from costly food self-sufficiency policies.

Transforming agriculture to ensure household, national, and world food security while protecting the natural resource base on which food and productivity depend involves enormous challenges. The challenges are technological and political. They require dramatic improvements in national and international policies, institutions, and public expenditures. The role of research is fundamental. The 1.3 billion poor who subsist on less than a dollar a day have waited far too long for us to help meet their needs. They are entitled to be impatient, now. Those of us with the capacity to help the poor begin their ascent from poverty are not entitled to ignore either their needs or their impatience. Let us get on with the job.

CHAPTER THREE

Food Security: An Assessment of the Global Condition and Review of New Initiatives

Jacques Diouf

In an oft-quoted phrase, author Tomasi di Lampedusa reminded us that "If you want things to stay as they are, things will have to change" (Lampedusa 1991, 41). Regrettably, however, the situation that the world now finds itself in demands that staying the same is not enough; we need to do much better and in earnest. Therefore, we not only need real change, we need major innovation. Our problems are not local, not even national or regional. They are global problems, requiring global changes to respond to them.

What changes are we referring to? At the most basic level, humanity must change the way that it feeds itself. We must develop new and improved modes of production, distribution, and consumption. We must galvanize the ingenuity and creativity of all sectors of society to identify new strategies.

Let us look briefly at what we mean by *change*. We want change because we want to progress, and because we want to alter and improve an existing situation. But at the same time, we must recognize that we do not begin from zero—there is no tabula rasa. As a result, we should respect the need for continuity. The historical process provides us with legacies and structures; it yields starting points and imposes constraints. Weaving a new fabric for global society requires that we combine threads of change with threads of continuity. I think that we can say that today we are beginning to see the emergence of a new consensus. We know what is wrong and unacceptable—starvation, malnutrition, deprivation, and inequality. We also understand that we can get over these problems through a delicate balance between change and continuity: change to adapt to new circumstances, scientific and technological innovations, geopolitical adjustments, economic and sociopolitical shifts, and redefined roles and responsibilities for government and civil society; continuity, on the other hand, to ensure the endurance of principles that retain a validity through time and to acknowledge the influence of the past in order to fashion the present and shape the future.

In advocating change, therefore, we should strive for balance. We must find a path shrouded in a rationality that understands the process of change fully, appreciating its roots and sensing its direction, its rhythms, and the risks involved. Through this, we can determine and influence the course of change, rather than having it imposed on us through the sheer force of events. And, in so doing we can ensure that it becomes *change with a human face.*

Role Models for Change

In seeking to follow these guidelines I have always sought inspiration from role models whose teaching and guidance can help illuminate the path forward. Many can be called heroes. When we talk of heroes, of course, we think of those people who have given or devoted their lives for the betterment of others. But we must also think of other heroes: the mothers who walk 20 miles a day, every day, so that their children may have water; the fathers who toil in parched fields until they drop from exhaustion so that their families might have food; the children who work 16-hour days in sweatshops, their fingers rubbed raw, so that they might augment their family's meager income; the pastoralists who wander barren landscapes in search of water and sustenance for their famished cattle; and the fishers who venture out onto stormy seas or dangerous inland waterways to provide income and food for their families. These are the real heroes of the modern age— millions of people, unknown to all but their families and friends—the people whose daily struggles define them as heroes.

Then again, there are others, heroes of more classic design. These are the people who have dedicated their lives to easing the suffering of others; there were indeed several in The World Food Prize Symposium audience. They are people who have sought to mobilize scientific research, learning, and technological innovation as tools for increased food production and food security. There are others who have assumed leadership roles in international development finance, and still others who have devoted themselves to furthering the appeal and effect of ecotechnologies and other sustainable natural resource management practices. I should like to focus, albeit briefly, on one man's achievements—they demonstrate to us the real value and purpose of applied learning and innovation.

Norman Borlaug was awarded the Nobel Peace Prize in 1971, an honor that attests to an individual's contribution to humankind. This award was not given for brokering peace or for ending conflict. And neither, for that matter, was it in recognition of scientific endeavor per se. It was given in recognition of his work directed at increasing food security, in improving the everyday existence of millions of people. And I think that this confirms a very important point. Peace can only be found through economic, social, and nutritional security. How many wars and conflicts and how much discrimination and prejudice have we seen that have emerged in contexts of increasing poverty, declining livelihoods, and rising insecurity? It is only through marshaling the energies of science and devoting them to the direct improvement of individual and collective livelihoods that we can hope to ensure peace at local, national, and international levels. And this is why it is so very right that the Nobel Peace Prize was awarded to a person who, perhaps more than any other individual, has set about the task of increasing food production, thereby heightening opportunities for peace and security.

At this point I also wish to highlight the importance of other initiatives such as The World Food Prize. Since its inception in 1987 it has celebrated, without regard to race, religion, nationality, or political beliefs, the achievements of individuals who have advanced human development by improving the quality, quantity, or availability of food in the world. Mr. Ruan and your fellow members of The World Food Prize Council, I, as director-general of the Food and Agriculture Organization (FAO), applaud your endeavors.

The Global Food Condition

I was asked to discuss food security and the current global condition, and to review some of the more strident new initiatives. I have already identified two core principles: the need to balance change and continuity, and the necessity of marshaling science and technology to improve food security in an appropriate and sustainable manner. The latter, we know, is of vital importance. This is why initiatives such as Global 2000, in which Mr. Borlaug is using his skills and experience to produce higher crop yields in selected African nations through the use of Production Test Plots (PTPs) and easily available technology, must be both commended and extended.

Stated simply, the contemporary global condition gives cause for grave concern. More than 800 million people currently suffer energy-deficient diets. Demographic growth rates mean that world population will grow by about one billion in the next 10 years—and I call upon you to remember that it took over a million years for the human population to reach one billion. It is true that the growth of food supplies has exceeded population growth rates, although there has been some leveling off of this trend in recent years. Production must continue to increase in order to ensure that enough food is available for the growing population and to absorb the ever larger backlog of food-deficit households. What does energy deficiency mean? It means constant gnawing hunger, desperation, reduced labor outputs, reduced tolerance to infection, and ultimately, it means a lingering death. The World Health Organization (WHO), for example, reported that 41.5 percent of deaths in the developing world result directly from infectious and parasitic diseases. This means that during the 30 minutes that I will speak at this symposium, approximately 900 people will have died as a result of infectious and parasitic disease; many were simply too weak and undernourished to withstand the attendant infections.

There are 82 Low-Income Food-Deficit Countries—countries that are unable to produce or import sufficient food to guarantee food security for their populations. In all, more than 800 million people suffer chronic malnutrition and more than 200 million children suffer acute or chronic protein and calorie deficiencies. We are then very far from the vision of FAO's founding fathers. More than 20 years after the World Food Conference of 1974, the goal to eradicate hunger, food insecurity, and malnutrition within a decade remains stubbornly beyond our grasp. And yet, the right to food is the first and foremost of the human rights, without which the others have no meaning. How can a hungry person be expected to exercise his or her right to education, work, and culture, or to participate fully in the political and social life of the community? It was the Roman philosopher Seneca who warned us that hungry people do not listen to reason, do not care for justice, and are not bent by any prayers. Let us not forget this warning.

For so long we assumed, mistakenly as it turned out, that global food reserves would provide security against national shortfalls. We now know that this is wrong. For the majority of Low-Income Food-Deficit Countries, food security must be derived from within, not without. And this is why I, as director-general of the Food and Agriculture Organization of the United Nations, recently launched two initiatives. The first was the Special Programme for Food Security in Low-Income Food-Deficit Countries; the second was the World Food Summit, held in November 1996 at our FAO headquarters in Rome.

Allow me to expand briefly on some of the trends and conditions that characterize the modern age and prompted these two initiatives. We have already referred to the burgeoning crisis of food security. In addition, there are some clear trends: slower but continuing population growth rates;

improved economic growth prospects for many developing countries but deteriorating prospects for others; a continued slowdown in agricultural growth rates; and, perhaps most hearteningly, clear signs of progress in food and nutrition conditions for a large part of the population. But importantly, and indeed sadly, the persistence of totally inadequate food and nutrition conditions appears to be the future facing many millions of the world's less privileged and most vulnerable inhabitants. These, of course, are only predictions based on current trends. They could, and I hope they will, all be modified with the correct application of vision and will.

Modern Agriculture

Let us now look briefly at some of the features characterizing modern agriculture. We cannot really generalize: the rich heterogeneity that defines the world's peoples, cultures, and regions prohibits standardized *descriptif*. Nonetheless, certain trends appear to be emerging. Agriculture is becoming more feminized. Not since the so-called Neolithic Revolution have we seen such a predominant role for women in agricultural production and decisionmaking. Likewise, agriculture is becoming increasingly urbanized. By this, I do not refer only to the growth of agricultural production systems within many of the world's large and small cities—São Paolo and Nairobi spring readily to mind—but also to the growing difficulties associated with separating the urban and rural sectors. We see that agriculture is becoming increasingly linked economically, demographically, and socially to urban areas. The once clear separation is becoming increasingly indistinct. Likewise, we can see that the retreat of the state from many aspects of social, political, and economic life is reconfiguring the agricultural policy environment. In many countries we witness two parallel processes: an emergent institutional vacuum resulting from the decline of state involvement, and an increasing importance attached to the participation of the private sector and civil society in decision- and policymaking processes. This, I believe, is the main hope for the future: the formation of new partnerships that combine individuals and groups from diverse sectors of society, bringing together the most modern and appropriate scientific innovations with time-proven indigenous knowledge systems in order to confront the policy problems of the modern age.

I do not wish to sound too optimistic, however. The end of the Cold War has not brought an end to war or conflict itself. And while conflict abounds, development cowers. Indeed, there can be no economic development, and especially no improvement in food security, unless there is peace. There can be no peace without justice, and no just relations between individuals and peoples, unless the inalienable rights of peoples and nations are respected. We have only to recall the wise words of U.S. President Dwight D. Eisenhower, delivered at "The Chance for Peace" address on April 16, 1953: "Every gun that is made, every warship launched, every rocket fired, signifies, in the final sense, a theft from those who hunger and are not fed, those who are cold and are not clothed" (Eisenhower 1960, 182).

FAO Initiatives

Let me now return to the World Food Summit and FAO's Special Programme for Food Security in Low-Income Food-Deficit Countries. The World Food Summit was conceived as a large-scale operation to enlist a solemn commitment at the top level to eliminate hunger and malnutrition and to undertake concerted action at the global, regional, and national levels to ensure food security for all. This has been the underlying aspiration behind the organization of the World Food Summit.

Any solution to the terrible problems of today inevitably entails changes in policies and measures on an unprecedented scale that can only be implemented after collective and profound reflection by all interested parties, including public authorities, universities and researchers, the private sector, nongovernment organizations (NGOs), and, more particularly, farmers' organizations, women, and the young.

Such clarity of purpose will inevitably be countered by skepticism, here and there, questioning the need for the Summit. What, one may well ask, was the point, after so many initiatives of all kinds? Was not concern to feed the world the springboard for the establishment of FAO 50 years ago, followed by the Freedom from Hunger Campaign, the two World Food Congresses of 1963 and 1970, the World Food Conference of 1974, and more recently, the International Conference on Nutrition in 1992? We can answer this on several levels. First, this was the first time in the 50 years since FAO was founded that a meeting on these issues was held at the level of heads of state and government. And the fact that the Summit proposal was unanimously approved by the Conference of FAO and the United Nations General Assembly clearly attests that the world food problem has now become so serious that it demands priority consideration at the highest levels. Second, while FAO's mandate has not changed from that laid down by the founding fathers in its constitution, the sheer size and the nature of the problems have evolved with a speed typical of this, the *short* century, as well-known historian Eric Hobsbawm recently described it (Hobsbawm 1994). And finally, it is FAO's fundamental responsibility to alert world opinion and world leaders to the deteriorating food situation before it reaches catastrophic proportions.

The focus of the Summit was on meaningful, sustainable action. In the spirit of the United Nations Conference on Environment and Development's (UNCED) Agenda 21, rather than relentlessly pushing out agricultural boundaries and jeopardizing fragile ecosystems, the Summit's efforts centered on high-potential areas where productivity can be increased by intensifying farming practices that employ the conservation, collection, and harnessing—and hence better management—of water. Where this is not a feasible option, marginal land will have to be developed sustainably without causing environmental damage. The aim in both cases is to increase output sufficiently to provide for population growth and raise nutritional levels where serious food deficiencies exist. However, increasing output is only part of the equation; we need to ensure that the benefits from national and international efforts reach all members of society and particularly its poorest members. Measures are therefore needed for more equitable access to food by all, more efficient distribution, and far fewer food losses.

Beyond the Summit itself, what is needed is a truly global campaign, with cooperation and consultation at all levels. Following in the footsteps of the Freedom from Hunger Campaign, its theme would be "Food for All," which is the slogan FAO has adopted for the future. The driving force for this Food for All Campaign would be national committees involving all segments of civil society: the private sector, NGOs, academic and research institutions, women's associations, and youth groups. To muster the support and mobilization necessary to ensure its success would demand long-term commitment and sustained resources. The mandate and objectives of the Food for All Campaign would be determined by the Summit, and its structure adapted to the specific situation of each country. The mechanism established would supplement the governmental National FAO Committees already in place.

How did this Summit differ from many previous attempts to combat hunger and malnutrition? Is this initiative any more likely to succeed than all its predecessors? In times of increasing urgency and growing budgetary restraint and accountability, FAO would not have launched such a large-scale initiative without being confident that the Summit would be worthwhile. In fact I believe that it was more than this. The World Food Summit was *vital*. It directed global attention at one of the most pressing problems facing humankind as we enter the new millennium. It identified concrete actions and strategies. It helped to foster global responses. It acted as a catalyst for galvanizing all sectors of society to strive for the goals of food security and "Food for All."

The World Food Summit does indeed differ in many respects from previous events addressing the problem of world food security. In contrast to recent high-level meetings, the Summit was convened by a body that was specifically set up to deal with food and agricultural development. This ensures both a solid base and the human and material resources to implement its programs. Furthermore, two key practical initiatives are already in place to achieve food security for all: the Special Programme for Food Security, and the Emergency Prevention System for Transboundary Animal and Plant Pests and Diseases (EMPRES).

These two programs demonstrate FAO's concern for preventive measures as well as reactive responses to existing problems. EMPRES is aimed at prevention. It was established in 1994 and has two components: animal pests and plant pests. The unifying theme, however, and one that I feel reflects some of FAO's unique comparative advantages, is that it is directed at pests that are both transboundary and fluid in nature. Locusts, the plant pest element of EMPRES, are in many ways the stealth terrorists of the animal world. Their ability to shift location rapidly and invisibly is both remarkable and one of the causes of their endurance. And locusts, we know, are not limited by national boundaries. Similarly, livestock pests such as rinderpest, lumpy skin disease, and contagious bovine pleuropneumonia are both unrestricted by national boundaries and adept at escaping emergency responses.

The underlying theme of the Special Programme on Food Production in Low-Income Food-Deficit Countries is similar, a proactive strategy directed at food production increases. Food and water loom prominently among the major world challenges as we enter the third millennium. The dimensions of the problem are ethical, political, and strategic, and could lead to extremely violent and serious conflicts unless we put things right. FAO is so keenly aware of the need for strong, immediate action that it has launched this special program directed, as the name implies, toward low-income food-deficit countries. It is now being implemented in its pilot phase in approximately fifteen countries.

The main thrust of the program is to work on a specific, day-to-day basis with farmers, livestock owners, forest workers, fishing communities, and fish farmers, so that they can sustainably increase their productivity and thus combat poverty. The program's activities include demonstration of improved techniques in the farmers' own fields. Identification, implementation, and evaluation are all done by those most directly involved: the farmers themselves.

Additionally, the program strongly emphasizes the absolute necessity of people's participation, particularly that of women. Women indeed play a predominant, multifaceted, and totally irreplaceable role in feeding the household and community. In many regions of the world, women are the main providers of food, which they grow, prepare, and store. They are responsible for children's education and for handing down cultural values and expertise related to food. Without broad-based participation, particularly of the feminine population, there would be no momentum or spillover effect, no continuity, and no universal adhesion to a joint undertaking.

Conclusion

We are in a crisis. We require new solutions. We need innovative strategies and new methods. Eight hundred and forty million malnourished people are a poor testament to the supposed achievements of modernity. As we move into the third millennium, it is these voices, and the millions more of those as yet unborn, who demand change. But in seeking to confront these problems, I do believe that we can draw hope from the words of Jawaharlal Nehru, the first leader of post-colonial India, who said that crises and deadlocks when they occur have at least the advantage that they force us to think.

Finally, I would emphasize one important point. Our efforts at the FAO are not the fruit of our labor alone. They embody the efforts of our member nations, governments, the private sector, farmers' organizations, community groups, civil society, and, of course, our sister international agencies. Whatever we do and whatever we propose is enriched with their vision and their will. Our actions are a synthesis of their attempts to decipher the symptoms of the contemporary malaise and their endeavors to identify the pace and rhythm of necessary change and transformation. It is only when we succeed in the construction and amplification of partnerships and broad-based coalitions, when we move away from words and embrace deeds, that we will really be in a position to claim that we are fighting these problems. I am appealing for a new covenant for rural development, a New Deal for the countryside: in short, "Food for All."

References

Eisenhower, Dwight D. 1960. *Public Papers of the Presidents of the United States, 1953*. Washington, D.C.: General Services Administration.
Hobsbawm, Eric J. 1994. *Age of Extremes: The Short Twentieth Century 1914–1991*. London: Abacus.
Lampedusa, G. di. 1991. *The Leopard*. New York: Random House.

CHAPTER FOUR

Commodity Supply and Use Balances and Food Availability

*Amani E. El Obeid, Stanley R. Johnson,
Lisa C. Smith, and Helen H. Jensen*

Introduction

Approximately 840 million people in the developing world do not consume enough food to meet minimum energy requirements. Another quarter of a billion people periodically experience inaccessibility to food as a result of weather, instability in prices and employment, drought, diseases, and civil strife (Tweeten et al. 1992). Although production of the major food crops grew at an annual rate of 2.4 percent during the period from 1960 to 1980, this growth rate has declined in recent years due to reductions in food production by the developed nations. The global capability to feed the growing population will depend heavily on our ability to increase food production in the future.

During the decade of the 1980s, global grain production increased by 2.1 percent per year while population grew by 1.7 percent per year (Mitchell and Ingco 1995). Food production growth continues to exceed world population growth. Per capita food production is currently about 18 percent higher than it was three decades ago. However, the distribution of available food is uneven. For example, average per capita food consumption is more than 50 percent higher in developed countries than in sub-Saharan Africa (USDA 1996a). As a result, there are large numbers of countries with insufficient supplies of food to feed their populations. Sub-Saharan Africa and South Asia, in particular, are areas where food availability problems are most prevalent.

Assessments of the future of the food balance situation in the developing world are not, in general, encouraging (Christensen et al. 1981; Angel and Johnson 1992; IDRC 1992; IFPRI 1995). Angel and Johnson (1992) examined projections for developing countries by country and region for the period from 1993 to 1997 (using data through 1990). They concluded that the per capita trends in consumption, production, and trade of the major food commodities (particularly wheat and coarse grains) implied major food security problems for selected African and Latin American countries. The situation was not assessed to be as critical for Asia as a whole, since with higher per capita income levels, grain consumption requirements could more likely be met by growth in imports.

According to Angel and Johnson (1992), food consumption levels, already low by dietary standards, were projected to decrease for Africa, the Middle East, and certain countries in Latin America. The inability of the nations in these regions to import food or increase food production to meet demand was identified as a possible indicator of a future food crisis. In short, Angel and Johnson predicted a continuing deterioration of food availability, particularly in the poorer, more heavily populated countries. The deterioration of food availability has, in fact, continued since their report was released, confirming, in the short term, these earlier projections.

Many countries have fulfilled their unmet demand for food through trade. The volume of trade in food has increased during the past three decades. During the period from the 1960s to the 1980s, the largest growth in agricultural trade was in the import of cereals by developing nations (Paulino 1986). However, the share of cereals in food imports declined during the late 1980s and early 1990s in most of the developing world with the exception of sub-Saharan Africa (Table 4.1). There, the share of cereals in food imports increased from 32.1 percent in the early 1960s to 48.4 percent in the early 1990s. This percentage was lower in Latin America and the Caribbean (45.8 percent) and in Asia and the Pacific (42.9 percent) but higher in the Near East and North Africa (52.6 percent) for the period from 1989 to 1991. Although the share of cereals in food imports declined in the developing world as a whole since the early 1960s, it still remains slightly over 49 percent (FAO 1995).

Food security is defined as the "access by all people at all times to enough food for an active, healthy life" (The World Bank 1986, 1). There are two dimensions to food security: food availability, i.e., the availability of adequate quantities and quality of food; and food access, i.e., food obtainable through adequate income and resources. This chapter deals primarily with food availability and provides quantitative information on food balance trends in the developing world. Reasons are proposed why these trends persist and why they have deteriorated for selected countries and regions.

Table 4.1.   Share of cereals in food imports, selected regions

Region	1961–63	1975–77	1989–91
		(percent)	
Sub-Saharan Africa	32.1	43.6	48.4
Latin America and the Caribbean	45.1	56.9	45.8
Asia and the Pacific	63.2	82.1	42.9
Near East and North Africa	52.1	57.8	52.6
Total	51.1	60.3	49.2

SOURCE: FAO 1995. (Reprinted with permission)

The next section reviews regional and country trends in per capita production and consumption of wheat, rice, and coarse grain. The following section presents an analysis of food availability in terms of per capita dietary energy supply and considers the distribution of dietary energy by nation. Dietary energy supply is the total food availability estimate provided by the Food and Agriculture Organization of the United Nations (FAO) and obtained by aggregating the energy values of all food products available for human consumption. Then, suggested alternative reasons for the observed food trends in developing countries are provided.

Commodity Supply and Use Balances

Trends in food availability, which is usually measured by the supply of food available from production, imports, and stocks, are presented by region and country in this section. Data on production and consumption (defined as total domestic use from production, imports, and stocks) of wheat, rice, and coarse grains were obtained from the U.S. Department of Agriculture (USDA 1996b) for the period from 1960 to 1995 for developing regions and for specific countries. Historical population data from 1960 to 1990 and medium-variant population projections from 1990 to 1996 were obtained from the United Nations (UN 1994). Per capita production and consumption over time were calculated for each region and country for the entire period (Appendix 4.A). Although Eastern Europe and the former Soviet Union are not classified as developing countries, they have been included in this review because of the major changes affecting these countries during their ongoing transition to market economies.

Wheat, rice, and maize are the most important cereal crops for developing countries. Wheat accounted for 23 percent of cereal production, rice 47 percent, and maize 19 percent in the developing countries as a whole in 1992. Coarse grains—namely, maize, sorghum, and millet—which are important in Africa, accounted for 33 percent, 18 percent, and 14 percent of

Africa's total cereal production, respectively (Pinstrup-Andersen 1994). Cereals also made up the highest share in total dietary energy supply, approximately 38 percent in Latin America and the Caribbean, 45 percent in sub-Saharan Africa, 57 percent in Near East and North Africa, and over 64 percent in Asia during the period from 1990 to 1992 (Table 4.2).

Table 4.2.   Share of major food groups in total dietary energy supply by region and economic group, 1990–92

Food group	Latin America and the Caribbean	Sub-Saharan Africa	Near East and North Africa	East and Southeast Asia	South Asia
			(percent)		
Vegetable products	82.6	93.4	90.4	89.1	92.6
Cereals	38.4	44.7	56.9	66.5	64.5
Roots and tubers	4.1	21.0	2.2	5.1	1.6
Pulses and nuts	4.8	7.1	4.0	3.4	6.3
Others[a]	35.3	20.6	27.3	14.1	20.2
Animal products[b]	17.4	6.6	9.6	10.9	7.4

[a] Others include sugar, vegetable oils and fats, vegetables and fruits, alcoholic beverages, and stimulants and spices.
[b] Animal products include meat and offal, milk, animal oils and fats, eggs, and fish.
SOURCE: FAO 1996.

Data in the Appendix show that both per capita production and consumption of grains, i.e., wheat, rice, and coarse grains, increased in most of the developing world, except in certain regions during the period from 1960 to 1995. However, in many countries the deficit in domestic production over domestic consumption of grains has widened significantly, and may indicate a food security problem for countries that cannot fill the gap through imports. The major problems in food availability today lie in Africa as a whole, and in sub-Saharan Africa in particular, where both per capita production and consumption of grains have declined significantly. The situation is also critical in South Asia and some countries in Latin America (particularly for wheat). Furthermore, per capita production and consumption of grains declined in Eastern Europe and the former Soviet Union, especially in recent years, reflecting the economic and political transitions these nations are experiencing.

It is important to point out that one contributing factor to falling per capita grain consumption in some developing nations may be the substitution of animal products for cereals. Although cereals and roots constitute the major food groups in developing countries, diversification of diets has been increasing in developing regions such as East and Southeast Asia. This has not been true, however, for sub-Saharan Africa. Furthermore, the share of animal products in the total food supply increased in many of these countries, again with the exception of most of Africa and the Near East. The share of animal products in total dietary energy supply fell from 6.7 percent in the late 1960s to 6.6 percent in the early 1990s in sub-Saharan Africa, and fell from 10.4 to 9.6 percent in North Africa and the Near East during the same period (FAO 1996).

Wheat

Wheat has been particularly important to the diet in the Near East and North Africa, where the share of wheat in total dietary energy supply was 42.8 percent in the early 1990s (Table 4.3). The share of wheat in total dietary energy supply is also high in the transition economies (Eastern Europe and the former Soviet Union)—approximately 33 percent. In East and Southeast Asia, the share of wheat in total dietary energy supply increased from 9.8 percent in the early 1970s to 17.1 percent in the early 1990s. The share of wheat also increased in South Asia from 16.8 to 21 percent during the same period. In Latin America, wheat made up the second largest share in total dietary energy supply (after maize) with 13.2 percent. Sub-Saharan Africa has the lowest share of wheat in the total dietary energy supply although this share increased over the 20-year period from 3.6 to 5.4 percent.

Table 4.3. Share of wheat in total dietary energy and supply by region, 1969–71 and 1990–92

Region	1969–71	1990–92
	(percent)	
World	17.5	19.5
Transition economies[a]	32.7	32.9
Latin America and the Caribbean	13.9	13.2
Sub-Saharan Africa	3.6	5.4
Near East and North Africa	41.7	42.8
East and Southeast Asia	9.8	17.1
South Asia	16.8	21.0

[a] Eastern Europe and the former Soviet Union.
SOURCE: FAO 1996.

Latin America. Consumption levels of wheat rose in Latin America during the period from 1960 to 1995, with consumption higher than production (Table 4.A.1). The production of wheat increased until the early 1980s and then started to decline, widening the gap between production and consumption. Population growth rates declined steadily for the region from 2.8 percent in the 1960s to 1.8 percent in 1995. On a per capita basis, wheat production in the region has declined since 1982 (Figure 4.1). Per capita consumption showed a gradual increase until the 1980s, after which it fell slightly. However, per capita consumption remained higher than per capita production for the historical period shown.

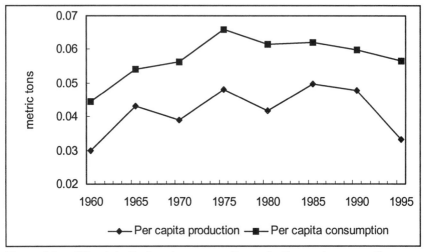

Figure 4.1.   Wheat: Latin America, per capita production and consumption
SOURCE: Based on data from USDA 1996b.

Central America has had low wheat production. However, per capita consumption of wheat rose in this region as population growth declined. The Caribbean, which does not produce wheat, showed an increase in per capita consumption of wheat before the mid-1980s and a decline in consumption per capita during the late 1980s and early 1990s. South America had rising levels of consumption and stagnant or slightly rising production of wheat. Consumption was higher than production for the entire period. Per capita production of wheat has increased slightly since the early 1980s following a decline during the 1960s and 1970s, while per capita consumption remained stable throughout the period.

Data for specific countries show somewhat different trends. Mexico has had increasing production and consumption of wheat, with consumption higher than production for the period. Consumption and production of wheat, measured on a per capita basis, rose during the period from the late 1960s to 1980, but both exhibited declining trends in the period after 1980. In Brazil, however, production, both in absolute and per capita terms, has declined significantly since the late 1980s, with slower increases in consumption and leveling off in per capita consumption. During the entire period from 1960 to 1995, Argentina produced higher levels of wheat than Mexico and other South American countries, and production was higher than consumption. Per capita consumption of wheat declined slightly during the same period.

Africa. Africa exhibited rising production and consumption levels of wheat for the period between 1960 and 1990, with consumption rising significantly faster than production (Table 4.A.1). Production declined slightly after 1990 while consumption stabilized. The gap between production and consumption increased throughout the entire period. For the continent as a whole, population growth increased from 2.5 percent in 1965 to 2.8 percent in 1995. Population growth rates in sub-Saharan Africa averaged approximately 3 percent during the past decade, while rates in North Africa declined, averaging 2.4 percent for the same period. Per capita production was low and stagnant in the African continent for the period from 1960 to 1995. Per capita consumption of wheat showed a rising trend but has declined since the mid-1980s.

In North Africa, consumption increased faster than production during the 1960 to 1995 period, and this resulted in a widening domestic gap. Per capita production of wheat was stagnant in the region during the same period while per capita consumption showed a slight decline in recent years (Figure 4.2). Although population growth in North Africa declined, particularly in Morocco, reductions in population growth rates were not consistent for Algeria, Egypt, and Tunisia. Algeria showed rising population growth until the 1980s when the rate started to decline. Algeria also exhibited declining per capita production of wheat for the entire period. Per capita consumption of wheat rose until the mid-1980s before stabilizing. Per capita wheat production in Egypt was somewhat stagnant, with slight increases after the mid-1980s, while per capita consumption continued to rise. However, the past decade saw a decline in per capita consumption of wheat in Egypt. The gap between production and consumption widened for the region as a whole, and particularly for Egypt and Algeria.

Sub-Saharan Africa showed rising trends in consumption levels of wheat, and also showed slightly increasing production levels. Consumption was higher than production, and the gap between consumption and production widened during the period from 1960 to the early 1980s. However, per capita consumption started to decline in the early 1980s and has continued this decline. Per capita production of wheat was stagnant throughout the period. Population growth rates are high in the region and have risen steadily.

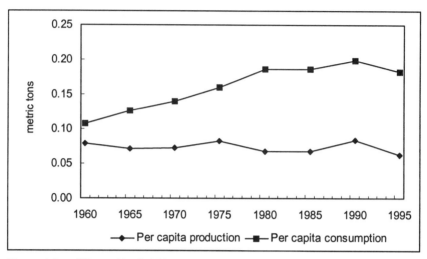

Figure 4.2. Wheat: North Africa, per capita production and consumption
SOURCE: Based on data from USDA 1996b.

West and Central Africa have negligible production of wheat and rising wheat consumption. In both regions, per capita consumption has declined since the mid-1980s. East Africa showed increasing consumption and production of wheat, with consumption rising faster than production. On a per capita basis, consumption continued to rise while production declined slightly or remained stagnant. However, per capita consumption of wheat declined in the 1990s. The situation was not much different for southern Africa where per capita production has been stagnant and a wide gap has existed between per capita consumption and production. Per capita consumption of wheat in southern Africa has declined since the early 1980s. In general, changes in production levels of wheat for all of Africa are not keeping pace with population growth.

Asia. Asia has shown rising wheat consumption and production, with consumption higher than production for the period from 1960 to 1995 (Table 4.A.1). With declining population growth rates, from 2.1 percent in 1960 to approximately 1.5 percent in 1995, the picture remained the same on a per capita basis, i.e., rising per capita consumption and production of wheat.

Most of the Asian countries have shown steadily declining population growth rates. In China, both per capita consumption and production of wheat rose; per capita consumption was slightly higher than per capita production. However, since 1985, China's per capita production and consumption appear to have leveled off. East Asia exhibited rising per capita consumption and low or stagnant production, with a wide gap between consumption and production. Per capita consumption in both Japan and Taiwan did not increase significantly, particularly in Japan, but the gap between consumption and production continued to be wide. Hong Kong's per capita wheat consumption showed a sharp rise after the mid-1980s.

Southeast Asia has very low production levels of wheat and increasingly high consumption levels. Per capita consumption of wheat in the Philippines has risen sharply since the mid-1980s. In South Asia and in countries like India, both per capita production and consumption of wheat have increased with consumption slightly above production. Per capita production and consumption of wheat also rose in Bangladesh, although per capita consumption has declined and per capita production has stabilized since 1980.

Other Regions. The Middle East experienced rising production and consumption of wheat with consumption slightly higher than production (Table 4.A.1). Population growth in the region started to decline only after the mid-1980s. Per capita consumption of wheat declined and per capita production increased in the region after 1980, reducing the wide gap between per capita production and consumption that existed during the mid-1970s to the early 1980s (Figure 4.3). Per capita production of wheat has increased significantly in Saudi Arabia since the 1980s, although the country experienced a sharp decline in production in recent years. Per capita consumption of wheat in the region increased, albeit slightly, during the entire period.

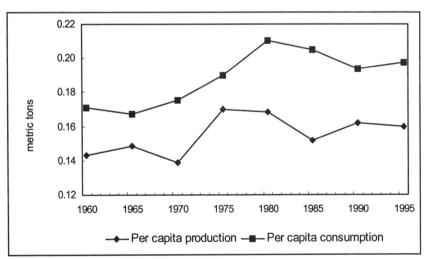

Figure 4.3.   Wheat: Middle East, per capita production and consumption
SOURCE: Based on data from USDA 1996b.

In Eastern Europe, per capita production and consumption of wheat increased between 1960 and the mid-1980s, with consumption not significantly different from production (Figure 4.4). However, the region experienced a sharp drop in per capita production and consumption of wheat in the late 1980s. Population growth rates in Eastern Europe declined during the period from 1960 to 1995. Despite declining population growth rates in the countries of the former Soviet Union, these countries have experienced declining per capita consumption and production of wheat since the late 1970s, with per capita consumption above per capita production.

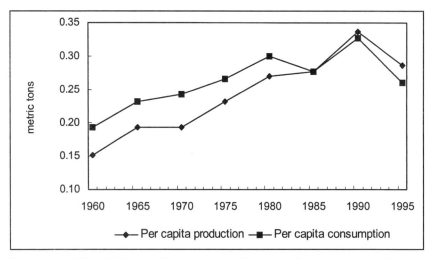

Figure 4.4.   Wheat: Eastern Europe, per capita production and consumption
SOURCE: Based on data from USDA 1996b.

Rice

The share of rice in total dietary energy supply was highest in Asia— 40.8 percent in East and Southeast Asia and 33.7 percent in South Asia during the period from 1990 to 1992 (Table 4.4). The share of rice in total dietary energy supply increased in sub-Saharan Africa from 4.8 to 7.8 percent over the period from the early 1970s to the early 1990s. Latin America had a slight increase in the share of rice in the total dietary energy supply from 9 percent during the period from 1969 to 1971 to 9.4 percent during the early 1990s. The share of rice remained unchanged at 6.2 percent for the Near East and North Africa during the periods between 1969 to 1971 and 1990 to 1992. Rice is relatively unimportant in Eastern Europe and the former Soviet Union compared to the developing regions.

Table 4.4. Share of rice in total dietary energy supply by region, 1969–71 and 1990–92

Region	1969–71	1990–92
	(percent)	
World	20.3	22.0
Transition economies[a]	1.1	1.3
Latin America and the Caribbean	9.0	9.4
Sub-Saharan Africa	4.8	7.8
Near East and North Africa	6.2	6.2
East and Southeast Asia	43.9	40.8
South Asia	35.4	33.7

[a] Eastern Europe and the former Soviet Union.
SOURCE: FAO 1996.

Latin America. Consumption and production of rice rose in Latin America during the period from 1960 to 1995, with consumption higher than production during the late 1980s and 1990s (Table 4.A.2). In per capita terms, consumption increased steadily after 1970, while per capita production exhibited a sharp decline in the late 1980s before increasing during the 1990s. Per capita consumption of rice in Latin America was above per capita production during the 1990s. In Central America, the gap between per capita production and consumption of rice increased significantly after 1985, with per capita consumption rising and per capita production declining. In the Caribbean countries, per capita consumption of rice increased while per capita production increased only from the mid-1960s to the mid-1980s and then declined in recent years, widening the gap between production and consumption. In South America, both per capita production and consumption of rice increased, with production slightly higher than consumption throughout the entire period from 1960 to 1995.

Although total consumption of rice increased steadily in Mexico, per capita consumption remained stagnant. Both production levels and per capita production of rice started to decline in Mexico in 1975. Argentina's per capita production of rice was higher than its per capita consumption, which remained somewhat stable throughout the three decades. Argentina has witnessed a sharp increase in rice production in the 1990s. In Brazil, per capita production and consumption of rice declined during the 1960s, followed by a slow increase in per capita consumption and a declining trend in per capita production. Both production and consumption increased in absolute terms. During the 1990s Brazil's per capita consumption of rice was stable and exceeded production.

Africa. The gap between production and consumption of rice has increased steadily in the African continent since the 1970s (Table 4.A.2). Although both production and consumption rose, consumption increased faster than production. In per capita terms, however, both production and consumption of rice declined after the late 1970s.

Production of rice in North Africa was higher than consumption until the early 1970s; thereafter, the gap narrowed significantly. Per capita production and consumption of rice declined during the early 1970s and the 1980s, then increased sharply in the 1990s. This was also true in Egypt where both per capita production and consumption of rice showed declining trends for the period prior to 1990 and rising trends during the 1990s. Production of rice was negligible in Algeria; therefore, per capita consumption was significantly higher than per capita production.

Sub-Saharan Africa experienced rising production and consumption of rice with consumption higher than production (Figure 4.5). However, per capita consumption has declined significantly since the 1970s while per capita production was below consumption and declined slightly or remained stagnant. The situation was the same for West Africa, especially Nigeria, which witnessed sharp increases in the consumption of rice after 1975, before consumption started to decline in the early to mid-1980s. Central Africa showed a rising trend in per capita production and consumption of rice, with consumption higher than production. However, the region witnessed a decline in both production and consumption during the 1970s and 1980s.

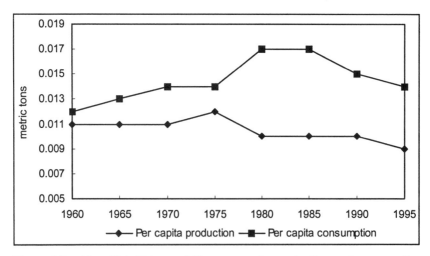

Figure 4.5.   Rice: Sub-Saharan Africa, per capita production and consumption
SOURCE: Based on data from USDA 1996b.

The gap between per capita consumption and production increased in East Africa, because consumption rose faster than production both on a per capita basis and in absolute terms. The region experienced declines in per capita production and consumption of rice in the late 1980s. In southern Africa, despite rising consumption and production levels with consumption higher than production, per capita rice consumption and production declined for the period from 1965 to 1995. The high growth in population rates in the region has been the major reason for these indicators of deterioration in the food balances.

Asia. Per capita consumption and production of rice increased in Asia with no significant gap between the two during the period from 1960 to 1995 (Table 4.A.2). This trend held for most of the region, particularly China. One exception is East Asia where both per capita production and consumption of rice declined steadily during the 30-year period (Figure 4.6). In Southeast Asia, per capita production was higher than per capita consumption after 1977 and has continued to rise (Figure 4.7). Thailand, on the other hand, witnessed a decline in per capita consumption of rice and a stagnant trend in per capita production, although production remained above consumption during the entire period. In South Asia, particularly in India and Bangladesh, production and consumption levels increased, with an insignificant gap between the two (Figure 4.8). In per capita terms, however, the increase was most significant after 1980.

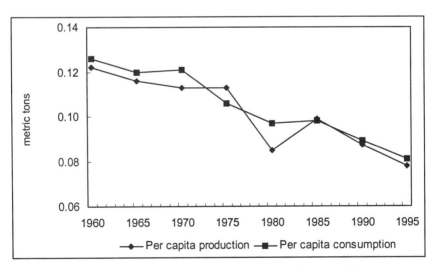

Figure 4.6.   Rice: East Asia, per capita production and consumption
SOURCE: Based on data from USDA 1996b.

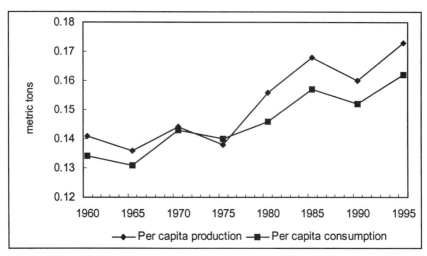

Figure 4.7.   Rice: Southeast Asia, per capita production and consumption
SOURCE: Based on data from USDA 1996b.

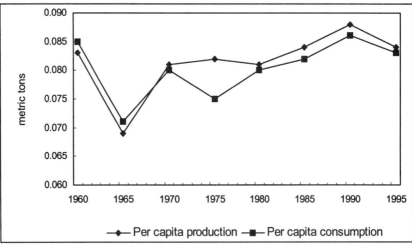

Figure 4.8.   Rice: South Asia, production and consumption
SOURCE: Based on data from USDA 1996b.

Other Regions. Per capita production of rice remained stagnant in the Middle East while per capita consumption rose steadily, and the gap between production and consumption widened throughout the entire period from 1960 to 1995 (Table 4.A.2). The gap between per capita production and consumption of rice in Eastern Europe was also wide, and consumption was higher than production. Both production and consumption have declined since the mid-1980s in per capita and absolute terms. In the early part of the period, per capita production and consumption of rice rose in the former Soviet Union with consumption slightly higher than production. However, both production and consumption per capita started to decline after the mid-1980s.

Coarse Grains

Maize, sorghum, and millet are the most important cereals in sub-Saharan Africa where the share in total dietary energy supply was 14.7 percent for maize and 14.6 percent for sorghum and millet in the early 1990s (Table 4.5). Cassava made up 14.9 percent of total dietary energy supply for the same region during the period 1990 to 1992. In Latin America, maize made up the largest share in total dietary energy supply at 15.3 percent during the 1990 to 1992 period. The share of coarse grains in total dietary energy supply is significantly lower in the rest of the developing regions and in the transition economies. In fact, this share has declined in most of these countries during the period from 1969 to 1992.

Table 4.5. Share of coarse grains in total dietary energy supply by region, 1969–71 and 1990–92

Region	Maize		Sorghum and Millet		Cassava	
	1969–71	1990–92	1969–71	1990–92	1969–71	1990–92
	(percent)					
World	5.4	6.1	4.4	2.6	1.7	1.6
Transition economies[a]	1.4	1.2	0.6	0.2	0.0	0.0
Latin America and the Caribbean	15.7	15.3	0.3	0.1	4.2	2.2
Sub-Saharan Africa	13.5	14.7	19.2	14.6	14.3	14.9
Near East and North Africa	6.1	4.7	2.6	0.8	0.0	0.0
East and Southeast Asia	6.8	6.8	4.6	0.9	1.1	0.9
South Asia	3.4	2.8	10.5	6.6	0.0	0.5

[a] Eastern Europe and the former Soviet Union.
SOURCE: FAO 1996.

Latin America. Per capita production and consumption of coarse grains increased in Latin America during the period from 1960 to 1980, then declined during the 1980s before increasing again in the 1990s (Table 4.A.3). Until the mid-1980s, consumption of coarse grains was below production, in per capita terms. However, particularly in the 1990s, per capita consumption rose above per capita production (Figure 4.9). In Central America, per capita consumption increased slightly and per capita production showed a decline, more noticeably after 1985. This significantly widened the gap between consumption and production of coarse grains. The gap was also wide in the Caribbean countries, where per capita production of coarse grains declined and per capita consumption rose, albeit slowly. However, the past decade showed stable per capita production and a decline in per capita consumption of coarse grains in the Caribbean countries. South America's per capita consumption of coarse grains rose faster than per capita production, which started to decline in the late 1980s.

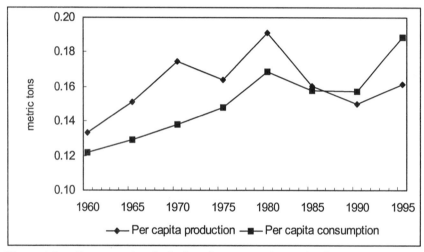

Figure 4.9.   Coarse grains: Latin America, per capita production and
              consumption
SOURCE: Based on data from USDA 1996b.

Mexico's per capita consumption of coarse grains rose steadily until 1980, and then declined until a significant increase in the 1990s. Per capita production of coarse grains declined during the 1970s and 1980s but rose sharply in the 1990s. Production of coarse grains in Mexico has remained below consumption since the 1970s. In Argentina, per capita production was higher than per capita consumption, with the latter exhibiting no notable changes during the entire period. Per capita production of coarse grains rose until the 1980s and then experienced a sharp decline. The gap between the rising per capita production and consumption of coarse grains was not significant in Brazil throughout the period.

Africa. Production and consumption of coarse grains rose in Africa during
the period from 1960 to 1995, but both per capita production and consumption declined (Table 4.A.3). Consumption was below production during the
1960s and 1970s but shifted above production during most of the 1980s and
the 1990s. In North Africa, per capita consumption of coarse grains rose
significantly above per capita production, which exhibited a declining trend
after 1975. However, per capita consumption showed a slight decline during the mid-1980s and early 1990s. Per capita production and consumption
of coarse grains also declined in Algeria after rising during the period from
the mid-1970s to the mid-1980s. Egypt's per capita coarse grain production remained stagnant, whereas per capita consumption rose sharply
particularly during the late 1970s to early 1980s and then again during
the 1990s.

In sub-Saharan Africa, per capita production and consumption of coarse
grains declined, although, in absolute terms, both production and consumption rose during the period from 1960 to 1995 (Figure 4.10). This was also
the case both in West and East Africa. In Central Africa, however, per
capita production and consumption rose; the gap between production and
consumption closed during the past decade. Per capita production and consumption of coarse grains in southern Africa declined slightly over the past
three decades, and the gap between production and consumption has not
been significant.

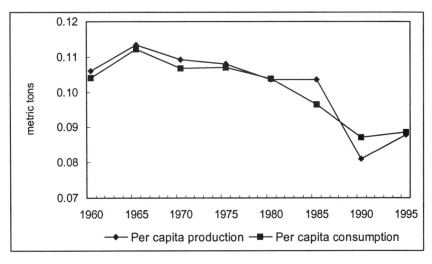

Figure 4.10. Coarse grains: Sub-Saharan Africa, per capita production and
            consumption
SOURCE: Based on data from USDA 1996b.

Asia. Asia's per capita consumption of coarse grains rose faster than per capita production during the 1960 to 1995 period (Table 4.A.3). In China, however, both per capita production and consumption increased at approximately the same rate. East Asia showed low and declining per capita production of coarse grains compared to per capita consumption, which rose sharply during the period, drastically increasing the gap between production and consumption (particularly in Japan and Taiwan). North Korea registered a significant decline in both per capita production and consumption after 1985, while South Korea experienced a sharp increase in per capita consumption and a decline in per capita production after 1970. Per capita consumption in Hong Kong has declined sharply since the late 1970s.

Both production and consumption of coarse grains rose in Southeast Asia. Per capita production was higher than per capita consumption until 1987, after which consumption exceeded production as per capita consumption increased and per capita production declined. The Philippines, however, experienced a decline in both per capita production and consumption of coarse grains in the 1990s. Singapore also saw a large decline in per capita consumption during the mid-1980s and the 1990s. Although production and consumption levels rose in South Asia, per capita production and consumption of coarse grains declined during the entire period. This was true for India and Bangladesh. Bangladesh witnessed a sharp decline in per capita consumption after the late 1970s.

Other Regions. Consumption levels of coarse grains in the Middle East rose sharply after 1975 (Table 4.A.3). Production also increased, but not as fast as consumption; therefore, there was a significant gap between production and consumption of coarse grains in the region. In per capita terms, there was a decline in production during the period from 1960 to 1995 while consumption rose sharply between the mid-1970s and the mid-1980s. Since then, the Middle East witnessed a decline in per capita consumption of coarse grains. Saudi Arabia saw a sharp increase in per capita consumption but, like the region as a whole, experienced a decline in consumption per capita since 1985.

Both per capita production and consumption of coarse grains in Eastern Europe showed an increase until the early 1980s and then a decline to 1960s levels by the 1990s (Table 4.A.3). Former Soviet Union countries also experienced sharp declines in per capita production and consumption of coarse grains in the 1990s. Prior to that, both per capita production and consumption rose; consumption was not significantly different from production in the early 1960s but was higher than production during the 1980s.

Food Availability and Distribution

We have considered the availability of food in terms of geographic per capita production and consumption of grains. Another aspect of food security relates to the distribution of the food available to the population in terms of per capita calories consumed (Mitchell, Ingco, and Duncan 1997). Although an overall increase in a country's supply of food through production and imports does lead, on average, to higher per capita calorie consumption, this does not imply that the whole population consumes an adequate diet. Thus, for a better understanding of the food situation, it is important to look at the proportion of the population that does not consume a sufficient amount of food. To do so, we utilize a measure of chronic undernourishment reported by the FAO. A person is defined to be chronically undernourished if his or her food intake is inadequate to meet dietary energy requirements[1] (FAO 1996).

The percentage of the total population in developing countries that is chronically undernourished dropped to approximately 20 percent in the 1990s, down from 35 percent in the 1970s. However, sub-Saharan Africa witnessed an increase from 38 percent in the 1970s to 43 percent in the early 1990s (FAO 1996). This makes sub-Saharan Africa the only region in the developing world that has a growing food security problem—exacerbated by unusual droughts and a series of wars and civil strife.

East and Southeast Asia have the largest numbers of undernourished people, although the proportion of the undernourished fell from 41 percent in the early 1970s to 16 percent by 1992, the largest improvement in the developing region. South Asia also experienced a rise in the proportion of undernourished people during the 1970s and the 1980s, from 33 to 34 percent, before the number fell to 22 percent in the 1990s.

During the decade from 1970 to 1980, the proportion of undernourished people fell in the Near East and North Africa from 27 to 12 percent. However, this percentage has not declined since then. In Latin America and the Caribbean, the percentage of chronically undernourished dropped from 19 percent in 1970 to 15 percent in 1992, although the actual numbers rose from 53 to 64 million during the same period due to population growth. The proportion of the population that is undernourished in Latin America and the Caribbean actually increased from 14 percent in the 1980s to 15 percent in the 1990s (FAO 1996).

In the 1990s, sub-Saharan Africa represented 26 percent of the total number of undernourished people in the developing world, up from 11 percent two decades earlier (Table 4.6). The share of the Near East and North Africa in the total number of undernourished people dropped from 5 to 4 percent during the same period. In the early 1990s, Asia as a whole (South, East, and Southeast) contained 62 percent of the undernourished people in the developing world. East Asia and Southeast Asia experienced a decline from 52 percent in the 1970s to 32 percent in the 1990s, while in South Asia the proportion rose from 26 to 30 percent. Latin America and the Caribbean also faced an increase from 6 to 8 percent over the 20-year period.

Table 4.6.   Distribution of undernourished by developing region, 1969–71 and 1990–92

Region	1969–71	1990–92
	(percent)	
Sub-Saharan Africa	11	26
Near East and North Africa	5	4
East and Southeast Asia	52	32
South Asia	26	30
Latin America and the Caribbean	6	8
	(million)	
Total of undernourished in the developing world	918	841

SOURCE: FAO 1996.

The average person in the developed world consumes one-third more calories than the average person in the developing world (Table 4.7). This number jumps to two-thirds when comparing industrialized countries to the least-developed countries. During the period from 1979 to 1981, the average person in sub-Saharan Africa consumed 2,080 kilocalories (kcal) per day. By 1992, sub-Saharan Africa's per capita dietary energy supply had dropped to 2,040 kcal per day, the lowest in the developing world.

In Latin America and the Caribbean, per capita dietary energy supply increased from 2,720 kcal per day during the period from 1979 to 1981 to 2,740 kcal per day during the early 1990s. The situation also improved in Asia where dietary energy supply per capita increased from 2,370 to 2,680 kcal per day in East Asia and Southeast Asia and from 2,070 to 2,290 kcal per day in South Asia during the same period. The Near East and North Africa currently have the highest per capita dietary energy supply in the developing world at 2,960 kcal per day, an increase from 2,850 kcal per day during the period from 1979 to 1981. However, these numbers are well below those in the developed and industrialized nations where, in 1992, the per capita dietary energy supply averaged 3,350 and 3,410 kcal per day, respectively (Table 4.7).

Table 4.7.   Per capita dietary energy supply by region and economic group,
              1979–81 and 1990–92

Region/economic group	Per capita dietary energy supply		Average annual rate of increase
	1979–81	1990–92	1979–81 to 1990–92
	(kcal/day)		(percent)
Developed countries[a]	3,280	3,350	0.2
Industrialized countries	3,220	3,410	0.5
Developing countries	2,330	2,520	0.7
Latin America and the Caribbean	2,720	2,740	0.0
Sub-Saharan Africa	2,080	2,040	-0.2
Near East and North Africa	2,850	2,960	0.3
East and Southeast Asia	2,370	2,680	1.1
South Asia	2,070	2,290	0.9
World	2,580	2,720	0.5

[a] Includes industrialized countries and transition economies.
NOTE: Per capita dietary energy supply is derived as a ratio of the total food supply to population size. The total food supply is based on information relating to food production, food products traded, wastage from the farm up to the retail level, stock changes, and nonfood uses of food products.
SOURCE: FAO 1996.

According to the FAO (1996), current levels of per capita dietary energy supplies in some developing countries have reached required levels and thus food inadequacy (i.e., energy deficiency relative to requirement standards) could be eliminated through redistribution. This is not the case, however, for the majority of developing countries, where increases in per capita food supplies are also necessary. The average per capita energy consumption of the undernourished population in sub-Saharan Africa was about 1,470 kcal per day for the period from 1990 to 1992, again the lowest in the developing region.

The undernourished in the rest of the developing region were consuming between 1,580 kcal per day (South Asia) and 1,660 kcal per day (East Asia and Southeast Asia, and Latin America and the Caribbean). In the Near East and North Africa, average per capita energy consumption for the undernourished population was 1,640 kcal per day in the early 1990s. The average per capita energy requirement for that period was approximately 2,170 kcal per day (FAO 1996).

Sources of Food Security Problems

A number of reasons have been advanced to explain why millions among the global population experience food insecurity. Rapid population and urban growth rates are among the major factors. High population growth rates have been associated with reduced per capita income and consumption, increased poverty, and food shortages. It has been argued that population growth may be the cause of the poverty facing many developing countries today and that a reduction in population growth rates may reduce poverty and, in turn, prevailing food problems (Dasgupta 1995). Furthermore, rising incomes and the dramatic increase in urban populations have resulted in increased demand for food from market sources and in dietary patterns that shift away from basic food staples toward higher-quality cereals (wheat and rice). It is estimated that by 2025 the populations of urban areas in the developing countries will increase from 1 billion to 4 billion (USDA 1996a).

Inadequate domestic agricultural production, uneven distribution of food, trade barriers, natural disasters, and war and civil strife are also factors contributing to food availability and access problems in many developing nations. Scarcity of land, low and declining investments in agricultural research and development, low investment in human capital, lack of adequate health and sanitation services, absence of productivity-enhancing technology, inadequate infrastructure, and high transportation costs have contributed to low productivity in many rural areas. This, in turn, has led to low and unstable levels of income (IDRC 1992). As a result of low productivity levels, food producers have become increasingly vulnerable to adverse weather conditions, pest infestations, and crop diseases. Marginal lands, environmental degradation, and the prevalence of subsistence farming have also contributed to low food production, particularly in sub-Saharan Africa (Christensen et al. 1981).

Certain economic policies, particularly structural adjustment policies implemented by developing countries, may have adversely affected the poor. One of the main criticisms of the structural adjustment policies is that they have not been designed to adapt fully to each individual country's economic situation. These policies have not targeted poverty alleviation or food security issues (Pinstrup-Andersen 1994). The effects and outcomes of development assistance in general and of food aid, in particular, on food security in developing regions have also been debated. Economists have argued that foreign aid to developing nations has resulted in the distortion of domestic markets and thus has exacerbated food problems.

However, the consensus is that for many developing nations food problems are closely associated with the persistence of poverty (see chapter 5). Poverty is the major cause of food insecurity for most of the world's at-risk populations. Some have even argued that it is poverty that is the major contributor to the high population growth rates prevalent in most developing countries and not high population growth rates that lead to poverty. In many cases, the poor do not have the purchasing power to buy food even when the country is not facing a national food supply problem. To improve human nutrition, it is crucial for developing countries to aim not only at increasing food supplies but also at implementing policies that increase the income levels of the poorer populations. In chapter 7, Muhammad Yunus, the 1994 World Food Prize laureate and founder of the Grameen Bank in Bangladesh, contends that providing access to credit for self-employment and private investment is essential for raising the levels of income of the poor and improving food security.

More than one-half of the 1.1 billion people living in poverty in developing countries in 1990 lived in conditions of *extreme poverty*. Fifty percent of the poor in the developing world live in South Asia, 15 percent in East Asia, 19 percent in sub-Saharan Africa, and 10 percent in Latin America and the Caribbean. In both South Asia and Africa, 50 percent of the population is poor. It is projected that there will be significant reductions in poverty in both East Asia and South Asia. In sub-Saharan Africa, however, the number of poor could rise to 304 million by 2000, an increase of almost 50 percent. By the same year, the percentage of the developing world's poor who reside in sub-Saharan Africa is predicted to rise from the current 19 percent to approximately 33 percent (Pinstrup-Andersen 1994).

It is therefore evident that sub-Saharan Africa requires special attention. The food situation in this region has been deteriorating for the past three decades, and continued deterioration is projected unless current policies are reformed. Most sub-Saharan African countries have low per capita calorie intake and declining per capita income levels. Low food production and rapid population and urban growth have been major causes of rising demand for food imports. Since the 1980s average population growth rates (at 3 percent per year) have outstripped the average growth rate in food production (at 2 percent per year) in the region (Mitchell, Ingco, and Duncan 1997). Regional population numbers are expected to increase by 160 percent between 1990 and 2025, compared to 107 percent in the Middle East and North Africa, 56 percent in Asia, 59 percent in Latin America, and 60 percent in the world (Bongaarts 1995).

Food imports in Africa have risen recently both to meet rising demand and as a result of declining world food prices. In addition to lower international prices for wheat and rice, imported food was cheaper relative to domestic staples, a consequence of appreciated real exchange rates. Furthermore, food was imported because the high domestic cost of transportation from production to consumption points often made it less expensive to import food than to encourage domestic production (Jaeger 1992).

Changes in dietary preferences, especially in urban areas, have contributed dramatically to higher demands for imported wheat and rice in Africa. These crops are more difficult to grow in the sub-Saharan region than other food crops, thus leading to higher import levels. The reasons for the growth in food demand, however, differ among regions of the developing world. In Latin America, for example, the growth in food consumption has been mainly the result of rising incomes.

The framework and stability of political institutions are important, but often overlooked, factors contributing to food insecurity in many developing countries. A number of studies show that political instability has a negative effect on economic growth, which is a necessary though insufficient condition for reducing poverty and hunger (Fosu 1992; Beghin and Kherallah 1994; McMillan, Rausser, and Johnson 1994). Democratic systems are seen as facilitating reform while nondemocratic systems have hindered development in many regions—for example, in Eastern Europe and South America as well as in countries such as Haiti and Zaire.

An analysis by Fosu (1992) concluded that political instability in developing countries may have contributed greatly to their stagnating economies and to the low or declining growth rates of their gross domestic product (GDP). The average per capita GDP in sub-Saharan Africa, a highly politically unstable region, grew by only $73 in relation to purchasing power parity between 1970 and 1992, compared to $420 in South Asia and $900 in East Asia (The World Bank 1996). Although an evaluation of economic reform based solely on differing political systems is incomplete, evidence has shown that there is a positive link between the level of democracy and economic development. Improvements in social indicators such as higher investments in education and human capital have also been associated with freer political systems.

In general, the agricultural sectors in developing countries have been heavily taxed while those of developed nations have been protected. There is a high political cost of subsidizing agriculture in developing countries because of an "urban bias" (urban areas being the main source of political power). In their study examining the influence of political systems and civil rights on the patterns of agricultural protection, Beghin and Kherallah (1994) found that the institutional setting (party system and civil liberties) influenced agricultural policy. In countries with more democratic systems, there was a high level of agricultural assistance. Democratic institutions were held politically accountable and therefore tended to dissipate rent. As the level of democracy increased (within a certain range), the agricultural sector was able to secure more protection.

Political instability affects production because the lack of both political stability and democracy has a negative effect on investment in technologies that lead to expansions in production. This negative effect has a significant impact on productivity in the agricultural sector and ultimately on food security. In terms of food, the extent of the power relationships exercised by the government, agro-industry, and market forces influences production decisions that, in turn, influence the entire food chain. Government policy decisions concerning food are affected not only by production, consumption, and trade considerations but also by the political power of producers, consumers, and other groups (van de Walle 1994).

According to Amartya Sen (1994), hunger must be viewed in terms of the failure of people's entitlements, i.e., people's inability to establish command over adequate quantities of food and other necessities. For this reason, hunger can and does exist in times of plentiful national supply of food. Famine prevention policies are imperative and, according to Sen, democratic governments are more apt to undertake these policies because they face reelections. For example, as a result of quick public action certain democratic African nations, such as Botswana and Zimbabwe, were able to avoid famines despite serious food problems caused by droughts during the early 1980s. On the other hand, countries that did not have democratic systems, such as Sudan and Ethiopia, were not as fortunate although they faced smaller declines in food output during the same period. Thus, to address the issue of food security, the political as well as the economic and scientific aspects must be considered.

Conclusion

Global food production is rising and is increasing faster than population growth. However, it is evident that in many developing countries food consumption exceeds food production. Some of these countries are unable to fill this gap with food imports due to budgetary problems and a scarcity of foreign exchange. Annual growth rates in global consumption of grains outpaced annual growth rates in production during the 1960s and 1970s but not during the 1980s (Table 4.8). It is expected that the annual growth rates in yields, although relatively high until recently, will fall during the 1990s and into the twenty-first century. This may pose a significant problem in terms of food security.

Historical data indicate that the "availability gaps" among the regions of the developing world have become more pronounced. The assembled data show that sub-Saharan Africa and Asia (South Asia in particular) are the two regions in the developing world that have the world's poorest

Table 4.8.   Actual and projected annual growth rates in grain production, consumption, and yields

Period	Production	Consumption	Yields
		(percent)	
1960–1970	2.7	3.1	2.4
1970–1980	2.8	2.7	2.0
1980–1990	2.1	1.7	2.5
1990–2000	1.1	1.4	0.9
2000–2010	1.4	1.4	1.2

SOURCE: Adapted from Mitchell, Ingco, and Duncan 1997.

populations and highest proportions of undernourished people (a combined total of 88 percent, Table 4.6). In both sub-Saharan Africa and South Asia, the proportion of undernourished increased during the past two decades. This is a clear indication that things are getting worse for a large segment of the developing world's population. Unless appropriate policies are implemented, poverty will continue to rise, mostly in the African continent, but especially in sub-Saharan Africa, and regional food crises seem inevitable.

According to Mitchell, Ingco, and Duncan (1997), net grain imports in developing countries are projected to increase from 87 million tons in 1990 to 210 million tons by 2010. Other projections show that there could be a net deficit of 70 million metric tons of grains in developing countries by 2000, if production patterns and per capita income trends remain unchanged. The largest shortfall is expected in North Africa and the Middle East (60 million metric tons), followed by sub-Saharan Africa (50 million metric tons) and Latin America (10 million metric tons). Asia, as a whole, is expected to have a net food surplus of 50 million metric tons (IDRC 1992).

The food shortage in the African continent has been estimated to total 250 million metric tons by 2020 (Pinstrup-Andersen 1994). The number of undernourished is expected to increase by 100 million in sub-Saharan Africa by 2010 (USDA 1996a). Unlike Asia, where higher consumption will likely be covered by an increase in imports, the African region may not be able to meet food needs through higher imports. To meet the expected rise in food requirements, sub-Saharan Africa will need to increase agricultural production from 2 percent per annum (during 1970 to 1990) to 4 percent per annum for the next 20 years (USDA 1996a). Furthermore, because of reductions in assistance programs, international food aid is not expected to fill the gap. The situation in Latin America, although critical in some areas, is expected to improve as growth in food production exceeds population growth in the future.

There are differing views on why a large number of people in many developing nations continue to face food security problems. Poverty, high population growth rates, urbanization, inappropriate economic policies, low agricultural productivity, and low investments in agricultural research and development are among the major reasons. Political rights and civil liberties as well as political instability have become increasingly important in discussions of solutions to the food security problems in the developing world. It is safe to suggest that a combination of all of these factors has contributed to the deteriorating food security situation in many food-deficit countries. However, the debate continues on the priority given to, and the extent of, the effects of each of these factors in the food security equation.

Many developing countries must make drastic policy changes in order to solve the problems of food availability and food access. Governments must address food problems through policies leading to poverty reduction; lower population and urban growth rates; increased investment for agricultural research and development; higher investments in human capital through education and better health and sanitation services; increased employment opportunities and access to credit; better infrastructure; and natural resource conservation. The shifting urban consumer tastes, the prevalence of a large subsistence sector, particularly in Africa, and the impact on the environment also need to be considered in effective policy decisions.

Political institutions cannot be ignored in addressing issues of food security in developing countries and in attempting to combat and alleviate both food insecurity and poverty. Developing countries require reforms that lead to economic growth that is not only sustainable but also targeted toward the poor. Only then can economic development that reaches the masses be achieved and, in turn, poverty and food insecurity drastically reduced or eliminated. Increases in food production are important to reduce domestic imbalances, but increased incomes for the poor are the real key to food security. Food imports and economic specialization can support food security in an environment of growing incomes for the poor.

Appendix

Table 4.A.1.   Wheat production, consumption, and trade

Year	Production	Consumption	Per capita production	Per capita consumption	Net imports	Ending stocks	Feed
Latin America							
1960	6,368	9,476	0.030	0.045	2,640	1,098	270
1965	10,527	13,211	0.043	0.054	-197	1,232	474
1970	10,869	15,630	0.039	0.056	4,355	1,576	232
1975	15,073	20,714	0.048	0.066	5,397	1,931	1,308
1980	14,735	21,680	0.042	0.061	7,231	2,363	472
1985	19,529	24,344	0.050	0.062	4,178	1,843	1,526
1990	20,726	25,916	0.048	0.060	5,645	1,651	887
1995	15,802	26,856	0.033	0.056	10,470	1,873	655
Central America							
1960	21	231	0.002	0.019	212	9	0
1965	27	324	0.002	0.023	307	31	0
1970	32	406	0.002	0.024	382	53	0
1975	41	430	0.002	0.022	378	34	0
1980	42	590	0.002	0.027	568	104	0
1985	53	666	0.002	0.027	623	94	3
1990	23	855	0.001	0.030	848	165	0
1995	22	1,030	0.001	0.032	1,000	167	110
South America (excluding Brazil)							
1960	847	2,001	0.016	0.037	1,168	31	102
1965	2,111	3,803	0.034	0.061	1,821	393	300
1970	2,034	4,212	0.029	0.059	2,100	429	151
1975	1,774	4,814	0.022	0.060	2,955	421	206
1980	1,587	5,303	0.018	0.059	3,626	752	222
1985	2,276	5,642	0.023	0.056	3,370	685	356
1990	2,603	6,047	0.023	0.054	3,267	398	87
1995	2,069	7,155	0.017	0.058	4,885	601	195
Caribbean							
1960	0	462	0	0.028	462	0	0
1965	0	1,350	0	0.073	1,350	0	0
1970	0	1,167	0	0.057	1,167	18	0
1975	0	1,502	0	0.067	1,504	28	0
1980	0	1,737	0	0.072	1,747	37	0
1985	0	2,196	0	0.084	2,199	46	0
1990	0	2,179	0	0.078	2,179	14	0
1995	0	1,705	0	0.057	1,705	20	50
Africa							
1960	6,039	8,318	0.023	0.032	2,476	288	102
1965	6,178	10,884	0.021	0.037	4,590	578	111
1970	7,980	14,508	0.024	0.044	6,695	829	146
1975	9,773	18,026	0.026	0.048	9,495	3,664	182
1980	8,919	24,096	0.021	0.056	14,874	1,891	245
1985	10,332	27,866	0.021	0.056	17,284	1,702	351
1990	13,737	32,834	0.024	0.057	19,777	3,004	1,496
1995	12,662	33,303	0.019	0.050	18,500	2,688	170

Table 4.A.1.   (continued)

Year	Production	Consumption	Per capita production	Per capita consumption	Net imports	Ending stocks	Feed
North Africa							
1960	4,405	5,981	0.079	0.107	1,773	288	83
1965	4,491	7,928	0.071	0.126	3,340	207	94
1970	5,174	10,003	0.072	0.140	4,741	160	116
1975	6,581	12,861	0.082	0.160	7,439	2,819	160
1980	6,129	16,956	0.067	0.186	10,772	1,086	140
1985	7,088	19,469	0.068	0.186	12,381	925	200
1990	9,952	23,493	0.084	0.199	14,178	2,178	1,350
1995	8,250	24,156	0.062	0.183	13,800	1,940	160
Sub-Saharan Africa							
1960	865	1,495	0.005	0.008	630	0	19
1965	1,031	2,099	0.005	0.010	1,088	93	17
1970	1,410	3,222	0.006	0.014	1,797	150	30
1975	1,400	3,421	0.005	0.013	2,087	327	19
1980	1,320	5,147	0.004	0.016	3,860	485	10
1985	1,564	6,197	0.004	0.017	4,675	349	45
1990	2,083	7,046	0.005	0.017	5,071	506	3
1995	2,287	6,597	0.005	0.014	4,250	290	0
Asia							
1960	39,447	52,747	0.025	0.033	12,416	7,280	821
1965	46,482	65,450	0.026	0.037	20,910	9,111	914
1970	60,077	76,456	0.030	0.038	17,959	14,541	1,078
1975	81,348	96,389	0.037	0.043	23,319	31,625	1,130
1980	103,381	141,104	0.042	0.058	28,620	39,884	2,172
1985	148,082	178,514	0.056	0.067	23,073	51,744	4,082
1990	167,845	196,621	0.058	0.067	29,181	37,125	5,332
1995	187,365	219,225	0.059	0.070	34,780	41,530	5,363
East Asia							
1960	1,815	5,459	0.013	0.038	3,769	805	401
1965	1,582	6,445	0.010	0.041	4,843	1,020	234
1970	762	8,017	0.004	0.047	7,298	1,285	338
1975	432	8,712	0.002	0.047	8,732	1,886	80
1980	776	9,454	0.004	0.048	8,936	2,024	152
1985	1,016	10,997	0.005	0.053	9,763	2,020	1,196
1990	1,073	11,776	0.005	0.055	11,341	2,459	1,737
1995	523	10,500	0.002	0.047	9,775	1,811	673
Southeast Asia							
1960	10	914	0.000	0.004	914	51	0
1965	97	1,037	0.000	0.004	985	156	0
1970	40	2,275	0.000	0.008	2,289	336	0
1975	57	2,498	0.000	0.008	2,452	217	0
1980	116	3,747	0.000	0.011	3,694	534	20
1985	190	3,296	0.000	0.008	3,049	465	86
1990	135	5,139	0.000	0.012	5,324	753	425
1995	150	8,825	0.000	0.019	8,675	525	590

Table 4.A.1.   (continued)

Year	Production	Consumption	Per capita production	Per capita consumption	Net imports	Ending stocks	Feed
South Asia							
1960	16,662	22,467	0.029	0.039	5,786	3,424	20
1965	19,292	26,613	0.030	0.041	8,738	3,735	30
1970	29,840	33,466	0.041	0.046	4,609	5,720	40
1975	35,183	41,288	0.043	0.051	9,920	7,822	100
1980	47,072	51,657	0.052	0.057	2,161	5,626	400
1985	60,377	63,122	0.059	0.062	3,661	18,559	500
1990	67,808	73,047	0.060	0.064	3,080	10,400	470
1995	86,192	86,400	0.068	0.069	3,350	17,499	900
Other Regions							
Middle East							
1960	11,050	13,181	0.143	0.171	2,369	1,332	158
1965	13,154	14,787	0.149	0.167	1,395	1,580	127
1970	14,131	17,839	0.139	0.175	3,772	2,004	297
1975	20,082	22,436	0.170	0.190	4,397	5,388	485
1980	23,125	28,833	0.169	0.210	6,040	9,339	1,355
1985	24,794	33,527	0.152	0.205	8,447	8,855	1,536
1990	30,977	36,999	0.162	0.194	8,163	13,018	1,435
1995	34,831	42,981	0.160	0.197	7,970	9,671	1,700
East Europe							
1960	15,047	19,191	0.151	0.193	3,455	1,581	2,850
1965	20,123	24,073	0.194	0.232	4,148	1,138	5,112
1970	20,908	26,089	0.194	0.242	4,723	1,687	7,355
1975	25,974	29,880	0.232	0.266	2,330	528	10,121
1980	31,504	34,930	0.270	0.299	3,530	1,343	11,852
1985	33,209	33,148	0.276	0.276	-291	1,898	11,464
1990	41,274	40,005	0.337	0.327	-451	4,612	16,100
1995	34,990	31,830	0.287	0.261	-2,950	6,125	11,300
Former Soviet Union							
1960	59,350	53,915	0.276	0.251	-4,435	3,000	12,452
1965	55,677	70,595	0.241	0.305	5,918	5,000	23,576
1970	92,601	93,882	0.381	0.387	-6,719	6,000	43,478
1975	61,826	81,426	0.243	0.320	9,600	3,000	33,478
1980	91,485	105,985	0.345	0.400	15,500	8,000	53,085
1985	72,575	86,075	0.262	0.311	15,200	19,200	39,447
1990	101,891	111,052	0.353	0.385	14,649	30,272	60,454
1995	59,553	74,295	0.204	0.254	5,490	9,864	28,182

NOTES: All production, trade, and use data are in 1,000 metric tons. Commodity years listed are the first year of the marketing year (e.g., 1980 represents 1980/81). Production is rough production. Consumption is total domestic use during the local marketing year and is the sum of feed, food, seed, and industrial uses. Imports and exports are the aggregate of local marketing year trade of the individual countries of the region. Stocks are at the end of the local marketing year.
SOURCE: USDA 1996b.

Table 4.A.2.   Rice production, consumption, and trade

Year	Production	Consumption	Per capita production	Per capita consumption	Net imports	Ending stocks	Rough production
Latin America							
1960	5,441	5,429	0.026	0.026	-9	682	8,134
1965	5,944	5,815	0.024	0.024	-147	747	8,880
1970	6,327	7,352	0.023	0.026	7	642	9,480
1975	9,830	8,598	0.031	0.027	-191	2,157	14,727
1980	10,251	10,932	0.029	0.031	155	1,711	15,390
1985	11,724	11,385	0.030	0.029	953	3,101	17,544
1990	11,509	12,994	0.027	0.030	1,486	2,099	17,184
1995	12,454	14,499	0.026	0.030	1,593	1,933	18,707
Central America							
1960	158	178	0.013	0.014	20	41	248
1965	214	229	0.015	0.016	15	41	333
1970	227	267	0.014	0.016	45	32	355
1975	396	344	0.021	0.018	6	117	619
1980	368	385	0.017	0.017	9	56	568
1985	486	461	0.019	0.018	-19	111	741
1990	452	503	0.016	0.018	66	81	692
1995	439	606	0.013	0.019	160	62	674
South America (excluding Brazil)							
1960	962	861	0.018	0.016	-89	249	1,499
1965	1,225	1,132	0.020	0.018	-77	304	1,894
1970	1,563	1,362	0.022	0.019	-184	409	2,404
1975	2,415	1,934	0.030	0.024	-222	582	3,720
1980	2,916	2,943	0.032	0.032	-85	801	4,514
1985	3,056	2,696	0.030	0.027	-432	695	4,679
1990	3,098	3,059	0.028	0.027	-156	372	4,708
1995	4,040	3,798	0.033	0.031	-393	527	6,215
Caribbean							
1960	338	563	0.020	0.034	225	22	518
1965	201	428	0.011	0.023	227	22	308
1970	412	764	0.020	0.038	352	79	634
1975	558	853	0.025	0.038	317	109	859
1980	647	878	0.027	0.037	227	69	994
1985	726	978	0.028	0.038	229	71	1,126
1990	660	1,192	0.024	0.043	510	87	1,024
1995	555	1,150	0.019	0.039	570	87	863
Africa							
1960	3,222	3,424	0.012	0.012	202	89	3,687
1965	3,716	4,072	0.012	0.013	356	89	5,662
1970	4,828	5,112	0.014	0.014	325	124	7,339
1975	5,172	5,721	0.013	0.014	618	280	7,876
1980	5,273	7,525	0.011	0.016	2,263	375	8,020
1985	5,824	8,543	0.011	0.016	2,735	234	8,973
1990	6,793	9,262	0.011	0.015	2,594	509	10,527
1995	7,210	10,364	0.010	0.015	2,972	556	11,436

Table 4.A.2.   (continued)

Year	Production	Consumption	Per capita production	Per capita consumption	Net imports	Ending stocks	Rough production
North Africa							
1960	1,015	826	0.020	0.016	-189	14	1,514
1965	1,216	881	0.021	0.015	-335	14	1,815
1970	1,772	1,284	0.027	0.019	-489	7	2,645
1975	1,643	1,488	0.022	0.020	-153	7	2,452
1980	1,616	1,554	0.019	0.018	-58	9	2,412
1985	1,551	1,628	0.016	0.017	76	7	2,315
1990	2,126	1,952	0.019	0.018	-25	152	3,173
1995	2,121	2,226	0.017	0.018	55	233	3,419
Sub-Saharan Africa							
1960	2,207	2,547	0.011	0.012	340	75	2,173
1965	2,500	3,134	0.011	0.013	634	75	3,847
1970	3,056	3,774	0.011	0.014	760	117	4,694
1975	3,529	4,154	0.012	0.014	692	273	5,424
1980	3,657	5,845	0.010	0.017	2,195	366	5,608
1985	4,273	6,737	0.010	0.017	2,481	227	6,658
1990	4,667	7,018	0.010	0.015	2,324	347	7,354
1995	5,089	7,718	0.009	0.014	2,517	258	8,017
Asia							
1960	138,598	143,643	0.087	0.090	-658	9,125	203,614
1965	158,516	157,121	0.090	0.089	-508	16,651	232,181
1970	195,898	191,719	0.098	0.096	-921	26,908	287,171
1975	220,279	211,334	0.099	0.095	-171	34,492	323,377
1980	244,887	245,701	0.100	0.100	-3,721	44,885	359,921
1985	291,188	288,621	0.109	0.108	-5,434	47,556	427,238
1990	321,577	310,768	0.110	0.106	-5,817	54,616	474,530
1995	339,248	332,124	0.107	0.105	-5,678	44,911	501,729
East Asia							
1960	17,588	18,186	0.122	0.126	465	1,917	24,378
1965	18,186	18,756	0.116	0.120	910	1,863	25,243
1970	19,246	20,602	0.113	0.121	260	6,421	26,832
1975	20,734	19,516	0.113	0.106	441	4,500	28,918
1980	16,699	19,111	0.085	0.097	1,291	6,235	23,591
1985	20,351	20,298	0.099	0.098	257	3,392	28,289
1990	18,622	19,113	0.087	0.089	507	3,944	25,750
1995	17,284	17,970	0.078	0.081	1,225	3,835	23,774
Southeast Asia							
1960	31,686	29,990	0.141	0.134	-1,766	1,127	48,176
1965	34,286	33,072	0.136	0.131	-1,167	1,707	52,200
1970	41,218	40,919	0.144	0.143	-636	2,715	62,626
1975	44,721	45,164	0.138	0.140	60	3,177	68,415
1980	56,227	52,510	0.156	0.146	-2,920	6,122	85,306
1985	67,066	62,710	0.168	0.157	-3,780	6,503	101,748
1990	70,678	66,986	0.160	0.152	-4,080	5,665	109,853
1995	83,784	78,134	0.173	0.162	-5,170	4,149	130,212

Table 4.A.2.   (continued)

Year	Production	Consumption	Per capita production	Per capita consumption	Net imports	Ending stocks	Rough production
South Asia							
1960	47,513	49,084	0.083	0.085	1,071	3,081	71,330
1965	44,639	45,875	0.069	0.071	1,236	7,081	67,017
1970	58,441	57,989	0.081	0.080	739	6,772	87,723
1975	66,932	61,524	0.082	0.075	90	7,315	100,484
1980	74,027	72,493	0.081	0.080	-1,745	7,528	111,118
1985	85,772	83,219	0.084	0.082	-1,306	10,161	128,631
1990	99,745	97,944	0.088	0.086	-1,622	16,787	149,596
1995	105,180	105,020	0.084	0.083	-2,283	13,385	157,743
Other Regions							
Middle East							
1960	663	1,000	0.009	0.013	327	39	999
1965	933	1,243	0.011	0.014	306	44	1,406
1970	986	1,541	0.010	0.015	596	101	1,486
1975	996	2,073	0.008	0.018	1,031	259	1,501
1980	1,248	2,970	0.009	0.022	1,807	312	1,894
1985	1,446	3,487	0.009	0.021	2,001	151	2,193
1990	1,626	4,248	0.009	0.022	2,252	129	2,465
1995	2,260	5,215	0.010	0.024	2,675	873	3,403
East Europe							
1960	94	263	0.001	0.003	169	11	148
1965	80	322	0.001	0.003	242	11	125
1970	133	393	0.001	0.004	242	33	209
1975	154	408	0.001	0.004	243	24	241
1980	109	369	0.001	0.003	254	33	171
1985	178	354	0.002	0.003	176	0	278
1990	106	266	0.001	0.002	160	0	164
1995	37	232	0.000	0.002	195	0	58
Former Soviet Union							
1960	97	116	0.000	0.001	19	0	149
1965	282	553	0.001	0.003	271	0	434
1970	747	1,005	0.003	0.004	258	0	1,149
1975	1,173	1,370	0.005	0.006	197	0	1,805
1980	1,629	2,786	0.007	0.011	1,157	0	2,506
1985	1,500	1,794	0.006	0.007	294	0	2,308
1990	1,406	1,640	0.005	0.006	234	0	2,163
1995	948	1,103	0.003	0.004	155	0	1,461

NOTES: All production, trade, and use data are in 1,000 metric tons. Commodity years listed are the first year of the marketing year (e.g., 1980 represents 1980/81). Production is on a milled basis. Rough production is paddy weight for rice. Consumption is total domestic use during the local marketing year and is the sum of feed, food, seed, and industrial uses. Imports and exports are the aggregate of local marketing year trade of the individual countries of the region. Stocks are at the end of the local marketing year.
SOURCE: USDA 1996b.

Table 4.A.3.   Coarse grains production, consumption, and trade

Year	Production	Consumption	Per capita production	Per capita consumption	Net imports	Ending stocks	Rough production
Latin America							
1960	28,339	25,962	0.133	0.122	-2,270	3,186	11,854
1965	36,883	31,598	0.151	0.129	-6,419	4,138	14,824
1970	48,554	38,447	0.175	0.138	-9,139	4,089	18,836
1975	51,606	46,480	0.164	0.148	-4,766	3,263	25,571
1980	67,374	59,484	0.191	0.169	-3,731	7,613	35,159
1985	62,835	61,970	0.160	0.158	-2,247	3,427	38,568
1990	64,879	68,088	0.150	0.157	3,190	5,251	41,788
1995	76,850	89,573	0.162	0.188	7,469	5,093	59,472
Central America							
1960	1,488	1,545	0.121	0.126	4	140	349
1965	1,707	1,731	0.119	0.121	46	209	424
1970	2,004	2,002	0.120	0.120	14	128	445
1975	2,436	2,346	0.127	0.122	86	221	534
1980	2,539	2,807	0.115	0.127	289	352	788
1985	2,963	3,122	0.119	0.125	146	334	839
1990	3,189	3,694	0.112	0.130	554	312	1,357
1995	3,390	4,167	0.104	0.128	790	143	1,689
South America (excluding Brazil)							
1960	2,874	2,897	0.054	0.054	43	85	845
1965	3,730	3,814	0.060	0.062	92	64	1,255
1970	4,221	5,148	0.060	0.073	810	245	1,873
1975	4,982	6,110	0.062	0.076	1,150	407	2,724
1980	5,503	7,278	0.061	0.081	2,225	1,137	3,111
1985	6,480	8,064	0.064	0.080	1,545	594	4,479
1990	7,146	8,952	0.064	0.080	1,467	643	4,713
1995	7,323	11,519	0.060	0.094	3,794	609	6,892
Caribbean							
1960	547	597	0.033	0.036	50	0	32
1965	563	836	0.031	0.046	273	0	36
1970	744	1,091	0.037	0.054	346	4	70
1975	420	1,013	0.019	0.045	593	18	268
1980	661	1,530	0.028	0.064	866	32	637
1985	347	1,306	0.013	0.051	916	14	501
1990	380	1,224	0.014	0.044	857	107	723
1995	381	1,346	0.013	0.045	955	90	926
Africa							
1960	32,683	30,145	0.118	0.109	-2,110	1,221	1,124
1965	37,115	36,198	0.118	0.115	-385	1,107	2,412
1970	44,135	39,952	0.123	0.112	-3,297	2,140	2,417
1975	47,542	46,792	0.117	0.115	-772	4,544	5,512
1980	59,034	53,597	0.126	0.115	-1,597	9,077	8,957
1985	60,173	58,758	0.112	0.109	1,029	7,352	14,518
1990	56,270	64,343	0.091	0.104	5,149	3,503	18,426
1995	67,365	71,694	0.094	0.100	3,117	4,040	17,179

Table 4.A.3.   (continued)

Year	Production	Consumption	Per capita production	Per capita consumption	Net imports	Ending stocks	Rough production
North Africa							
1960	5,336	5,390	0.096	0.097	54	0	0
1965	5,299	5,430	0.084	0.086	190	169	0
1970	6,035	6,289	0.084	0.088	115	116	0
1975	7,330	7,828	0.091	0.098	781	2,346	1,794
1980	7,852	9,795	0.086	0.107	2,075	2,226	4,271
1985	9,246	13,059	0.088	0.125	4,037	737	10,035
1990	9,418	14,821	0.080	0.125	4,983	704	12,776
1995	8,526	15,163	0.064	0.115	5,275	645	11,713
Sub-Saharan Africa							
1960	21,595	21,175	0.106	0.104	-430	108	94
1965	26,255	25,999	0.113	0.112	-171	195	724
1970	28,798	28,118	0.109	0.107	-564	305	410
1975	32,482	32,166	0.108	0.107	-23	1,088	568
1980	35,879	36,006	0.104	0.104	1,258	2,081	949
1985	41,550	38,647	0.104	0.096	-50	5,326	1,269
1990	37,692	40,637	0.081	0.087	641	1,799	1,124
1995	47,546	47,918	0.088	0.089	322	2,295	1,094
Asia							
1960	63,979	69,773	0.040	0.044	2,374	8,484	11,060
1965	78,184	86,119	0.044	0.049	5,150	12,020	21,930
1970	100,709	109,368	0.050	0.055	10,062	22,576	24,066
1975	117,237	126,168	0.053	0.057	14,853	38,292	33,199
1980	127,209	156,348	0.052	0.064	25,015	39,388	59,224
1985	128,946	160,541	0.048	0.060	21,035	29,464	83,910
1990	168,041	186,006	0.058	0.064	27,943	31,925	99,812
1995	173,353	218,773	0.055	0.069	45,175	34,194	142,827
East Asia							
1960	5,037	7,158	0.035	0.050	2,100	565	2,781
1965	4,433	9,608	0.028	0.061	5,351	1,166	5,083
1970	3,994	15,750	0.024	0.093	11,822	2,169	10,422
1975	4,260	20,950	0.023	0.114	16,589	3,525	14,330
1980	3,928	29,418	0.020	0.150	25,325	3,919	22,059
1985	4,485	33,864	0.022	0.164	29,905	2,958	25,768
1990	3,926	37,274	0.018	0.174	33,573	4,479	27,914
1995	2,936	40,706	0.013	0.184	37,285	4,317	31,852
Southeast Asia							
1960	4,534	3,932	0.020	0.018	-580	454	570
1965	5,115	4,008	0.020	0.016	-1,193	243	598
1970	7,148	5,352	0.025	0.019	-1,736	795	998
1975	9,297	6,888	0.029	0.022	-2,053	660	2,099
1980	11,391	10,459	0.032	0.029	-942	449	4,370
1985	14,896	12,260	0.038	0.031	-2,444	1,061	5,584
1990	15,436	15,760	0.035	0.036	590	1,176	8,936
1995	14,815	20,070	0.031	0.042	5,095	1,094	13,620

Table 4.A.3.   (continued)

Year	Production	Consumption	Per capita production	Per capita consumption	Net imports	Ending stocks	Rough production
South Asia							
1960	26,892	26,592	0.047	0.046	192	4,952	404
1965	24,993	27,054	0.039	0.042	1,163	4,670	443
1970	34,051	36,482	0.047	0.050	7	5,095	951
1975	34,041	32,536	0.042	0.040	667	4,300	1,544
1980	32,133	32,064	0.035	0.035	-19	1,700	1,728
1985	29,655	30,445	0.029	0.030	-10	600	2,372
1990	36,994	37,224	0.033	0.033	0	1,020	4,060
1995	33,962	33,957	0.027	0.027	-5	420	5,835
Other Regions							
Middle East							
1960	7,563	7,996	0.099	0.105	649	572	4,594
1965	9,384	9,734	0.107	0.111	166	611	4,856
1970	9,006	10,825	0.089	0.107	1,273	357	5,683
1975	11,929	13,642	0.102	0.117	2,063	916	7,370
1980	12,600	17,764	0.093	0.131	5,694	2,168	13,311
1985	14,587	26,166	0.090	0.162	11,839	2,850	21,071
1990	17,136	26,116	0.091	0.138	9,407	5,716	20,652
1995	18,444	28,752	0.086	0.134	9,015	3,213	23,120
East Europe							
1960	38,028	38,194	0.383	0.384	22	1,290	25,692
1965	37,069	38,163	0.357	0.367	1,152	1,536	26,715
1970	39,429	42,803	0.366	0.397	2,182	1,476	32,303
1975	53,212	55,753	0.474	0.497	3,327	2,273	44,325
1980	55,730	62,832	0.477	0.538	6,263	3,511	50,208
1985	57,305	60,298	0.477	0.502	1,872	5,229	48,579
1990	51,361	56,063	0.419	0.458	2,673	5,446	42,311
1995	50,989	48,470	0.418	0.397	-2,270	3,556	37,891
Former Soviet Union							
1960	52,683	50,875	0.245	0.237	-1,808	5,000	33,216
1965	48,698	50,472	0.211	0.218	-2,226	8,000	31,837
1970	70,357	69,548	0.290	0.286	-809	6,000	50,186
1975	60,203	78,753	0.237	0.309	15,550	5,000	57,853
1980	73,678	92,678	0.278	0.350	18,000	3,000	65,319
1985	91,685	105,085	0.331	0.380	13,700	9,200	81,966
1990	103,329	113,699	0.358	0.394	11,405	11,743	88,984
1995	59,688	66,804	0.204	0.229	975	6,350	39,917

NOTES: Coarse grains include corn, barley, sorghum, millet, rye, oats, and mixed grains. All production, trade, and use data are in 1,000 metric tons. Commodity years listed are the first year of the marketing year (e.g., 1980 represents 1980/81). Production is rough production. Consumption is total domestic use during the local marketing year and is the sum of feed, food, seed, and industrial uses. Imports and exports are the aggregate of local marketing year trade of the individual countries of the region. Stocks are at the end of the local marketing year.
SOURCE: USDA 1996b.

Note

1. The FAO measure of chronic undernourishment is based on actual household survey data on food consumption or expenditures for only 18 of the 99 countries for which it is reported. Due to the methodology employed for estimating undernourishment rates for the remaining 81 countries, the measure is highly correlated with countries' per capita dietary energy supplies; it does not fully capture the effects of household incomes on people's ability to acquire adequate calories to meet their requirements. Note, in particular, that the estimated changes in prevalence rates over time are driven solely by changes in countries' per capita dietary energy supplies (Smith 1997).

References

Angel, Bruna, and Stanley R. Johnson. 1992. Changing International Food Markets in the 1990s: Implications for Developing Countries. In *World Food in the 1990s: Production, Trade and Aid,* edited by Lehman B. Fletcher. Boulder, CO: Westview Press, 43–129.

Beghin, John C., and Mylene Kherallah. 1994. Political Institutions and International Patterns in Agricultural Protection. *Review of Economics and Statistics* 76(3):482–89.

Bongaarts, John. 1995. Global and Regional Population Projections to 2025. In *Population and Food in the Early Twenty-First Century: Meeting Future Food Demand of an Increasing Population,* edited by Nurul Islam. Washington, D.C.: International Food Policy Research Institute, 7–16.

Christensen, C., A. Dommen, N. Horenstein, S. Pryor, P. Riley, S. Shapouri, and H. Steiner. 1981. *Food Problems and Prospects in Sub-Saharan Africa: The Decade of the 1980's.* Foreign Agricultural Economic Report Number 166 (August). Washington, D.C.: United States Department of Agriculture, Economic Research Service.

Dasgupta, Partha. 1995. The Population Problem: Theory and Evidence. *Journal of Economic Literature* 33(December):1879–1902.

Food and Agriculture Organization of the United Nations (FAO). 1995. *The State of Food and Agriculture 1995.* Rome: FAO.

_____. 1996. *The Sixth World Food Survey 1996.* Rome: FAO.

Fosu, Augustin Kwasi. 1992. Political Instability and Economic Growth: Evidence from Sub-Saharan Africa. *Economic Development and Cultural Change* 40(4):829–41.

International Development Research Centre (IDRC). 1992. *Our Common Bowl: Global Food Interdependence.* Ottawa, Ontario: IDRC.

International Food Policy Research Institute (IFPRI). 1995. *Population and Food in the Early Twenty-First Century: Meeting Future Food Demand of an Increasing Population,* edited by Nurul Islam. Washington, D.C.: IFPRI.

Jaeger, William K. 1992. The Causes of Africa's Food Crisis. *World Development* 20(11):1631–45.

McMillan, John C., Gordon C. Rausser, and Stanley R Johnson. 1994. *Economic Growth, Political and Civil Liberties.* Occasional Paper Number 53. San Francisco, CA: International Center for Economic Growth.

Mitchell, Donald O., and Merlinda D. Ingco. 1995. Global and Regional Food Demand and Supply Prospects. In *Population and Food in the Early Twenty-First Century: Meeting Future Food Demand of an Increasing Population,* edited by Nurul Islam. Washington, D.C.: International Food Policy Research Institute, 49–60.

Mitchell, Donald O., Merlinda D. Ingco, and Ronald C. Duncan. 1997. *The World Food Outlook.* Cambridge: Cambridge University Press.

Paulino, Leonardo A. 1986. *Food in the Third World: Past Trends and Projections to 2000.* Research Report 52 (June). Washington, D.C.: International Food Policy Research Institute.

Pinstrup-Andersen, Per. 1994. *World Food Trends and Future Food Security.* Food Policy Report (March). Washington, D.C.: International Food Policy Research Institute.

Sen, Amartya K. 1994. The Political Economy of Hunger. In *Overcoming Global Hunger,* edited by Ismail Serageldin and Pierre Landell-Mills. Washington, D. C.: The World Bank, 85–90.

Smith, Lisa C. 1997. World Food Summit Follow-up: How Useful is the FAO Measure of Chronic Undernourishment? *Food Forum* 37:1–7.

Tweeten, L., J. Mellor, S. Reutlinger, and J. Pines. 1992. *Food Security.* Abridged and revised version of Food Security Discussion Paper (PN-ABK-883) released in May, prepared for U.S. Agency for International Development under supervision of the International Science and Technology Institute.

United Nations (UN). 1994. *Annual Populations 1950–2050* (The 1994 Revision) (diskette version). New York: United Nations Population Division.

United States Department of Agriculture (USDA). 1996a. *World Food Summit—Food Security Situation and Issues: A North American Perspective.* Washington, D.C.: USDA.

_____. 1996b. Production, Supply, and Distribution (PS&D) Database. Washington, D. C.: Foreign Agricultural Service and Economic Research Service.

van de Walle, Nicolas. 1994. Political Liberation and Economic Policy Reform in Africa. *World Development* 22(4):483–500.

World Bank. 1986. *Poverty and Hunger: Issues and Options for Food Security in Developing Countries.* A World Bank policy study. Washington, D.C.: The International Bank for Reconstruction and Development, The World Bank.

_____. 1996. Poverty in Sub-Saharan Africa: Issues and Recommendations. *Findings,* Africa Region, Number 73 (October).

CHAPTER FIVE

The Geography and Causes of Food Insecurity in Developing Countries

Lisa C. Smith, Amani E. El Obeid,
Helen H. Jensen, and Stanley R. Johnson

Introduction

Global food supply is currently sufficient to meet the food needs of the world's population. It is expected to continue to be sufficient well into the next century (Islam 1995). Despite this "abundance" of food, over 840 million people (20 percent) in the developing world today are estimated to suffer from chronic undernourishment (FAO 1996a); many more suffer from deficiencies of protein and essential micronutrients, such as iodine, vitamin A, and iron. Moreover, even those who consume sufficient food to meet their dietary needs can suffer from malnutrition due to poor health and inadequate or inappropriate caring behaviors.

The most vulnerable of the world's citizens are the first to suffer from such food and nutrition insecurity: 167 million developing-country children under age five, or about one in three, suffer from malnutrition. It is estimated that over 50 percent of child deaths in developing countries are related to malnutrition (Pelletier et al. 1995). As so vividly portrayed by Haddad (1995, 93), malnutrition has far-reaching consequences for the ability of those children who survive to contribute to their countries' development as adults:

> For nearly half of the children in the ... least developed countries, being born is a shock from which they will never recover. In these countries, nearly all of which are in South Asia or sub-Saharan Africa, out of every 1,000 children born alive, 112 will die before their first birthday. Another 48 will die before their fifth birthday. Of the remaining 840, 300 will be significantly underweight. As

school-aged children, they will be less able to learn in school. As adults, they will earn less income and accumulate less wealth. Only the remaining 540 children will emerge relatively unscathed.

In some countries and for some historical periods the inadequate food consumption and malnutrition that are partially responsible for such innocent suffering have been due partly to insufficient supplies of food at a national level. In fact, adequacy of global and national food supplies was the main concern of the first world global summit on food security, the World Food Conference of 1974, which coincided with rising grain prices worldwide (FAO 1996c).

While food availability is still a problem for some countries, the root cause of food insecurity in developing countries today is believed to be the inability of people to gain *access* to food due to poverty (Maxwell 1996a; Serageldin 1995; Alexandratos 1995; Von Braun et al. 1992; Foster 1992; see also Adelman's discussion in chapter 11 of this book). The problems of food access and poverty were, appropriately, a focal point at the second world summit on food security, the World Food Summit of 1996, at which 186 countries adopted a *Plan of Action* with the following commitment:[1]

> We will implement policies aimed at eradicating poverty and in
> equality and improving physical and economic access by all, at all
> times, to sufficient, nutritionally adequate and safe food and its
> effective utilization (FAO 1996b, 13).

At the Summit, these countries set a goal of reducing the number of chronically undernourished people by half by the year 2015 (FAO 1996b).

In order to formulate effective policies directed at improving food security, a thorough understanding of the location and causes of food insecurity is needed. This chapter provides a broad overview of the current character of food insecurity in developing countries. Focusing specifically on chronic, rather than transitory, food insecurity, this chapter addresses two questions: Where are the food insecure? and Why are they food insecure? We first lay out a conceptual framework for understanding the causes of food insecurity and introduce the empirical measures employed in the chapter. We then give an overview of the geographic distribution of food insecurity in the 1990s, broadly consider its basic causes in the developing-country regions, and discuss where it is likely to be located in the early decades of the twenty-first century. Next we employ data from 58 developing countries with high prevalences of food insecurity to examine the relative importance of national food availability and poverty in determining food insecurity in the 1990s. Finally, we consider the implications of the analysis for appropriate geographical and policy targeting to improve food and nutrition security for the greatest numbers of people at the fastest pace, now and into the twenty-first century.

The Causes of Food Insecurity

A person is food secure when he or she has access at all times to enough food for an active, healthy life. Accordingly, people are food secure when their consumption of food is sufficient, secure (not vulnerable to consumption shortfalls), and sustainable (Maxwell 1996). The list of causes of food insecurity is long and multifaceted: they range from political instability, war and civil strife, and macroeconomic imbalances and trade dislocations to environmental degradation, poverty, population growth, gender inequality, inadequate education, and poor health. All, however, can be related in some fashion to two basic causes: insufficient national food availability and insufficient access to food by households and individuals. A broad conceptual framework for food security is given in Figure 5.1, in which it is seen to be part of an overall process linking global and national food availability, households' and individuals' access to food, and individuals' nutrition security.[2]

Global and national food availability stand at the most macro level of the food security "equation." Global food availability is determined by total world food production. In any given year, national food availability is determined by a country's own food production, its stocks of food, and its net imports of food (imports minus exports), including food aid.

Still closer to food security is household and individual access to food which, in addition to national food availability, is determined by a household's "full" income. Along with cash income, full income includes the value of goods produced (such as food) and services provided (such as child care) in households that do not enter the market, as well as in-kind transfers of goods and services. Access to food may be gained through (1) production or gathering of food, (2) purchase of food on the market with cash income, and/ or (3) receipt of in-kind transfers of food (whether from other private citizens, national or foreign governments, or international institutions). Within households, individual food access is influenced by intrahousehold food distribution decisions. As shown in Figure 5.1, a household's expenditures of full income for achieving food security compete with expenditures on other basic needs (e.g., health care, housing, and basic education) as well as nonnecessities. A household or individual unable to meet all basic needs can be considered to be in absolute poverty (Frankenberger 1996).

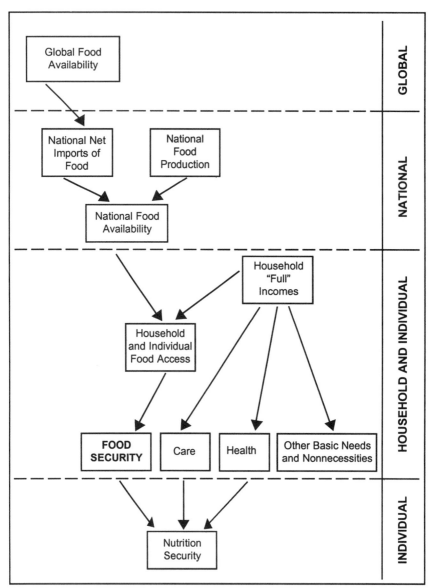

Figure 5.1.   Conceptual framework for food security: from global food
              availability to people's nutrition security
SOURCE: Adapted from UNICEF 1990 and Frankenberger et. al. 1997.

Finally, at the most micro level of the food security equation is nutrition security, without which food security is not translated into an active, healthy life. Nutrition security is defined as follows: "An individual is nutritionally secure when she or he has a nutritionally adequate diet and the food consumed is biologically utilized such that adequate performance is maintained in growth, resisting or recovering from disease, pregnancy, lactation, and physical work" (Frankenberger, Oshaug, and Smith 1997, 1). In addition to food security, nutrition security has two other determinants. The first is "care," or "the provision in the household and the community of time, attention, and support to meet the physical, mental, and social needs of the growing child and other household members" (ICN 1992, 1). Examples of important child care behaviors are the timing and frequency of breast-feeding, the degree of stimulation and interaction with parents, investments in disease prevention and domestic hygiene, the use of health services, and regular growth monitoring. The second nonfood determinant of nutrition security is health. Poor health, or illness, affects nutrition security by depressing appetite, inhibiting the absorption of the nutrients in food, and consuming calories and other nutrients while fighting off and recovering from illness, leaving less energy and nutrients available for growth and weight maintenance (Ramalingaswami, Johnsson, and Rohde 1996).

Indicators of Food and Nutrition Security

In the remainder of the chapter, to examine where the food insecure are located and the basic causes of their food insecurity, we rely on three indicators (reported for 99 developing countries in Appendix Table 5.A.1).

The first indicator, daily per capita dietary energy balance (DEB), is a measure of national food availability. It gives the sufficiency of countries' dietary energy supplies (DES) for meeting the dietary energy requirements (DER) of their populations *if* such dietary energy were distributed among people according to their requirements. Countries with daily per capita DES that are greater than their daily per capita DER are classified "dietary energy surplus"; otherwise they are classified "dietary energy deficit." National per capita DES are derived from the United Nations Food and Agricultural Organization (FAO) food balance sheets and population statistics. Food production and trade (including food aid)[3] data and information on seed rates, wastage, stock changes, and types of food use for each country are used to arrive at the totals of each commodity available for human consumption each year. Estimates of total energy availability for human consumption are then derived by aggregating the energy values (in kilocalories, or "kcals"). The DER for each country takes into account the country's demographic composition (sex and age distribution) and allowances for physical activity. The DERs employed in this chapter correspond to body mass indexes (BMI) observed among healthy, active people.[4]

The second indicator we employ is a measure of income-based absolute poverty, a proxy for people's ability to access food.[5] A country's absolute poverty rate is defined as the proportion of people whose income is less than one dollar per day. Such people are unlikely to be able to meet their food needs sustainably. Where possible, the measure is based on (1985) purchasing power parity (PPP) prices to ensure international comparability. It includes the imputed value of nonmarket goods (including production of food for own consumption, but excluding the value of people's time) (The World Bank 1997).

The third indicator, child malnutrition, is a measure of nutrition security. Specifically, we employ the prevalence of underweight children under five measured anthropometrically by weight-for-age, a summary measure capturing both short stature (stunting) and thinness (wasting) (Martorell and Mason 1996). Child malnutrition is determined by health and care as well as food security; all three determinants may act independently and interact synergistically in influencing child malnutrition (Haddad et al. 1996). Nevertheless, it is considered a good proxy measure for food insecurity (Maxwell and Frankenberger 1992; Rogers 1997). In the analysis below, in order to isolate the influence of food insecurity on child malnutrition from the influence of health, we employ an indicator of the quality of countries' health environments. This is an index composed of the percentage of each country's population with access to health services, safe water, and sanitation.

The Geography of Food Insecurity in Developing Countries

Employing child malnutrition, absolute poverty, and dietary energy availability data, it is possible to construct a broad geographic distribution of food insecurity and to identify its basic causes in the developing-country regions. South Asia and sub-Saharan Africa are the regions with the severest food insecurity problems (see Table 5.1 and Figures 5.2 and 5.3).

South Asia has by far the highest rate of child malnutrition, at 52 percent of all children under five. It also has the highest numbers of malnourished children (see dashed line in Figure 5.2). At 82 million, this is almost one-half of the entire developing-country total, which indicates a massive food and nutrition insecurity problem. Although the region has a dietary energy surplus of 180 kcals per capita per day, nearly one out of

Table 5.1. Child malnutrition, absolute poverty, and dietary energy availability by developing-country region

Region	Child malnutrition (underweight) (1989–92)		Absolute poverty (1993)		Dietary energy supply (1990–92)	Dietary energy supply/balance (1990–92) Dietary energy balance
	Prevalence	Number of Children	Prevalence	Number of People	(per capita kcals/day)	(per capita kcals/day)
	(percent)	(millions)	(percent)	(millions)		
South Asia	52	82	43	515	2,290	+180
Sub-Saharan Africa	30	27	39	219	2,040	−60
East and Southeast Asia	25	44	26	446	2,680	+460
Near East and North Africa	19	9	4	11	2,960	+810
Latin America and Caribbean	10	5	24	110	2,740	+540
All developing countries	32	167	29	1,301	2,520	+350

NOTE: The countries included in each region are given in Appendix Table 5.A.1.
SOURCES: Data columns 1 and 2: Calculated from data in Appendix Table 5.A.1, column 3 and data on population of children under five in 1990 in United Nations (1994). Due to lack of data, North and South Korea, Hong Kong, Suriname, and South Africa are left out of the malnutrition calculations. Data columns 3 and 4: Table 5, Ravallion and Chen (1996). Data column 5: Table 1, FAO (1996a). Data column 6: Calculated as column 5 minus regional per capita "average" energy requirements (see Endnote 4).

every two South Asian people is estimated to be living in absolute poverty. This suggests that South Asia's food insecurity is associated mainly with a food access, rather than food availability, problem. In addition to food insecurity, South Asian countries' high child malnutrition prevalences are also caused by their poor health environments (only 46 percent of its population has access to health services, safe water, and sanitation) and inadequate care for women and children, factors that are linked in turn to the very low social and health status of South Asian women (Ramalingaswami, Johnsson, and Rohde 1996; Osmani 1997). Within the South Asian region, the severest problems are in Bangladesh, which has the highest child malnutrition rate in the world (67 percent), India (53 percent), and Nepal (51 percent). India is the country with the highest number of malnourished children in the world, at 60 million, or 35 percent of the developing-country total.

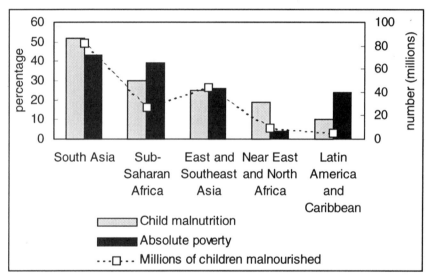

Figure 5.2.   Child malnutrition and absolute poverty in developing-country regions
NOTE: Percentages of children's malnutrition and of people in absolute poverty are measured on the left-hand vertical axis. The numbers of malnourished children are measured on the right-hand vertical axis.
SOURCE: Table 5.1.

Sub-Saharan Africa's child malnutrition rate of 30 percent is much lower than South Asia's. Due to the region's current low population density, the number of malnourished children in sub-Saharan Africa is also fairly low, at 27 million. Its poverty prevalence, however, is almost as high as South

Asia's, suggesting severe food access problems. In addition, it has a major food availability problem, being the only region with a dietary energy deficit in the period from 1990 to 1992. Indeed, of the 26 developing countries that were dietary energy deficit in this period, 17 were in sub-Saharan Africa. Contributing to child malnutrition is the region's poor health environment (with only 42 percent of the population having access to health services, safe water, and sanitation). The countries in the region with the severest food insecurity problems are Ethiopia and Mozambique, with child malnutrition rates of 40 and 47 percent and dietary energy deficits of 480 and 360 kcals per capita per day, respectively. Somalia has the worst food availability problem of all developing countries, with a dietary energy deficit of 510 kcals per capita per day.

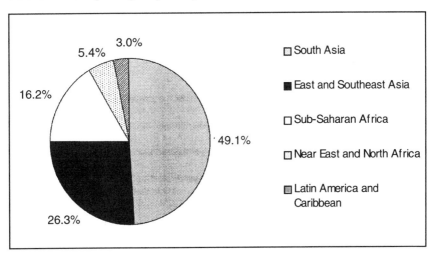

Figure 5.3.   Regional proportions of total developing-country child malnutrition, in percent
SOURCE: Table 5.1.

Although the other developing-country regions have less severe food insecurity and much better health environments than do South Asia and sub-Saharan Africa,[6] some countries within these regions have major food insecurity problems. While the East and Southeast Asia region has a fairly high dietary energy surplus of 460 kcals per capita per day, it maintains relatively high poverty (26 percent) and child malnutrition rates (25 percent). It is the home of a large number of malnourished children, 44 million, which is over one-quarter of all developing-country malnourished children. Countries with particularly high child malnutrition rates are Vietnam (42 percent) and Indonesia and Cambodia (both at 38 percent).

The Near East and North Africa region has the highest dietary energy surplus of the developing-country regions, at 810 kcals. It also has a low poverty prevalence, at 4 percent, suggesting that food insecurity is not a major problem. Nevertheless, the region's child malnutrition rate is 19 percent, suggesting that nutrition insecurity remains a problem due to poor health or care. Within the region, Afghanistan has an extremely high dietary energy deficit of 490 kcals and a 40 percent child malnutrition rate. Iran also has a relatively high child malnutrition rate of 39 percent.

The Latin America and Caribbean region has the lowest child malnutrition rate of all the developing-country regions. Although it has a high dietary energy surplus (at 540 kcals), its poverty rate (24 percent) is almost on par with that of East and Southeast Asia. This suggests that the main problem underlying food and nutrition insecurity in Latin American and Caribbean countries is probably one of food access. Compared to other countries in the region, Haiti and Guatemala, for which one-quarter of all children under five are estimated to be malnourished, have the severest food insecurity problems.

Some developing countries have relatively low overall prevalences of food insecurity but nevertheless contain large numbers of food insecure people, due either to large populations or in-country regional differences that cancel each other out (Von Braun et al. 1992). For example, Brazil and Mexico, with dietary energy surpluses and child malnutrition rates of 7 and 14 percent, respectively, nevertheless contain pockets of extreme absolute poverty and food insecurity. China has a dietary energy surplus and a child malnutrition rate of 21 percent, which is low compared to other developing countries. Yet it is the home of the second largest number of developing-country malnourished children, at 25 million, or 15 percent of the total.

Across developing countries, the prevalence of food insecurity is generally higher in rural areas than in urban areas (Von Braun et al. 1992). An International Food Policy Research Institute (IFPRI) study has found that the prevalence of children's (preschoolers') malnutrition is consistently lower in urban samples compared to rural samples (see Appendix Table 5.A.2). In Peru, for example, only 6.4 percent of children living in urban areas are malnourished, compared to 17.7 percent of children living in rural areas (Ruel et al. 1996). However, urban food insecurity and malnutrition will become increasingly important problems in the future as rates of urbanization increase and urban sanitation, diet quality, and food safety problems grow (FAO 1996d).

Of course future food insecurity is dependent on what actions are taken now to avert it. Nevertheless, current regional trends can give us some idea of where food insecurity may be located in the early decades of the twenty-first century. By most projections, both food and nutrition insecurity are on the decline in developing countries as a whole. Child malnutrition in Southeast Asia is falling rapidly, at around one percentage point per year. It is declining more slowly in South Asia (by one-half a percentage point per year), where it is expected to remain high relative to other regions well into the twenty-first century. Without accelerated preventative action, the region's poverty, poor health conditions, and poor quality of care could persist for some time. Despite a declining child malnutrition rate, the Latin American region is also likely, without accelerated action, to experience a persistence of absolute poverty, and thus problems of food access, into the twenty-first century (IFPRI 1995; ACC/SCN 1992).

The only region in which food insecurity is expected to *increase* over the next 20 years is sub-Saharan Africa. Trends show that child malnutrition has either increased or remained static for most countries. Absolute poverty is expected to rise; predictions show the region will face continued food availability problems and probably will be unable to import the difference between its food needs and production without increased investment (IFPRI 1995; ACC/SCN 1992).

National Food Availability, Poverty, and Food Security

We have seen that inadequate food availability at a national level and inadequate access to food due to poverty are the two most basic causes of food insecurity. Although both are crucial factors, for policy purposes it is important to know *which* is relatively more important in determining food insecurity in each developing country. In this section we give some quantitative evidence on this issue for the developing world as a whole using data from the early 1990s.

As noted in Figure 5.1, enough food available in a country to meet the food needs of all of its population is a necessary—but not sufficient—condition for food security. This point is made clear by the fact that in many countries food supply is ample to meet the needs of all people but food insecurity nevertheless remains high. Using child malnutrition as a proxy for food insecurity, Figure 5.4 illustrates how the severity of food insecurity changes across six groups of developing countries differentiated by their food availability positions. The groups range from one containing high dietary energy-deficit countries through one made up of high dietary energy-surplus countries.[7]

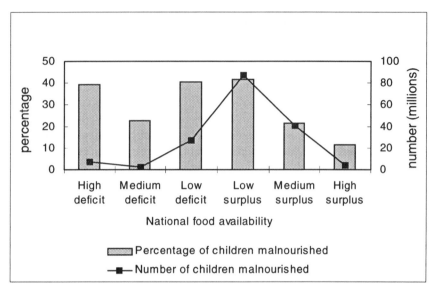

Figure 5.4.   Child malnutrition by dietary energy availability country groupings
NOTE: The horizontal axis gives six groups of countries differentiated by their dietary energy balances. Countries are defined as dietary energy deficit if their daily per capita dietary energy supply is insufficient to meet their per capita dietary energy requirement; otherwise they are dietary energy surplus. The dietary energy balance cut-offs for each group are given in endnote 7. The percentage of child malnutrition is measured on the left-hand vertical axis. The percentages are calculated as the total number of malnourished children under five in the group divided by the total number of children under five in the group. The number of malnourished children is measured on the right-hand vertical axis.
SOURCE: Calculated from data in Appendix Table 5.A.1.

We find that the probability that a child will suffer from malnutrition, rather than decreasing steadily with increasing food availability, is likely to be slightly higher in countries with low dietary energy surpluses than in those with high dietary energy deficits. Countries with medium dietary energy deficits show a tendency to have approximately the same child malnutrition rates as those with medium surpluses. Overall, the dietary energy-surplus countries have a lower incidence of child malnutrition (31 percent) than the dietary energy-deficit countries (38 percent). The difference is small, however. Figure 5.4 suggests a weak relationship between national food availabilities and people's food security in developing countries.[8]

Figure 5.4 also shows the *numbers* of malnourished children for the same six groupings (see dashed line). It is apparent that in any given year a larger number of malnourished children may live in dietary energy-surplus countries than in dietary energy-deficit countries. In fact, in 1990 to 1992, approximately 131 million of the 167 million children under five who were estimated to be malnourished (78 percent, or about three in four) lived in countries with dietary energy surpluses. This suggests that in the 1990s, for the large majority of food insecure people, the underlying problem probably is not one of insufficient food availability in their countries.

To examine the relative extent to which poverty and national food availability are sources of food insecurity, we focus on a subset of 58 developing countries with high degrees of food insecurity. This group includes all countries with dietary energy deficits and/or prevalences of child malnutrition of at least 15 percent. To differentiate them by their food availability positions, the countries are first grouped into dietary energy deficit and surplus groups. To differentiate them by their poverty positions, within each of these groups are embedded a "high poverty" group and a "low poverty" group. The criteria for falling into the high poverty group are that the country has either a poverty rate of 40 percent or higher and/or an estimated 1992 per capita gross national product (GNP) of less than 780 dollars.[9]

Table 5.2 lists the countries in each group, giving a snapshot of their food security situations for the 1990 to 1992 period. The columns of the table represent the countries' food availability positions. The rows represent their (absolute) poverty positions. Each country's child malnutrition rate is given in parentheses. In addition, at the bottom of each quadrant is the total number of malnourished children in the group. The highest number fall into the food surplus/high poverty quadrant (72 million), mainly due to the presence of India. Overall, as was the case for the entire sample of 99 developing countries (see Figure 5.4), the majority of the malnourished children living in these highly food-insecure countries (78 percent) live in food-surplus countries. In addition, a majority (64 percent) fall into the high poverty category.

To see how the severity of food insecurity differs across the groups, Table 5.3 gives estimates of the prevalences of child malnutrition for each group. The prevalence is the greatest for the dietary energy-surplus/high poverty group, at 48 percent. Consider the differences in the child malnutrition rates across the dietary energy-deficit and surplus groups. The incidence of child malnutrition is slightly lower in the dietary energy-surplus group than in the deficit group for the low poverty group; for the high poverty group, however, the incidence is *higher* in the dietary energy-surplus group than in the deficit group by seven percentage points.[10]

Next consider the differences in the prevalences of child malnutrition across the high and low poverty groups. The prevalence is greater in the high poverty group than in the low poverty group across both dietary

Table 5.2.   Classification of highly food-insecure countries into national food availability and poverty groups

	National food availability (1990 to 1992)	
	Dietary energy deficit	Dietary energy surplus
High poverty	Bangladesh (67)	India (53)
	Mozambique (47)	Nepal (51)
	Vietnam (42)	Niger (44)
	Ethiopia (40)	Iran (39)
	Afghanistan (40)	Madagascar (38)
	Somalia (39)	Sudan (34)
	Cambodia (38)	Burkina Faso (27)
	Angola (35)	Guatemala (25)
	Laos (34)	Tanzania (25)
	Rwanda (32)	Uganda (23)
	Chad (31)	Mali (22)
	Burundi (29)	Senegal (20)
	Sierra Leone (26)	Honduras (20)
	Haiti (24)	Nicaragua (19)
	Malawi (24)	El Salvador (19)
	Kenya (17)	Guyana (18)
	Cameroon (14)	Lesotho (18)
	Zimbabwe (14)	Gambia (17)
	Peru (11)	
	(26 million children)	(72 million children)
Low poverty	Nigeria (36)	Sri Lanka (42)
	Ghana (27)	Pakistan (40)
	Zambia (25)	Indonesia (38)
	Mongolia (12)	Philippines (34)
	Bolivia (11)	Yemen (30)
		Congo (28)
		Botswana (27)
		Namibia (26)
		Guinea (24)
		Benin (24)
		China (21)
		Togo (18)
		Malaysia (18)
		Mauritius (17)
		Mauritania (16)
		Gabon (15)
	(8 million children)	(48 million children)

NOTES AND SOURCE: After each country name, the percentage of children malnourished (Appendix Table 5.A.1, column 1) is given in parentheses. A country is dietary energy deficit if its per capita dietary energy supply is insufficient to meet its average per capita dietary energy requirement; otherwise it is dietary energy surplus. Countries with either poverty prevalences of greater than 40 percent or per capita GNPs less than $780 (both measured in 1985 PPP-adjusted dollars) are classified into the high poverty group; otherwise they are in the low poverty group.

Table 5.3.   Prevalence of child malnutrition by national food availability and
poverty positions of highly food-insecure countries, in percent

	National food availability (1990-92)	
	Dietary energy deficit	Dietary energy surplus
High poverty	41	48
Low poverty	32	26

NOTES AND SOURCE: The countries in each group are given in Table 5.2. Child
malnutrition data are from Appendix Table 5.A.1, columns 1 and 2. Malnutrition rates are
calculated as the total number of underweight children in the countries falling into each
quadrant divided by the total number of children. See Table 5.2 notes for classification
criteria.

energy-surplus and deficit groups—consistent with the widely held view
that poverty is the most widespread cause of food insecurity. The difference is particularly pronounced within the dietary energy-surplus
group, with the prevalence in the high poverty group being almost double
that of the low poverty group.[11] This difference may partially reflect large
differences in the groups' health environment.[12] Overall, the numbers suggest that for most countries with high food and nutrition insecurity and
for most people who are food and nutrition insecure, the underlying cause
is likely to be poverty rather than national food availability.

Policy to Improve Food Security

The causes of food insecurity are complex and interrelated. A myriad
of interventions could be useful in reducing or eliminating it. The key policy
goal, however, is not simply to improve food security, but to improve it in a
sustainable fashion for the *most* people at the *fastest* pace. To reach this
goal, given financial and institutional capabilities, policies should be targeted at the underlying causes that pose the most limiting or "binding"
constraint to improving food security in each specific context.

The above analysis suggests that, in the 1990s, poverty is the most
binding constraint to improving people's food security in developing
countries. In most settings, policies that improve people's access to food
by reducing poverty are likely to have the greatest gain in food security
improvements. A powerful means of reducing poverty is through equitable
economic growth. In order for growth to benefit poor people as well as
those better off, special efforts are often needed to enable the poor to
participate in the growth process, such as efforts to enhance their access
to financial and productive resources, information and training, and
physical and market infrastructure. A second and complementary means

of reducing poverty is investment in the human capital of the poor, mainly their health and educational attainment. Developing people's skills and capacities equips them to take advantage of new opportunities placed before them by economic growth (The World Bank 1991). A final means of reducing poverty is investing in the social capital of the poor. Social capital is the networks, norms, and trust among members of communities that enable them to coordinate and cooperate for their members' mutual benefit (Moser 1996). It is thus a further means of enabling poor people to take advantage of economic opportunities and gain access to services that enhance their human capital.

In addition to poverty, many countries still face problems of national food availability (particularly sub-Saharan African countries). Most countries also face nutrition security problems linked to problems of health and care (particularly South Asia and sub-Saharan Africa). To make good the gains of food security in terms of actual human physical well-being, attention to these problems must continue as well.

How then do governments and international organizations committed to reducing food insecurity go about choosing appropriate interventions? Choice of priority interventions requires, above all, identification of the people who are food and/or nutrition insecure and of the primary causes of their insecurity. In this way, interventions that are appropriate for the given context can be chosen to alleviate the constraints posed by these causes (Frankenberger 1997; FAO 1996d). Because there are many factors that are necessary but not sufficient for bringing about food and nutrition security, all possible causes—whether related to problems of food availability, food access, or the nonfood determinants of nutrition security (care and health)—must be considered simultaneously for an optimal choice and sequencing of interventions. Beyond identification alone, priority setting requires knowledge of the *relative* importance of each cause in determining food and nutrition insecurity. Emphasis can then be placed first on those causes that pose the most binding constraint.

In general, when more than one cause is being addressed at a time, policy objectives should be linked so that multiple causes can be tackled with single interventions and synergies can be obtained from multiple interventions (Frankenberger 1997). In countries where food production is the main income source of the poor, policies that increase agricultural productivity can, when appropriately formulated, improve food access at the same time as they improve food availability because they can raise incomes and, in the long run, lower real food prices. Similarly, productivity-enhancing plant breeding that enriches staple foods with scarce micronutrients ameliorates not only food availability problems but micronutrient malnutrition as well (Bouis 1996). An example of a case where multiple interventions can have a synergistic effect is combining income generation (especially of women) with nutrition education (Von Braun et al. 1993). Combining human capital development and food assistance in low-potential agricultural areas where high food insecurity exists with investments in increased food production in adjacent high-potential areas is another example. In this case, jobs that are created in the high-potential areas through increased productivity can be filled by people from the low-potential areas, and both food availability and food access problems are addressed (Frankenberger 1997).

With an eye both to solving immediate problems and reaching long-term goals, choice of an optimal mix of interventions should consider interventions that have an impact in the long run as well as those having an impact only in the short run (Von Braun et al. 1992). The benefits of investments in agricultural research, population stabilization, basic education, and economy-wide economic growth may only accrue after several years. Yet they can have powerful positive impacts on a large number of people's food and nutrition security. To avert or respond to crisis situations, some interventions are needed to tackle immediate problems. Examples of interventions that have their impact mainly in the short run are supplementary feeding, food stamp, and food rationing programs. Sometimes it is possible to contribute to both short- and long-term goals simultaneously. For example, programs to combine supplementary feeding with nutrition education, as well as food-for-work with investment in rural transport and health infrastructure, can improve food and nutrition security both immediately and in the future.

In many cases there will be trade-offs in reaching policy goals directed at different problems, and an optimal balance between competing goals must be achieved. For example, a focus on food availability leads in the direction of investment in new technologies to increase food (or export crops) production in high-potential areas. A focus on poverty reduction, on the other hand, leads in the direction of investment in programs to raise incomes of poor people who often live in resource-poor, low-potential areas where obtaining large food surpluses may not be possible (Maxwell 1996b). Beyond cost considerations, the optimal investment balance should ultimately depend on the relative importance of food availability and food access in determining food and nutrition insecurity in the particular context.

Finally, in formulating efficient food security policies targeted at the underlying causal problems, care must be taken to account for possible negative side effects so that new problems are not created. Actions that focus on improving food availability and/or access may have negative consequences for nutrition security by affecting its nonfood determinants. For example, interventions that lead to increases in household members' time in income-generating activities may lead to a decrease in the time they spend in child care or health enhancement, slowing down improvements in children's nutrition that accompany increases in food access (Smith 1995). Macroeconomic and trade policies that are designed to promote food-security-enhancing economic growth may have harmful impacts on some people's food security. For example, reductions in food subsidies raise food costs for net food consumers, reductions in public expenditures on health services can lead to a deterioration in health environments, and currency devaluations can reduce the incomes of those relying on imported productive inputs. Increased food production that takes place through expansion onto marginal lands (or through shorter fallow periods) causes soil degradation, compromising future food production and income. As a final example, influxes of food aid, without precautions, can lead to dependence and production disincentives, undermining long-term food production and poverty-reduction efforts. Such possible adverse consequences should be taken into account in policy formulation and either averted or compensated for through complementary interventions.

Conclusion

To improve food security for the greatest numbers of people at the fastest pace, policies must be targeted, first, on the regions, countries, and people who are most food insecure. Second, they must be targeted on alleviating the underlying problems causing food insecurity. To ensure that food security results in actual improvements in human physical well-being, nutrition security must be improved as well. Thus it is not only food itself that is important in efforts to improve people's food security, but also health and care (especially of children), both of which are necessary for nutrition security.

Geographically, South Asia and sub-Saharan Africa are the developing country regions with the most severe food and nutrition insecurity. These regions, as well as countries in other regions with dietary energy deficits and high levels of poverty and malnutrition, should be the focus of international efforts to improve food and nutrition security. Whereas poverty is the leading cause of food insecurity for most developing countries, including those that are in dietary energy *surplus,* many countries also have problems related to national food availability (especially sub-Saharan African countries) and to health and care problems (especially South Asian and sub-Saharan African countries) that must be resolved as well.

In general, policy combinations that improve food and nutrition security for the greatest number of people at the fastest pace have the following qualities. First, they focus on the most binding constraint to their improvement. Second, they are based on linked policy objectives so that single interventions can be used to solve multiple problems and synergistic impacts can be obtained from multiple interventions. Third, they contain, as needed, both long-run and short-run instruments. Fourth, they achieve an optimal balance between competing policy goals. Finally, they avoid or compensate for any possible negative side effects of interventions.

Improving people's physical well-being through improving their food and nutrition security will have two beneficial effects that enhance the long-run sustainability of interventions. The first is to increase the capacity of the poor to pull themselves out of poverty, thus improving their access to food in a sustainable manner. The second is to provide the human capital basis for accelerated economic growth and national development (Martorell 1996; Behrman 1992), in turn enabling countries as a whole to obtain sufficient food for their populations over time, whether through domestic production or imports.

Appendix

Commitments Made at the 1996 World Food Summit

1. We will ensure an enabling political, social, and economic environment designed to create the best conditions for the eradication of poverty and for durable peace, based on full and equal participation of women and men, which is most conducive to achieving sustainable food security for all;

2. We will implement policies aimed at eradicating poverty and inequality and improving physical and economic access by all, at all times, to sufficient, nutritionally adequate and safe food and its effective utilization;

3. We will pursue participatory and sustainable food, agriculture, fisheries, forestry, and rural development policies and practices in high- and low-potential areas, which are essential to adequate and reliable food supplies at the household, national, regional, and global levels, and combat pests, drought, and desertification, considering the multifunctional character of agriculture;

4. We will strive to ensure that food, agricultural trade, and overall trade policies are conducive to fostering food security for all through a fair and market-oriented world trade system;

5. We will endeavor to prevent and be prepared for natural disasters and man-made emergencies and to meet transitory and emergency food requirements in ways that encourage recovery, rehabilitation, development, and a capacity to satisfy future needs;

6. We will promote optimal allocation and use of public and private investments to foster human resources, sustainable food, agriculture, fisheries and forestry systems, and rural development in high and low-potential areas;

7. We will implement, monitor, and follow up this *Plan of Action* at all levels in cooperation with the international community.

SOURCE: FAO 1996b.

Appendix Table 5.A.1.   Food and nutrition security data for developing countries

Country	Predicted prevalence of child malnutrition (underweight) (1990) (percent) (1)	Predicted number of children malnourished (underweight) (1990) (thousands) (2)	Health environment (access to health services, safe water, and sanitation) (percent) (3)	Prevalence of absolute poverty (1993) (percent) (4)	Dietary energy supply (1990–92) (per capita kcals/day) (5)	Dietary energy balance (1990–92) (per capita kcals/day) (6)
Sub-Saharan Africa						
Angola	35.3	641	28	a	1,840	-260
Benin	23.5	213	47	33.0	2,520	420
Botswana	26.8	58	60	34.7	2,320	220
Burkina Faso	27.1	443	49	83.6	2,140	40
Burundi	29.1	301	58	36.2	1,950	-150
Cameroon	13.6[b]	272	24	40.0	2,040	-60
Central Afr. Rep.	31.9	347	15	a	1,720	-380
Chad	30.6	349	26	a	1,810	-290
Congo	27.5	112	20	a	2,210	110
Côte d'Ivoire	12.3	297	60	a	2,460	360
Ethiopia	39.8	3,577	30	a	1,620	-480
Gabon	15.1	26	79	34.9	2,490	390
Gambia	17.1	28	83	64.0	2,320	220
Ghana	26.7	733	54	31.4	2,090	-10
Guinea	24.0	268	32	26.3	2,400	300
Kenya	17.4	788	37	50.2	1,970	-130
Lesotho	17.5	52	49	50.4	2,260	160
Liberia	20.1	95	32	a	1,780	-320
Madagascar	38.1	899	65	72.3	2,160	60
Malawi	23.5	428	66	a	1,910	-190
Mali	21.6	389	36	55.2	2,230	130
Mauritania	15.7	52	48	31.4	2,610	510
Mauritius	17.0	16	69	10.6	2,780	680

Appendix Table 5.A.1.   (continued)

Country	Predicted Prevalence of child malnutrition (underweight) (1990) (percent) (1)	Predicted number of children malnourished (underweight) (1990) (thousands) (2)	Health environment (access to health services, safe water, and sanitation) (percent) (3)	Prevalence of absolute poverty (1993) (percent) (4)	Dietary energy supply (1990–92) (per capita kcals/day) (5)	Dietary energy balance (1990–92) (per capita kcals/day) (6)
Mozambique	46.8	1,184	24	a	1,740	-360
Namibia	26.2[b]	58	a	a	2,190	90
Niger	44.0	676	33	61.5	2,190	90
Nigeria	35.7[b]	6,350	42	28.9	2,100	0
Rwanda	31.7	401	62	a	1,860	-240
Senegal	19.6	259	46	54.0	2,310	210
Sierra Leone	25.9	186	41	a	1,820	-280
Somalia	38.8[a]	656	29	a	1,590	-510
South Africa			a	a	2,810	710
Sudan	33.7	1,438	41	a	2,150	50
Swaziland	8.8	11	39	16.4	2,680	580
Tanzania	25.2[b]	1,197	74	32.3	2,110	190
Togo	18.4	119	47	50.0	2,290	120
Uganda	23.3[b]	845	33	a	2,220	10
Zaire	33.2	2,413	36	84.6	2,090	-10
Zambia	25.1[b]	393	63	41.0	2,020	-80
Zimbabwe	14.1	243	50	a	2,080	-20
Near East and North Africa						
Afghanistan	40.3	1,078	35	a	1,660	-490
Algeria	9.2[b]	338	64	a	2,900	750
Egypt	10.4[b]	869	80	a	3,340	1190
Iran	39.0	4019	78	a	2760	610
Iraq	11.9	373	88	a	2270	120
Jordan	6.4[b]	46	96	a	2900	750
Kuwait	5.0	15	99	a	2460	310
Lebanon	8.9	29	90	a	3260	1110

Libya	4.0	33	98	a	3,290	1,140
Morocco	12.0	396	93	a	2,730	850
Saudi Arabia	12.6	323	93	a	2,730	580
Syria	12.5	294	80		3,220	1,070
Tunisia	8.9	95	67	a	3,260	1,110
Turkey	10.5	766	92	a	3,510	1,360
United Arab Em.	7.0	15	95	a	3,370	1,220
Yemen	30.0[b]	660	a	19.1	2,160	10
Latin America and Caribbean						
Argentina	1.2	40	89	a	2,950	750
Bolivia	11.4[b]	116	40	7.1	2,030	-170
Brazil	7.1[b]	1,249	87	a	2,790	590
Chile	2.0	29	88	a	2,540	340
Colombia	10.1[b]	387	82	a	2,630	430
Costa Rica	8.1	33	96	a	2,870	670
Cuba	8.4	73	100	a	3,000	800
Dom. Republic	10.4[b]	99	61	a	2,270	70
Ecuador	13.0	182	65	30.4	2,540	340
El Salvador	19.4	153	51	48.3	2,530	330
Guatemala	25.0	402	59	53.3	2,280	80
Guyana	18.0	17	89	84.2	2,350	150
Haiti	24.4	241	36	65.0	1,740	-460
Honduras	19.8	166	49	46.5	2,310	110
Jamaica	7.2[b]	19	81	a	2,580	380
Mexico	13.9	1,593	81	a	3,190	990
Nicaragua	18.7	128	40	43.8	2,290	90
Panama	11.0	33	83	a	2,240	40
Paraguay	3.7[b]	25	45	a	2,620	420
Peru	10.8[b]	300	53	49.4	1,880	-320
Suriname	a	a	a	a	2,510	310
Trin. & Tobago	9.0	13	98	a	2,630	430
Uruguay	7.0	18	84	a	2,680	480
Venezuela	5.9	159	90	a	2,590	390

(continued on next page)

Appendix Table 5.A.1.   (continued)

Country	Predicted Prevalence of child malnutrition (underweight) (1990) (percent) (1)	Predicted number of children malnourished (underweight) (1990) (thousands) (2)	Health environment (access to health services, safe water, and sanitation) (percent) (3)	Prevalence of absolute poverty (1993) (percent) (4)	Dietary energy supply (1990–92) (per capita kcals/day) (5)	Dietary energy balance (1990–92) (per capita kcals/day) (6)
South Asia						
Bangladesh	66.5[b]	10,602	55	47.5	1,990	-120
India	53.0[b]	59,774	44	52.5	2,330	220
Nepal	50.5	1,577	21	53.1	2,140	30
Pakistan	40.4[b]	8,877	52	11.6	2,340	230
Sri Lanka	42.0	770	67	4.0	2,230	120
East and Southeast Asia						
Cambodia	37.7	614	13	[a]	2,100	-120
China	21.0 [a]	24,870 [a]	83.5	29.4 [a]	2,710	490
Hong Kong					3,150	930
Indonesia	38.0 [a]	8,441 [a]	43	14.5	2,700	480
Korea-DPR	[a]	[a]	100	[a]	2,930	710
Korea-Republic			92		3,270	1050
Laos	34.0	255	35	46.1	2,210	-10
Malaysia	17.6	459	87	5.6	2,830	610
Mongolia	12.3[b]	40	81	36.3	2,100	-120
Myanmar	32.4[b]	1,950	39	[a]	2,580	360
Philippines	33.5[b]	2,934	75	27.5	2,290	70
Thailand	13.0[b]	761	64	[a]	2,380	160
Vietnam	41.9[b]	3,860	67	50.9	2,200	-20
Oceania						
Papua New Guinea	[a]	[a]	[a]	[a]	2,610	730

Appendix Table 5.A.1.   (continued)

NOTES AND SOURCES: The superscript " ª " means data are not available or (in the case of the poverty data) not necessary for the chapter's analysis. The superscript " ᵇ " is explained in note 1.

1. *Predicted prevalence of children underweight.* A child is considered underweight when his or her weight falls below -2 standard deviations of the expected weight of healthy children of her or his age using National Center for Health Statistics norms. The underweight rates with superscript "b" are from nationally representative surveys that took place over the period 1989 to 1992. The source for these data is from United Nations Administrative Committee on Coordination/Subcommittee on Nutrition (ACC/SCN 1993) Table 2.1, with the following exceptions: the Namibia and Mongolia rates are for 1992 from FAO (1996a) Appendix 2, Table 8; the rate for India is for 1992–93 and reported in Gillespie, Mason, and Martorell (1996).

For the remaining 74 countries, data close to 1990 are not available. For these countries, ACC/SCN-predicted 1990 underweight rates are employed. These are estimated using data from nationally representative surveys that took place from 1975 to 1991 using two multivariate regression models: 1) a global model estimated with data from 100 surveys in 66 countries; and 2) a sub-Saharan Africa model using data from 20 surveys in 20 countries. The independent variables employed were per capita dietary energy supply, infant mortality rates, percent government fiscal expenditures on health, education, and social security, education of females, and the under-five child population. The global and sub-Saharan African models had (adjusted) R²s of .9 and .82, respectively. The predicted values are given in ACC/SCN (1993) Table 2.6. See ACC/SCN (1993) Chapter 2 for further explanation.

2. *Predicted number of children underweight.* Calculated from column (1) and data on the estimated numbers of children under five found in United Nations (1994).

3. *Health environment.* This is an average of three indicators: the percentage of the population with access to health services, safe water, and sanitation (each indicator is given equal weight). Note that data on all three indicators are not available for a small number of countries. Source: UNDP (1993), Table 12.

4. *Prevalence of absolute poverty.* Data are only given for countries defined to be "highly food insecure" (see Table 5.2). The majority of the data are from the World Bank (1997) Table 2.5. They are from nationally representative survey samples for 1987 to 1995 (except for Botswana and Cameroon, for which data are for 1985–86 and 1984, respectively). For Benin, Burundi, Cameroon, El Salvador, Gambia, Ghana, Mauritius, Togo, Yemen, Haiti, Bangladesh, Laos, Mongolia, and Vietnam, nationally defined poverty lines were employed. For the rest, an "international poverty line" of one dollar per day in 1985 international prices was employed. To enhance cross-country comparability, purchasing power parity (PPP) exchange rates from the latest version of the Penn World Table (5.6) were employed. Most surveys included the imputed value of income earned from nonmarket goods (including production of food for own consumption).

Data for Gabon, Burkina Faso, and Mali are for 1988 and are derived from rural poverty prevalence rates reported in Stewart (1995) Table 6.7. An urban-rural poverty prevalence ratio of .69 (estimated from data in Stewart for which both urban and rural poverty rates were

(continued on next page)

Appendix Table 5.A.1.   (continued)

reported) was used to estimate an urban poverty rate for each country. The final poverty prevalences were calculated as weighted averages of the urban and rural poverty prevalence rates using rural and urban population proportions as weights. The rate for Guyana is for 1989 and is the average of the urban and rural rates reported in Stewart (1995) Table 7.10

5. *Per capita dietary energy supply (daily).* Data given are averages of the daily per capita supplies (DES) for human consumption for 1990, 1991, and 1992. Source: FAO (1996a) Appendix 2, Table 1.

6. *Per capita dietary energy balance (daily).* Calculated as per capita DES (using the data in Table 5.A.1, column 5) minus average per capita dietary energy requirements. A negative national food energy deficit indicates that a country's DES is insufficient for meeting the requirements of its population. A positive national food energy deficit indicates sufficiency. Regional requirements, which are used as proxies for national requirements, are given in FAO (1996a) Table 16.

Appendix Table 5.A.2.   Urban-rural differences in prevalences of underweight children under five

Country	Underweight children Urban	Rural	Percent difference
	(percent)		
Sub-Saharan Africa			
Burkina Faso	20.0	31.3	-36.10
Burundi	20.2	38.9	-48.07
Ghana	17.5	31.4	-44.27
Kenya	12.6	23.5	-46.38
Madagascar	33.4	40.0	-16.50
Malawi	15.4	28.6	-46.15
Namibia	17.8	29.8	-40.27
Rwanda	17.8	29.8	-40.27
Senegal	13.3	24.4	-45.49
Togo	11.9	20.9	-43.06
Uganda	12.8	24.3	-47.33
Zambia	20.8	29.0	-28.28
Zimbabwe	12.5	16.6	-24.70
Latin America and the Caribbean			
Bolivia	11.6	20.4	-43.14
Brazil	9.8	15.5	-36.77
Dom. Repub.	7.7	14.3	-46.15
Guatemala	25.7	36.5	-29.59
Paraguay	2.8	4.3	-34.88
Peru	6.4	17.7	-63.84
Trin & Tobago	5.0	8.2	-39.02
South Asia			
Pakistan	32.5	44.6	-27.13

SOURCE: Table 3 in Ruel et al. 1996.

Notes

1. A full list of the World Food Summit commitments can be found in the Appendix.

2. Examples of more elaborate conceptual frameworks can be found in UNICEF (1990), Oshaug (1994), and FAO (1996d).

3. Note that since food aid is included in per capita DES figures, the figures are not an indicator of national food self-reliance—or the ability of countries to meet their food needs on their own—for all countries. In the absence of food aid, countries that are classified as dietary energy surplus may fall into the dietary energy deficit category (Ramakrishnan 1997). In 1994–95, the developing-country region with the highest percentage of its food supply met through food aid was Latin America (6 percent), followed by sub-Saharan Africa (5.6 percent) (USDA 1995).

4. The per capita DER corresponding to BMI observed among healthy, active people are termed "average" (rather than "minimum") per capita DER (FAO 1996a). For the developing-country regions these are estimated to be (in kcals): sub-Saharan Africa 2,100, Near East and North Africa 2,150, East and Southeast Asia 2,220, South Asia 2,110, and Latin America and the Caribbean 2,200 (FAO 1996a, Table 16).

5. Nationally representative data on direct measures of food access, such as dietary intakes and food expenditures, are not available for a large number of countries. To give a rough estimate of the numbers of food insecure in each developing country, the FAO uses per capita DES and estimates of the distribution of food intake within countries to calculate the numbers of "chronically undernourished." While the FAO measure gives a broad idea of the prevalence and numbers of food insecure in the developing world, because it is methodologically biased toward national food availabilities, it can not be employed for making accurate cross-country comparisons and in undertaking analyses of the causes of food insecurity (Smith 1997).

6. The approximate percent of people having access to health services, safe water, and sanitation in East Asia is 76. For the Near East and North Africa region the percentage is 78. For the Latin American and Caribbean region the percentage is 80.

7. The six groups are defined by upper and lower bounds on the countries' per capita dietary energy balances as follows:

Category	Daily per capita dietary energy balance bounds (kcals)		Number of countries
	Lower	Upper	
High deficit	-510	-340	6
Medium deficit	-339	-170	8
Low deficit	-450	0	13
Low surplus	1	450	43
Medium surplus	451	900	19
High surplus	701	1,360	10

8. Note that the quality of countries' health environments improves over the six groups. The percentages of the groups' total populations with access to health services, safe water, and sanitation (presented in the order given in Figure 5.4) are 30, 47, 49, 50, 79, and 85. This indicates that the child malnutrition prevalence differences across the groups are more likely to be driven by variations in food security than by variations in health.

9. Where possible, all poverty rates and GNPs are based on PPP-converted 1985 international dollars (see Appendix Table 5.A.1). The countries for whom 1992 estimated per capita GNPs of less than 780 alone classify them as high poverty are (with per capita GNPs in parentheses) Ethiopia (340), Tanzania (630), Burundi (750), Mozambique (570), Chad (710), and Sierra Leone (770) (The World Bank 1994, Table 30). For Somalia, Sudan, Afghanistan, Cambodia, Angola, and Iran, GNP and recent poverty data are not available; we assume these countries fall into the high poverty group because of their very poor social indicators. While poverty data are not available, we assume that Namibia (with a per capita GNP of 3,040) and Congo (with a per capita GNP of 3,450) fall into the low poverty category. It was not possible to classify the following highly food-insecure countries: Liberia, Zaire, the Central African Republic, and Myanmar.

10. Since the health environments of the groups hardly differ (the percentages of population with access to health services, safe water, and sanitation for the high poverty/deficit group is 46 percent; for the high poverty/surplus group it is 46.2 percent), the differences in the groups' child malnutrition rates probably do not reflect differences in health status.

11. The prevalences given in Table 5.3 are based on population-weighted means of the countries' prevalence weights. A two-tailed test for the difference in the unweighted means of the dietary energy-deficit group (29.7 percent) and the dietary energy-surplus group (27.3 percent) reveals no significant statistical difference ($p > 0.1$). The difference

between the unweighted means for the high poverty group (30.1 percent) and the low poverty group (25.1 percent) is statistically significant at the 10 percent level.

12. The percentage of population with access to health services, safe water, and sanitation for the low poverty/deficit group is 45 percent; for the low poverty/surplus group it is 75 percent.

References

Administrative Committee on Coordination/Subcommittee on Nutrition, United Nations (ACC/SCN). 1992. *Second Report on the World Nutrition Situation. Vol. I: Global and Regional Results.* Geneva: ACC/SCN.
_____. 1993. *Second Report on the World Nutrition Situation. Vol. II: Country Trends, Methods and Statistics.* Geneva: ACC/SCN.
Alexandratos, Nikos, ed. 1995. *World Agriculture: Towards 2010.* Rome: Food and Agricultural Organization of the United Nations; New York: John Wiley and Sons.
Behrman, Jere R. 1992. *The Economic Rationale for Investing in Nutrition in Developing Countries.* Washington, D.C.: United States Agency for International Development.
Bouis, Howarth. 1996. Enrichment of Food Staples through Plant Breeding: A New Strategy for Fighting Micronutrient Malnutrition. *Nutrition Reviews* 54(5):131–37.
Food and Agriculture Organization of the United Nations (FAO). 1996a. *The Sixth World Food Survey 1996.* Rome: FAO.
_____. 1996b. *Rome Declaration on World Food Security and World Food Summit Plan of Action.* Rome: FAO.
_____. 1996c. Food, Agriculture and Food Security: Developments since the World Food Conference and Prospects. In *World Food Summit Technical Background Documents 1–5.* Technical Background Document No. 1. Rome: FAO.
_____. 1996d. Food Security and Nutrition. In *World Food Summit Technical Background Documents 1–5.* Technical Background Document No. 5. Rome: FAO.
Foster, Phillips. 1992. *The World Food Problem: Tackling the Causes of Undernutrition in the Third World.* Boulder, CO: Lynne Rienner Publishers.
Frankenberger, Timothy. 1996. Measuring Household Livelihood Security: An Approach for Reducing Absolute Poverty. *Food Forum* 34 (November–December):1–6. Washington, D.C.: Food Aid Management.
_____. 1997. Food Security Indicators: Issues of Targeting, Monitoring and Evaluation. In *Proceedings of the USAID Workshop on Performance Measurement for Food Security.* December 11–12, 1995. Arlington, VA: United States Agency for International Development.

Frankenberger, Timothy, Larry Frankel, Susan Ross, Marshall Burke, Carol Cardenas, Debra Clark, Anne Goddard, Kevin Henry, Maurice Middleberg, Dan O'Brien, Carlos Perez, Rand Robinson, and Jeannie Zielinski. 1997. Household Livelihood Security: A Unifying Conceptual Framework for CARE Programs. In *Proceedings of the USAID Workshop on Performance Measurement for Food Security.* December 11–12, 1995. Arlington, VA: United States Agency for International Development.

Frankenberger, Timothy, Arne Oshaug, and Lisa Smith. 1997. A Definition of Nutrition Security. CARE memo. Atlanta, GA: CARE.

Gillespie, Stuart, John Mason, and Reynaldo Martorell. 1996. *How Nutrition Improves.* ACC/SCN State-of-the-Art Series Nutrition Policy Discussion Paper No. 15. Geneva: United Nations Administrative Committee on Coordination/Sub-Committee on Nutrition.

Haddad, Lawrence. 1995. The March of Malnutrition to 2020: Where Are the Solutions? *Speeches Made at an International Conference.* 2020 Vision, June 13–15, 1995. Washington, D.C.: International Food Policy Research Institute.

Haddad, Lawrence, Saroj Bhattarai, Maarten Immink, and Shubh Kumar. 1996. *Managing Interactions between Household Food Security and Preschooler Health.* Food, Agriculture and the Environment Discussion Paper No. 16, 2020 Vision. Washington, D.C.: International Food Policy Research Institute.

International Conference on Nutrition (ICN). 1992. *Caring for the Socioeconomically Deprived and Nutritionally Vulnerable.* Theme paper No. 3 in Major Issues for Nutritional Strategies. Geneva: Food and Agriculture Organization of the United Nations/World Health Organizaiton.

International Food Policy Research Institute (IFPRI). 1995. *A 2020 Vision for Food, Agriculture, and the Environment: The Vision, Challenge, and Recommended Action.* Washington, D.C.: IFPRI.

Islam, Nurul, ed. 1995. *Population and Food in the Early Twenty-First Century: Meeting Future Food Demand of an Increasing Population.* Washington, D.C.: International Food Policy Research Institute.

Martorell, Reynaldo. 1996. The Role of Nutrition in Economic Development. *Nutrition Reviews.* 54(4):S66–71.

Martorell, Reynaldo, and John B. Mason. 1996. Use of Growth Data at the Global Level: Examples from the Subcommittee on Nutrition of the UN System (ACC/SCN). In *Maternal and Extrauterine Factors: Their Influence on Fetal and Infant Growth,* edited by F. Battaglia et al. Madrid: Ediciones Ergon, 309–18.

Maxwell, Daniel G. 1996. Measuring Food Insecurity: The Frequency and Severity of 'Coping Strategies'. *Food Policy* 21(33):291–303.

Maxwell, Simon. 1996a. Food Security: A Post-Modern Perspective. *Food Policy* 21(2):155–70.

_____. 1996b. Perspectives on a New World Food Crisis. *Journal of International Development* 8(6):859–67.

Maxwell, Simon, and Timothy R. Frankenberger. 1992. *Household Food Security: Concepts, Indicators, Measurements: A Technical Review*. New York: United Nations Children's Fund; Rome: International Fund for Agricultural Development.

Moser, Caroline. 1996. *Confronting Crisis: A Comparative Study of Household Responses to Poverty and Vulnerability in Four Poor Urban Communities*. ESD Monographs Series No. 8. Washington, D.C.: The World Bank.

Oshaug, Arne. 1994. Nutrition Security in Norway? A Situation Analysis. *Scandinavian Journal of Nutrition* 38(S28):1–68.

Osmani, S. R. 1997. Poverty and Nutrition in South Asia. Abraham Horwitz lecture delivered at the symposium on Nutrition and Poverty, United Nations ACC/SCN meeting, March 17–18, 1997, Katmandu.

Pelletier, David, E. A. Frongillo, D. G. Schroeder, and J. P. Habicht. 1995. The Effects of Malnutrition on Child Mortality in Developing Countries. *WHO Bulletin* 73(4):443–48.

Ramakrishnan, Usha. 1997. Personal communication.

Ramalingaswami, Vulimiri, Urban Johnsson, and Jon Rohde. 1996. The Asian Enigma. In *Progress of Nations*. New York: United Nations Children's Fund.

Ravallion, Martin, and Shaohua Chen. 1996. What Can New Survey Data Tell Us About Recent Changes in Distribution and Poverty? The World Bank Policy Research Department, The World Bank, Washington, D.C. Mimeographed.

Rogers, Beatrice. 1997. Indicators of Improved Food Utilization as an Element of Food Security. In *Proceedings of the USAID Workshop on Performance Measurement for Food Security*. December 11–12, 1995. Arlington, VA: United States Agency for International Development.

Ruel, Marie T., James L. Garrett, Patrice Engle, Lawrence Haddad, Dan Maxwell, Purnima Menon, Saul Morris, and Arne Oshaug. 1996. Urban Challenges to Nutrition Security: A Review of Food Security, Health and Care in the Cities. International Food Policy Research Institute, Washington, D.C. Mimeographed.

Serageldin, Ismael. 1995. *Nurturing Development: Aid and Cooperation in Today's Changing World*. Washington, D.C.: The World Bank.

Smith, Lisa C. 1995. *The Impact of Agricultural Price Liberalization on Human Well-Being in West Africa: Implications of Intrahousehold Preference Heterogeneity*. Ph.D. diss., University of Wisconsin, Madison. Ann Arbor: University Microfilms International.

_____. 1997. World Food Summit Follow-up: How Useful is the FAO Measure of Chronic Undernourishment? *Food Forum* 37 (June–July):1–7. Washington, D.C.: Food Aid Management.

Stewart, Frances. 1995. *Adjustment and Poverty: Options and Choices*. New York: Routledge.

United Nations. 1994. *The Age and Sex Distribution of the World's Population: 1994 Revision*. New York: United Nations.

United Nations Children's Fund (UNICEF). 1990. *Strategy for Improved Nutrition of Children and Women in Developing Countries.* UNICEF policy review paper. New York: UNICEF.

United Nations Development Programme (UNDP). 1993. *Human Development Report.* New York: Oxford University Press.

United States Department of Agriculture (USDA). 1995. *Food Aid Needs Assessment.* International Agriculture and Trade Reports, Situation and Outlook Series. Washington, D.C.: USDA, Economic Research Service.

Von Braun, Joachim, Howarth Bouis, Shubh Kumar, and Rajul Pandya-Lorch. 1992. *Improving Food Security of the Poor: Concept, Policy, and Programs.* Washington, D.C.: International Food Policy Research Institute.

Von Braun, Joachim, John McComb, Ben Fred-Mensah, and Rajul Pandya-Lorch. 1993. *Urban Food Insecurity and Malnutrition in Developing Countries: Trends, Policies, and Research Implications.* Washington, D.C.: International Food Policy Research Institute.

World Bank. 1991. *World Development Report 1991.* Washington, D.C.: The World Bank.

_____. 1994. *World Development Report 1994.* New York: Oxford University Press.

_____. 1997. *World Development Indicators 1997.* Washington, D.C.: The World Bank.

CHAPTER SIX

Toward a Food-Secure World

M. S. Swaminathan

Introduction

Since the advent of the industrial and technological revolutions, economic indicators have been used as the principal criteria for measuring sustainability. Population expansion, rapid industrialization, commercialization of agriculture, and quantum jumps in economic activity have been some of the results of the development paradigm adopted after World War II. Technological progress in several areas such as space, information technologies, biotechnology, energy, and new materials has been impressive. At the same time there is a growing understanding of the ecological and social costs of such progress.

The UN Conference on the Human Environment held at Stockholm in 1972 and the UN Conference on Environment and Development held at Rio de Janeiro in 1992 helped to articulate the serious environmental repercussions of contemporary development pathways. The UN Conference on Social Development held at Copenhagen in March 1995 warned that development which is not socially equitable will lead to social disintegration and conflicts. Jobless economic growth and feminization of poverty are the other consequences of the current pattern of development. Thus, the concept of sustainable development has now to be viewed in terms of ecology, social and gender equity, employment, and economics. How to achieve such a synthesis in developmental thinking, planning, and implementation, thereby enabling humankind to take to the pathway of green productivity, is the task facing us today.

In its report titled "Changing Course" the International Business Council pointed out that where there is a will there is a way (Schmidheiny 1992). If technology has so far been a major cause of ecological damage, it can be a leader in finding methods to ensure that development is sustainable. In a recent study, Repetto et al. (1996) of the World Resources Institute have

shown that environmental protection not only need not reduce productivity growth but can in fact stimulate growth without accompanying
ecological damage.

Ecotechnology and Sustainable Livelihoods

The Twentieth Century: A Balance Sheet

As we approach a new century we can look back and draw a balance
sheet of our achievements and failures (Table 6.1). Spectacular progress in
science and technology ranks first among our major accomplishments.
Recent advances in biotechnology and genetic engineering, space technology, information technology, and new materials have opened up uncommon
opportunities for a world where every individual can lead a healthy and
productive life. The spread of democratic systems of governance, the breakdown of skin color–based apartheid and the advent of the information age
have created the sociopolitical substrate essential for integrating the principles of intra- and intergenerational equity in public policy. The power of
a right blend of technology and public policy is strikingly evident from the
progress made in recent decades to keep the growth rate in food production
above the rate of growth in population, thereby ensuring that the Malthusian prediction of population overtaking our ability to produce adequate
food does not come true.

Table 6.1. Contemporary development: a 50-year balance sheet (1945–1995)

Achievements	Concerns
Impressive progress in science and technology	Scientific dimensions of sustainable development
Information Age: Global Village	Era of unfulfilled expectations
Unprecedented economic growth	Damage to life support systems and potential adverse changes in climate and sea level
Uncommon opportunities for food, health, literacy, and jobs for all	Population growth, jobless economic growth, and social disruption; gross economic and gender inequity
Peace Dividend: from defense to development	Growing violence in the human heart: from AIDS to ethnic strife
End of skin color–based apartheid: spread of democratic tradition and renewed faith in UN and other multilateral institutions	Debt burden and emerging technological and economic apartheid

While the positive achievements are many and make us proud of the power of the human intellect, we will be entering the new millennium with some of the greatest social and scientific challenges humankind has ever faced. Several of these challenges have been articulated with great clarity in the "Human Development Report" series published by the United Nations Development Programme (UNDP) in recent years.

Environmental degradation and increasing economic and gender inequality are among the most serious problems we face today. The rich/poor divide is increasing at an alarming rate. The pattern of development adopted by rich societies is leading to jobless economic growth, pollution, and potential changes in climate. Unsustainable lifestyles on the part of the rich billion and unacceptable poverty on the part of another billion coexist. The absence of an educational and health environment that is conducive to every child achieving his or her innate genetic potential for physical and mental development leads to the spread of poverty in capability. UNDP has proposed indicators for measuring both human development and human capability.

The U.S. National Academy of Sciences, the Royal Society of London, the Indian National Science Academy, and 55 other scientific bodies in a statement made in 1993 pointed out that stress on the environment is caused by four interacting factors: population growth, consumption habits, technology, and social organization (Graham-Smith 1994). Concurrent attention is needed on all these four factors to promote sustainable development and sustainable societies. The report "Sustainable America" indicates what an affluent society should do. In poor nations, the social sustainability of the development process is as important as ecological and economic sustainability (The President's Council on Sustainable Development 1996). Also, if the current pace of damage to the ecological foundations essential for sustainable advances in biological productivity—land, water, flora, fauna, forests, oceans, and the atmosphere—continues, sustainable food and nutrition security cannot be achieved. Therefore, as we approach the new millennium, we need a broader concept of sustainability which encompasses environmental, economic, and social parameters. Among social factors, gross economic and gender inequity need priority attention. If such a paradigm shift in developmental thinking and pathways does not occur, the successes achieved in the twentieth century in abolishing skin color–based apartheid, in conquering space, and in splicing genes will be overshadowed by the spread of technological and economic apartheid. If these forms of apartheid are allowed to grow and spread, social disintegration and ecological genocide will be the result.

Ecotechnology: The Emerging Solution

Technologies rooted in the principles of ecology, economics, and equity are now referred to as ecotechnologies. The United Nations Educational, Scientific, and Cultural Organization (UNESCO) and the Cousteau Foundation established by Commandant Jacques Cousteau are promoting ecotechnology networks in different parts of the world. The M. S. Swaminathan Research Foundation (MSSRF) at Madras is the coordinating center for the Asian Ecotechnology Network. A major purpose of this network is the creation of *ecojobs,* which are economically viable, environmentally benign, and socially equitable. A multimedia database on opportunities for ecojobs is being developed, because the dissemination of information on ecojobs is essential for creating opportunities for sustainable livelihoods in rural and urban areas.

The most serious manifestation of poverty is hunger. It is now recognized that endemic hunger is largely the result of inadequate livelihood opportunities which in turn lead to inadequate purchasing power. "Hidden hunger" results from both micronutrient deficiencies and poor environmental hygiene, which impair the biological absorption and retention of food.

A Science Academies Summit on uncommon opportunities for a food-secure world held at MSSRF in July 1996 stressed that national policies for sustainable food and nutrition security should ensure two things: first, that every individual have the *physical, economic, social, and environmental access* to a balanced diet that includes the necessary macro- and micro-nutrients, safe drinking water, sanitation, environmental hygiene, primary health care, and education necessary to lead a healthy and productive life; second, that food originate from efficient and *environmentally benign production technologies* that conserve and enhance the natural resource base of crops, animal husbandry, forestry, and inland and marine fisheries.

During the next three decades, population is expected to increase by another 2.5 billion people. Food requirements will grow due to increases in both population and per capita purchasing power. World grain production has grown from 631 million tonnes (metric tons) in 1950 to nearly 1900 million tonnes in 1995. Such a phenomenal growth has had its environmental cost in terms of soil degradation, aquifer depletion, genetic erosion, and pesticide pollution. This is why we have to produce more in the coming decades but produce it differently. To achieve such a shift, the following ground rules must be followed in technology development and dissemination as well as in public policy.

First, production advances must be based on linking the ecological security of an area with the livelihood security of the people in a symbiotic manner. Second, steps must be taken to create widespread awareness of the human and animal population-supporting capacity of different ecosystems. Sustainable systems of management of soil, water, biodiversity, and forests should be internalized in rural societies.

Third, since the poor remain poor because they have no productive assets and there is no value to their time, *asset creation and value addition to time should receive high priority in poverty alleviation programs.* Women belonging to the economically underprivileged sections of the society, in particular, are often overworked and underpaid. What they need is a reduction in the number of hours of work and an increase in the economic value of each hour of work. This will call for massive efforts in information and skill empowerment of the poor, particularly women. The emerging technologies are largely knowledge intensive and *hence the transfer of knowledge and market-driven skills can become the most powerful instrument.* Modern information technology affords opportunities for reaching the unreached and thereby achieving a learning revolution within a short span of time.

Fourth, equal attention is needed to the problems of the rural and urban poor. Lack of livelihood opportunities in rural areas leads to the proliferation of urban slums. Damage to common property resources in villages results in the growth of environmental refugees. In many developing countries, agriculture, including crop and animal husbandry, forestry, fisheries, and agro-processing, provides most of the jobs and income in rural areas. Therefore the triple challenge of producing more food, income, and jobs from diminishing per capita land, water, and nonrenewable energy sources can be met only through agricultural intensification, diversification, and value addition. Integrated, intensive farming systems that are ecologically sustainable are needed for this purpose.

Fifth, microcredit and microenterprises should be linked in a symbiotic manner, as is being done successfully in Bangladesh through the Grameen Bank and other innovative efforts. Finally, an "Ever-green Revolution" of the kind described above can be given a self-propelling and self-replicating momentum only if it is based on the self-mobilization of the people. *In all externally funded and introduced development projects, there should be a built-in withdrawal strategy, so that the program does not collapse when the external inputs are withdrawn.*

Meeting the Multidimensional Challenges: Response of the M. S. Swaminathan Research Foundation

I will describe briefly each of the responses being developed and field tested by MSSRF to identify implementable approaches at the micro and policy levels to meet the challenges outlined earlier.

Linking the ecological security of an area with the livelihood security of the local community: Creating an economic stake in conservation. The community biodiversity program of MSSRF illustrates how such mutually beneficial linkages can be fostered in biodiversity-rich areas. It is a sad fact that the tribal and rural families who have conserved and enhanced biodiversity remain poor, while those who are utilizing the products of their efforts become rich. When the conservers have no social or economic stake in conservation, denudation of natural ecosystems becomes more rapid. The MSSRF has adopted a three-pronged strategy for creating an economic stake in biodiversity conservation.

First, a transparent and implementable methodology has been developed for incorporating in sui generis systems of plant variety protection procedures for recognizing and rewarding informal innovations in genetic resources conservation and enhancement.

Second, a symbiotic social contract between commercial companies and tribal and rural families is being fostered for the purpose of promoting the cultivation by local communities of genetic material of interest to the companies on the basis of buy-back arrangements. Such a linkage will prevent the primary material being unsustainably exploited.

Third, local women and men are trained in the compilation of biodiversity inventories and in biomonitoring, so that they themselves become custodians of their intellectual property. Such trained women and men constitute an agrobiodiversity conservation corps and will be able to help their respective communities to deal with issues such as "prior informed consent" in the use of genetic resources.

For assisting the community biodiversity movement, MSSRF has established a technical resource center for the implementation of the equity provisions of the Convention on Biological Diversity. This is the first technical resource center of its kind in the world and contains six major components.

1. Chronicling the contributions of tribal and rural families to the conservation and enhancement of agrobiodiversity through primary data collection in the states of Tamil Nadu, Kerala, Andhra Pradesh, and Orissa, as well as in the Lakshadweep and Great Nicobar group of islands.

2. Organization of an agrobiodiversity conservation corps of young tribal and rural women and men who have a social stake in living in their respective villages and who, with appropriate training, can undertake tasks such as compilation of local biodiversity inventories, revitalization of the in situ genetic conservation traditions of their respective communities, monitoring of ecosystem health with the help of appropriate bio-indicators, and restoration of degraded sacred groves. The members of the corps will be able to assist their respective communities in dealing with the prior informed consent provision of the Convention on Biological Diversity in the use of genetic resources.
3. Development of multimedia databases documenting the contributions of tribal and rural families in the conservation and improvement of agrobiodiversity, for the purpose of enabling them to secure their entitlements from the national and global community gene funds.
4. Maintenance of a community gene bank and herbarium. A community gene bank with facilities for medium-term storage has been established to conserve farmer-preserved and developed seeds from the tribal areas of South India. The material will be cataloged and linked to the technical resource center database. The herbarium serves as a reference center for the identification of landraces, traditional cultivars, and medicinal plants conserved by tribal and rural families.
5. Revitalization of genetic conservation traditions of tribal and rural families through social recognition of their contributions and the creation of an economic stake in conservation. For this purpose, replicable models of private sector engagement in contract cultivation of commercially valuable plants by tribal and rural families are being developed.
6. Establishment of a Legal Advice Cell to make available to tribal and rural families appropriate legal advice relating to intellectual property rights and plant variety protection.

The Population-supporting capacity of ecosystems: Local level sociodemographic charter. In order to help internalize an understanding of the vital need to restrict population growth within the supporting capacity of land, water, forests, and the other components of the ecosystem, training modules have been developed to enable the women and men members of village-level democratic institutions to prepare sociodemographic charters for their respective villages. A gender code is an important component of the charter. Such sociodemographic charters will help local communities to view population issues in the context of social development and to ensure that children are born for happiness and not just for existence.

Information and skill empowerment. For this purpose, the concept of information villages has been developed. Trained rural women and men will operate information shops where generic information on the meteorological, management, and marketing factors relevant to rural livelihoods will be converted into location-specific information. Trained farm women and men themselves will become trainers. The computerized extension system adopted in the information shops also helps to sensitize local families on their entitlements from government and other programs. Information technologies provide numerous opportunities for value-added jobs in rural areas. While new technologies are important, folk media are often even more effective in reaching the unreached. Hence, folk plays, folk arts, and theater are fully mobilized for achieving information empowerment. For ensuring the success of information empowerment programs, the information disseminated should be demand driven and locale specific.

Agricultural intensification, diversification, and value addition. These approaches are achieved through participatory research with farm families. Ecotechnologies such as integrated pest management and integrated nutrient supply are used. Ecotechnology development involves blending the best in frontier technologies with traditional wisdom and practices. Modern science and the ecological prudence of the past can thus be combined.

Ecotechnologies are also practiced in aquaculture. Integrated agriculture and aquaculture techniques enhance both farm income and the nutrition security of the household. Whole villages are being enabled to adopt such Integrated, Intensive Farming Systems (IIFS). This approach is essential for meeting the triple goals of more food, income, and jobs from the available land and water resources. There are seven basic principles guiding the IIFS movement.

1. Soil health care. This is fundamental to sustainable intensification. The IIFS approach fosters the inclusion of stem-nodulating legumes like *Sesbania rostrata*; the incorporation of *Azolla*, blue-green algae, and other sources of symbiotic and nonsymbiotic nitrogen fixation; and the promotion of cereal-legume rotation in the farming system. In addition, vermiculture composting and organic recycling constitute essential IIFS components. Farmers using IIFS are trained to maintain a soil health card to monitor the impact of farming systems on the physical, chemical, and microbiological components of soil fertility.

2. Water harvesting and management. The IIFS farm families include in their agronomic practices measures to harvest and conserve rainwater so that it can be used in a conjunctive

manner with other sources of water. Where water is the major constraint, technologies that can help to optimize income and jobs from every liter of water are chosen and adopted. Maximum emphasis is placed on on-farm water use efficiency and on the use of techniques such as drip irrigation that help to optimize the benefits from the available water.

3. Crop and pest management. Integrated nutrient supply (INS) and integrated pest management (IPM) systems form important components of IIFS. The precise composition of the INS and IPM systems depends on the components of a farming system as well as on the agro-ecological and soil conditions of the area. Computer-aided extension systems will provide farm families with timely and precise information on all aspects of land, water, pest, and postharvest management.

4. Energy management. Energy is an important and essential input. Besides the energy efficient systems of land, water, and pest management described earlier, every effort will be made to harness biogas, biomass, solar, and wind energies to the maximum extent possible. Solar and wind energy will be used in hybrid combinations with biogas for farm activities like pumping water and drying grains and other agricultural produce.

5. Postharvest management. Farmers using IIFS will not only adopt the best available threshing, storage, and processing measures, but will also try to produce value-added products from every part of the plant or animal. Postharvest technology assumes particular importance in the case of perishable commodities like fruit, vegetables, milk, meat, eggs, fish, and other animal products and processed foods. A mismatch between production and postharvest technologies adversely affects both producers and consumers. Growing urbanization leads to a diversification of food habits. Therefore there will be increasing demand for animal products like milk, cheese, eggs, and processed food. Agro-processing industries can be justified on the basis of an assessment of consumer demand and should be promoted in villages to increase employment opportunities for rural youth. In addition, these industries can help to mitigate micronutrient deficiencies in the diet. Investment in sanitary and phytosanitary measures is important for providing quality food, both for domestic consumers and for export. To assist the spread of IIFS, government should make major investments in storage, roads, and transportation, as well as in sanitary and phytosanitary measures.

6. Choice of the crop and animal components of farming systems. In IIFS, it is important to give very careful consideration to the

composition of the farming system. Soil conditions, water availability, agro-climatic features, home needs, and above all, marketing opportunities will have to determine the choice of crops, farm animals, and aquaculture systems. Small and large ruminants will have a particular advantage among farm animals, because they can live largely on crop biomass. Backyard poultry farming can help to provide supplementary income and nutrition.

7. Information, skill, organization, management, and marketing empowerment. The IIFS is based on the principle of precision farming. Hence, for its success the IIFS needs a meaningful and effective information and skill empowerment system. Decentralized production systems will have to be supported by a few key, centralized services, such as the supply of credit, seeds, biopesticides, and animal disease diagnostics. Ideally, an information shop will have to be set up by trained local youth to give farm families timely information on their entitlements as well as on meteorological, management, and marketing factors. Organization and management are key elements, and depending on the area and farming system, steps will have to be taken to provide small producers with the advantages of scale in processing and marketing.

The IIFS is best developed through participatory research between scientists and farm families. This will help to ensure economic viability, environmental sustainability, and social and gender equity in IIFS villages. The starting point is to learn from families who have already developed successful IIFS procedures.

It should be emphasized that IIFS will succeed only if *it is a human-centered rather than a mere technology-driven program*. The essence of IIFS is the symbiotic partnership between farm families and their natural resource endowments of land, water, forests, flora, fauna, and sunlight. Without appropriate public policy support in areas like land reform, security of tenure, credit supply, rural infrastructure, input and output pricing, and marketing, small farm families will find it difficult to adopt IIFS.

Increasing farm and nonfarm employment. The biovillage program addresses three key areas: preventing resource degradation, improvement of crop and animal productivity, and alleviation of poverty. The biovillage program in progress in villages in the Pondicherry region of India places equal emphasis on off-farm livelihood opportunities and on-farm jobs. This program avoids a patronage approach to poverty alleviation. It regards the poor as producers and innovators and helps to build their assets through value addition to time and labor. The basic approach is on asset building and sustainable human development leading to the growth of entrepreneurship.

The programs are designed on a pro-nature, pro-poor, and pro-women foundation. By placing emphasis on strengthening the livelihood security of the poor, the biovillage model of sustainable development revolves around the welfare of the economically and socially underprivileged. It is thus a human-centered pattern of development. The enterprises chosen are based on marketing opportunities. The technological and skill empowerment of the poor is the major approach. Because of the market-driven nature of the enterprises, the economic viability of the biovillage approach is assured. Production and postharvest technologies and farm and nonfarm occupations are brought together in a manner that benefits both producers and consumers.

Biovillages around biosphere reserves would help provide alternative sources for meeting the day-to-day needs for food, fuel, fodder, and other commodities of the families living near such biodiversity-rich areas. Also, biovillages near urban areas would help to link the rural producer and the urban consumer in a mutually beneficial partnership.

Producing the processed and semiprocessed food products required in urban areas in the villages around towns and cities minimizes the need for the rural poor to migrate to urban centers for livelihood opportunities. Also, food processing can be used as a method of providing the needed micronutrients by including millets and grain legumes in the food.

Toward an Ever-green Revolution in Agriculture

The term "Green Revolution," coined by Dr. William Gaud of the U.S. Department of Agriculture in 1968, has come to be associated not only with higher production through enhanced productivity, but also with several negative ecological and social consequences. There is also frequent reference to the "fatigue of the Green Revolution," due to stagnation in yield levels and to a larger quantity of nutrients required to produce the same yield as in the early 1970s. Experts such as Lester Brown of the Worldwatch Institute have been warning about an impending global food crisis due to increasing population, increasing purchasing power leading to the consumption of more animal products, increasing damage to the ecological foundations of agriculture, declining per capita availability of land and water, and the absence of technologies that can further enhance the yield potential of major food crops.

Should we therefore assume that as we enter a new millennium we will not have the benefit of new technologies which can help farmers to produce more food and other agricultural commodities from less land and water?

I believe we are now in a position to launch an Ever-green Revolution that can help to increase yield, income, and livelihoods per unit of land and water, if we bring about a paradigm shift in our agricultural research and development strategies. The Green Revolution was triggered by the genetic manipulation of yields in crops such as rice, wheat, and maize. The Ever-green Revolution will be triggered by farming systems that can help to produce more from the available land, water, and labor resources without either ecological or social harm. Thus, progress can be achieved if we shift our mind-set from a commodity-centered approach to an entire cropping or farming system. This does not mean that we should decelerate our efforts in crop improvement research. But such research should be tailored to enhancing the performance and productivity of entire production systems. The transition from the fatigue of the Green Revolution to an Ever-green Revolution involves a shift from a crop-centered to a systems-based approach to technology development and dissemination.

Let us take for example the prospects for "super-rice," capable of yielding over 10 tonnes of rice per hectare. Such a rice plant will need a minimum of 200 kg of nitrogen per hectare, together with other major and micro-nutrients. Addition of such nutrients solely through mineral fertilizers will lead to serious environmental problems; hence, the introduction of legumes in the rotation becomes important.

Scientists now have unique opportunities for designing farming systems to achieve the triple goals of *more food, more income, and more livelihoods* per hectare of land provided we harness the tools of ecotechnologies resulting from a blend of traditional knowledge with frontier technologies. Such tools include biotechnology, Geographic Information Systems (GIS) mapping, space technology, renewable energy technologies (solar, wind, biomass, and biogas), and management and marketing technologies.

We can enter a millennium of hope only if we abandon the old concept of a crop-centered Green Revolution and replace it with a farming systems– and frontier technologies–centered Ever-green Revolution. Ecotechnologies should have pride of place among frontier technologies.

Industrial countries are responsible for many global environmental problems such as potential changes in temperature, precipitation, sea level, and incidence of ultraviolet B radiation. While further agricultural intensification in industrialized countries will be ecologically disastrous, the failure to achieve agricultural intensification and diversification in developing countries where farming provides most of the jobs will be socially disastrous. This is because agriculture, including crop and animal husbandry, forestry and agro-forestry, fisheries, and agro-industries, provides livelihood to over 70 percent of India's population. The smaller the farm, the greater is the need for higher marketable surplus to increase income. Eleven million new livelihoods will have to be created every year in India and these must come largely from the farm and rural industries sectors. Importing food and other agricultural commodities will thus have the same impact as importing unemployment. What we need now is an environmentally sustainable and socially equitable Green Revolution or what may be termed an Ever-green Revolution.

Those who advocate going back to the old methods of farming ignore the fact that only a century ago when the population of undivided India was 281 million, famines claimed 30 million lives between 1870 and 1900. The famine eradication strategy comprising the following steps is perhaps the most important achievement of post-independence India:

- enhanced production and productivity;

- better distribution through the public distribution system;

- adequate grain reserves;

- purchasing power enhancement through various employment generation and guarantee schemes; and

- special intervention programs for children, pregnant and nursing mothers, and old and infirm persons.

While famines have been prevented, widespread undernutrition prevails among the economically underprivileged. Since nonfood factors such as health care, environmental hygiene, and literacy play an important role in promoting sustainable food security at the level of the individual, we should redefine security and the steps for achieving sustainable food security as follows. First, sustained physical access to food will involve a transition from chemical- and machinery-intensive to ecological farming technologies. Second, the emphasis on economic access underlines the need for promoting sustainable livelihoods through multiple income-earning opportunities. Third, environmental access involves on the one hand attention to soil health care, water harvesting management, and the conservation of forests and biodiversity, and on the other to sanitation, environmental hygiene, primary health care, and primary education.

If the political vision to implement this mission is forthcoming, population stabilization can be more readily achieved. The prediction of the French philosopher, Marquis de Condorcet, made in 1795, that population will stabilize itself if children are born for happiness and not just for existence will then come true.

The emphasis on the individual is important, since the household often is not a homogeneous unit. Women and girls tend to suffer more from undernutrition than men and boys. The UNDP's 1995 "Human Development Report" (UNDP 1995) contains distressing data on the growing feminization of poverty. To give operational content to such a concept of food security, we should initiate a Hunger-Free Area Programme (HFAP) with the following objectives.

- Ensure sustainable availability of food by maintaining the growth in food production over population growth through the development and dissemination of ecotechnologies, supported by appropriate packages of services and public policies. Ecotechnology involves blending the ecological prudence and technologies of the past with the best in frontier technologies, particularly biotechnology, information technology, space technology, renewable energy technology, and management technology. Without ecotechnological empowerment, farmers will not be able to produce more food and other agricultural commodities on an environmentally sustainable basis from fewer land, water, and energy resources.

- Sustain the productivity of the natural resource base by conserving and improving the ecological foundations essential for continuous advances in crop and animal productivity.

- Ensure adequate household incomes through promotional social security, such as accessing assets, employment, and organizational and marketing empowerment. Agricultural programs should aim concurrently at more food, more jobs, and more income. Integrated attention to farm and nonfarm employment and value-addition to primary agricultural commodities will be necessary to enhance income and rural livelihood security.

- Provide entitlement to food for the vulnerable groups through protective social security measures such as employment guarantees and food for nutrition programs.

- Introduce a National Food and Livelihood Security Act with the concurrence of the National Development Council for the purpose of paying integrated attention to important issues. These include conserving land, water, forests, and biodiversity, and protecting the atmosphere; enhancing productivity through ecotechnologies; improving distribution to eliminate endemic hunger; maintaining adequate food security reserves; strengthening the techno-infrastructure for better postharvest technology; expanding the coverage of sanitary and phytosanitary measures; and developing efficient research, education, extension, and marketing systems to take full advantage of emerging opportunities in international trade and to ensure that research and extension designed to promote public good receive adequate support.

Potential Future Challenges: Climate Change

In 1995 the Intergovernmental Panel on Climate Change (IPCC) published a second assessment of climate change and its impact. The study confirmed the potential for climate change due to the injection of greenhouse gases such as carbon dioxide (CO_2), methane, nitrous oxide, and others. Further refinement of general circulation models (GCM) and coupling of them with ocean models and negative feedback from aerosols led IPCC to project an increase in global mean surface temperature of 2°C, which is lower than that projected in 1990. The low value of climate change projects a change of 1°C by 2100. However, changes at the regional level continue to remain difficult to project. A mean change of 2°C, which is the "best estimate," could result in changes of from less than 0°C to more than 4°C in different regions. It is these changes that are important for biota, particularly the biodiversity and agriculture production system. The projected changes for this part of the world (India) show a change of 1°C with increased cloudiness and approximately 5 percent change in rainfall.

Sinha and Swaminathan (1991) showed that in India a temperature change of 1.7°C in a large grid caused a change from -0.4 to 4.5°C from coastal to interior areas. Recent studies in the United States also show that an increase of 0.4°C has resulted in changes from 0 to 2°C or more.

These changes would have an effect on major crops, especially grain crops. An increase of 1 to 2°C combined with reduced radiation causes enhanced sterility of spikelets in rice. Fortunately there is already evidence of genetic variability in rice for tolerance to both higher temperature and reduced radiation. Crops such as pigeon pea promote vegetative growth with enhanced humidity and reduced radiation. We do not know the response of most vegetable and fruit crops.

The general view has been that enhanced CO_2 concentration would benefit crops through an improved photosynthesis rate. At the doubling of CO_2 equilibrium, the actual concentration of CO_2 could be only around 450 parts per million (ppm) in the atmosphere. Therefore, we should be cautious in extrapolating data from a 600 ppm or higher concentration of CO_2. We need more studies on the interaction of CO_2, temperature, and radiation for projecting the impacts of climate change on individual crop and agricultural systems.

The projected changes in climate, particularly temperature, humidity, and radiation could have a profound effect on diseases, pests, and microorganisms. Incorporation of disease and pest resistance at higher temperature levels should be an important objective in crop improvement programs. It is known that a change of 1°C changes the virulence of some races of rush infecting wheat. The whole life cycle of insects can be affected by a small change in temperature.

The impact of change in precipitation and temperature on crops has received much less attention than it deserves. Records show that the rainfall pattern has changed during the last 75 to 80 years. In some areas there has been either an overall decrease in rainfall or the total number of rainy days has decreased while maintaining the same total rainfall. This causes increased intensity of rain and results in greater soil erosion and runoff. Thus new crops and varieties will have to adapt to a greater degree of soil and water stress. In this respect traditional agricultural systems such as mixed cropping may prove advantageous. Consequently, crop improvement for intercropping/mixed cropping should possibly receive a greater emphasis.

The Final Milestone: A Hunger-free World

In his budget speech delivered on July 17, 1996, the Chief Minister of Tamil Nadu, India, made the following announcement.

> "Feed the people who are hungry,
> Educate the people to uplift the World"

To fulfill this dream of Mahakavi Bharatiyar (a Tamil poet who fought for India's independence), this Government will launch a new "Hunger-Free Area Programme" with an aim to eradicate poverty-induced hunger. A number of schemes are already under implementation to alleviate poverty and to cater to the nutrition requirements of different groups of the population. Gaps in this coverage will be identified which can then be specifically targeted under the Hunger-Free Area Programme. Provision has been made in the Budget for preparing a detailed strategy to implement this programme in association with Dr. M. S. Swaminathan (Karunanidhi 1996).

Studies at MSSRF have shown that by adding a horizontal dimension to numerous vertically structured programs and by promoting a coalition of all concerned with ending hunger and deprivation, it is now possible to provide opportunities for a healthy and productive life for all.

Designing a Hunger-free Area Programme

The problem of food and nutrition security at the level of the individual has to be viewed in three dimensions. First, inadequate purchasing power leads to calorie-protein undernutrition. Second, the lack of the needed quantity and variety of micronutrients and vitamins in the diet leads to several nutritional disorders, including blindness caused by vitamin A deficiency. This kind of problem is referred to as hidden hunger, which affects more than 2 billion people in the world today. Third, lack of environmental hygiene and sanitation leads to a low biological absorption and retention of food due to intestinal infection and diarrhea. Thus, both food and nonfood factors assume importance in determining the nutrition security of an individual. Concurrent action at all levels is necessary in a Hunger-Free Area Programme. The steps involved in this process are the following.

1. Identify the basic underlying reasons for chronic under- and malnutrition.

2. Collate information on available programs and opportunities for the sustainable end of hunger.

3. Articulate the steps needed to provide the "missing elements" in achieving the end of hunger.

4. Mobilize local level community action and commitment for achieving the end of endemic hunger by the end of the Ninth Five Year Plan and enlist mass media support for this purpose.

5. Assess the extent of financial resources required, technical resources necessary for technological empowerment, and managerial and organizational resources needed.

6. Foster the organization of a grassroots-level coalition for ending hunger in each HFAP region. Each coalition would include delegates elected by the people as well as representatives from government agencies, nongovernmental organizations, voluntary agencies, service organizations (Rotary clubs and Lions clubs), academia (research, training institutions, and universities), the corporate sector, financial institutions, and the mass media.

Global Action for Sustainable Food Security

The commitment of the international community to ensure food and nutrition security to every child, woman, and man on our planet has been reiterated at several international conferences held during this decade.

1990   World Summit for Children, New York

1992   International Conference on Nutrition, Rome

1992   UN Conference on Environment and Development, Rio de Janeiro

1993   World Conference on Human Rights, Vienna

1994   International Conference on Population and Development, Cairo

1995   UN Conference on Social Development, Copenhagen

1995   Fourth World Conference on Women, Beijing

1996   World Food Summit, Rome

Despite such high-level political reaffirmation of the need to achieve rapidly the goal of "Food for All," over 800 million human beings now suffer from hunger and malnutrition. Hidden hunger, arising from the deficiency of essential micronutrients, is widespread. Endemic hunger is now due more to the lack of adequate purchasing power at the household level than to the lack of availability of food in the market. Both poverty and nonfood factors such as environmental hygiene, sanitation, and lack of safe drinking water are becoming major contributors of food insecurity at the level of communities and individuals. A majority of those who are food insecure live in South and Southeast Asia.

There is also increasing concern about the earth's capacity to produce adequate food for the growing population. The increase in human population, now confined mostly to the developing countries, coupled with enhanced purchasing power leading to greater capability to buy food as well as increasing urbanization will lead to a higher growth rate in demand for food as well as for more diversified food products in the coming millennium. Thus, on the one hand, there will be a need for intensification and diversification of agriculture, particularly in the developing countries. On the other hand, the ecological foundations essential for sustainable advances in crop and animal productivity are being eroded. This situation has led to the ringing of alarm bells by experts concerning the earth's capacity to produce adequate food to meet growing needs (Brown 1995). Lester Brown has concluded that the growth in world food production has been hindered by constraints imposed by the earth's natural systems, the environmental degradation of land and water resources, and the diminishing backlog of yield-raising agricultural technologies. This raises questions about the earth's population-carrying capacity. The overriding environmental issue for the twenty-first century is the earth's capacity to produce enough food to satisfy expanding demand (Brown and Kane 1994).

The year 1998 will mark the bicentenary of Malthus's essay on population. The onward march of science and technology, supported by appropriate public policies, has so far helped humankind to keep Malthusian forebodings on the population–food supply equation at bay. It is, however, becoming increasingly clear that without concerted national and global action to arrest environmental degradation, promote economic and gender equity, and check population growth, the Malthusian specter may still come true in the next century. The Food and Agriculture Organization of the United Nations (FAO), therefore, organized a World Food Summit in Rome from November 13 to 17, 1996. At this summit, a Rome Declaration on World Food Security and a World Food Summit Plan of Action were adopted. Following are highlights of these resolutions.

Rome Declaration on World Food Security

The Declaration makes a commitment to reduce the number of undernourished people to one-half their present level no later than 2015. It also states: "We will pursue participatory and sustainable food, agriculture, fisheries, forestry and rural development policies and practices in high and low potential areas which are essential to adequate and reliable food supplies . . .We will strive to ensure that food, agricultural trade, and overall trade policies are conducive to fostering food security for all through a fair and market-oriented world trade system" (FAO 1996, 3).

World Food Summit Plan of Action

The Plan of Action includes seven commitments.

1. Create an enabling political, social, and economic environment designed to create the best conditions for the eradication of poverty, and for durable peace.
2. Implement policies aimed at eradicating poverty and inequality and improving physical and economic access to sufficient, nutritionally adequate, safe food and its effective utilization.
3. Pursue participatory and sustainable food, agriculture, fisheries, forestry, and rural development policies and practices in high- and low-potential areas.
4. Ensure that food, agricultural trade, and overall trade policies are conducive to fostering food security for all.
5. Prevent and be prepared for national disasters and man-made emergencies.
6. Promote optimal allocation and use of public and private investments to foster human resources, sustainable food, agriculture, fisheries and forestry systems, and rural development, in high- and low-potential areas.

7. Implement, monitor, and follow up the Rome Plan of Action at all levels in cooperation with the international community.

The above plan of action integrates three principles: *conservation* of the ecological foundations essential for sustainable advances in agricultural productivity and production; *sustainable production* through ecotechnologies; and *equity in access* to food through an appropriate blend of national and international action in the area of entitlements and employment. Food and livelihood security must be viewed in an inter-linked manner.

Conclusion

We are facing a battle against time in safeguarding our natural resources. In his book, *The Diversity of Life,* Edward O. Wilson has warned that *Homo sapiens* are in imminent danger of precipitating a biological disaster of a greater magnitude than anything we have witnessed so far in our evolutionary history (Wilson 1992). There is no time to relax if we are to ensure that the Malthusian prophecy of famine and pestilence does not come true in the coming millennium. Legal, educational, and participatory measures of program implementation and benefit sharing will all be needed for promoting a peoples' movement for conservation.

It is clear that the concept of sustainable development should be broad based to incorporate considerations of ecology, equity, employment, and energy, in addition to those of economics. This will call for a systems approach in project design and implementation. Both unsustainable lifestyles and unacceptable poverty have to be eliminated. Factors that influence climate and sea level have to be addressed with the seriousness they deserve and need. Sustainable development will become a reality if we keep in mind that the greatest responsibility of our generation, as suggested by Dr. Jonas Salk, is to be good ancestors.

The first and foremost responsibility of governments, the corporate sector, and civil society is to give every child an opportunity for the full expression of her or his innate genetic potential for physical and mental development. This is an achievable goal provided the rich regard themselves as trustees of their financial and intellectual wealth and not just as owners. That there is adequate surplus wealth in the world is clear from 20 percent of the human population earning over 80 percent of the global annual income. According to the UNDP's 1996 "Human Development Report," the world's 358 billionaires are worth more than the combined income of the poorest 2.5 billion people (UNDP 1996). If the concept of trusteeship of financial and intellectual wealth becomes the twenty-first century human ethic, we can lay the foundation for a socially, ecologically, and economically sustainable future for humankind. This is the task as well as the hope for the new millennium.

It is appropriate to conclude with a statement from a document by the U.S. President's Council on Sustainable Development titled "Sustainable America."

Prosperity, fairness, and a healthy environment are interrelated elements of the human dream of a better future. Sustainable development is a way to pursue that dream through choice and policy. (The President's Council 1996, 4).

References

Brown, Lester R. 1995. *Who Will Feed China? Wake-up Call for a Small Planet.* New York: W. W. Norton.

Brown, Lester R., and Hal Kane. 1994. *Full House: Reassessing the Earth's Population Carrying Capacity.* New York: W. W. Norton.

Food and Agriculture Organization of the United Nations (FAO). 1996. *Rome Declaration on World Food Security and World Food Summit Plan of Action.* A report from the World Food Summit held in Rome, November 1996.

Graham-Smith, F., ed. 1994. *Population—the Complex Reality: A Report of the Population Summit of the World's Scientific Academies.* London: The Royal Society.

Karunanidhi, M. 1996. *Budget 1996–97.* Government of Tamil Nadu. July 17.

The President's Council on Sustainable Development. 1996. Sustainable America: A New Consensus for Prosperity, Opportunity, and a Healthy Environment for the Future. Washington, D.C.: The President's Council.

Repetto, Robert, and the World Resources Institute. 1996. *Has Environmental Protection Really Reduced Productivity Growth: We Need Unbiased Measures.* New York: World Resources Institute.

Schmidheiny, Stephan. 1992. Changing Course: A Global Business Perspective on Development and the Environment. Cambridge: MIT Press.

Sinha, S. K., and M. S. Swaminathan. 1991. Deforestation, Climate Change and Sustainable Nutrition Security: A Case Study for India. *Climate Change* 19:201–9.

United Nations Development Programme (UNDP). 1995. Human Development Report. Delhi: Oxford University Press.

_____. 1996. Human Development Report. Delhi: Oxford University Press.

Wilson, Edward O. 1992. *The Diversity of Life.* New York: W. W. Norton.

Background References

International Commission on Peace and Food. 1994. *Uncommon Opportunities: An Agenda for Peace and Equitable Development.* London: Zed Books.

M. S. Swaminathan Research Foundation. 1996. *Uncommon Opportunities for Achieving Sustainable Food and Nutrition Security.* Conference Proceedings of the Science Academies Summit.

Myers, Norman, and Jennifer Kent. 1995. *Environmental Exodus—An Emergent Crisis in the Global Arena.* Washington, D.C.: Climate Institute.

Swaminathan, M. S., ed. 1993. *Reaching the Unreached: Information Technology—A Dialogue.* Bangalore: Macmillan India Ltd.

_____. 1994. *Reaching the Unreached: Ecotechnology and Rural Employment—A Dialogue.* Bangalore: Macmillan India Ltd.

_____. 1996. *Agrobiodiversity and Farmers' Rights.* Madras: M. S. Swaminathan Research Foundation.

CHAPTER SEVEN

Food Security for the Poor: Eliminating the Apartheid Practiced by Financial Institutions

Muhammad Yunus

My efforts to empower and ensure food security for the poor began, appropriately enough, with an independence movement—the independence of Bangladesh. In 1971, my homeland was confronted with a war, with bloodshed, and with a tremendous amount of misery. During that war, I was teaching in the United States. After nine months of fighting, Bangladesh became independent and I went back, thinking I would join the people in the rebuilding process and help create the nation of our dreams. As the days went by, however, the situation in Bangladesh did not improve. It started sliding down quickly and we ended up with a famine in 1974.

By this time, I was teaching economics at Chittagong University in the outskirts of Chittagong City, Bangladesh, and I felt terrible. In the classroom I taught economics with all the enthusiasm of a brand-new Ph.D. from the United States. I felt as if I knew everything, that I had all the solutions. But then I would walk out of the classroom and see skeletons all around me, people waiting to die. There are many, many ways to die, but none is as cruel as to die of hunger; death inches toward you, you see it, and you feel so helpless because you cannot find one handful of food to put in your mouth. And the world moves on.

I could not cope with this daily tragedy. It made me realize that whatever I had learned, whatever I was teaching, was all make-believe; it had no meaning for people's lives. So I started trying to find out why the people in the village next door to the university were dying of hunger. Was there anything I could do as a human being to delay the process, to stop it, even for one person?

I traveled around the village and talked with its people. Soon, all my academic arrogance disappeared. I realized that, as an academic, I was not really solving global problems; I was not even solving national problems. I decided to abandon my bird's-eye view of the world, which allowed me to look at problems from above, from my ivory tower in the sky. I assumed, instead, the worm's-eye view and tried to probe whatever came right in front of me—smelling it, touching it, seeing if I could do something to improve it. Trying to involve myself in whatever capacity I could, I learned many things in my travels.

But it was one particular incident that pointed me in the right direction. I met a woman who was making bamboo stools. After a long discussion, I discovered that she made those stools for the equivalent of two American pennies a day. I could not believe anyone could work so hard, create such beautiful bamboo stools, and make such a tiny profit. She explained that she did not have the money to buy the bamboo that went into the stools; she had to borrow from the trader who required that she sell the product only to him and at a price he decided. She was virtually a bonded laborer for that person. And how much did the bamboo cost? She said about 20 cents; if it was good bamboo, 25 cents. I said, "My god, people suffer for 20 cents and there's nothing anyone can do about it?"

Thinking it over, I wondered whether I should just give her the 20 cents. But then, I came up with a better idea: I could make a list of people in the village who needed that kind of money. After several more days of traveling, a student of mine and I came up with a list of 42 such people. When I added up the total dollars they needed, I felt the biggest shock of my life. The amounts added up to $27! I felt ashamed to be a member of a society that could not provide $27 to 42 hardworking, skilled human beings. To escape my shame I took the $27 out of my pocket, gave it to my student, and said, "Take this money. Give it to those 42 people we met and tell them this is a loan, which they can pay back whenever they are ready. In the meantime, they can sell their products wherever they get a good price."

The people of the village were quite excited to receive the money; such a thing had never happened to them before. Seeing such excitement made me wonder what more I could do to help them. Should I continue providing them money or should I arrange for them to secure their own funds? I thought of the bank located on campus. In a meeting with the manager, I suggested that his bank lend money to the 42 people I had met.

"You're crazy!" he said. "It's impossible! How can you lend money to the poor people? They're not creditworthy."

"At least give it a try, find out," I pleaded with him. "It's only a small amount of money."

"No," he insisted, "our rules don't permit it. We cannot offer them funds, and such a tiny amount isn't worth giving." He suggested that I, instead, meet with higher officials in the Bangladesh banking establishment; maybe they could find a way to provide the loans.

I took his advice, and was told the same thing. Finally, after several weeks of getting the runaround, I offered myself as a guarantor of the loans; I would sign whatever papers the bank wanted, but then I would give the money to whomever I chose. They warned me that such poor people would never pay me back, but I took the chance. As it turned out, the people paid back every cent; nothing was owed. Excited, I went to the bank manager and declared, "Look! They paid me back, and there are no problems."

"No," one high-ranking banking official countered, "they're just fooling you. Soon they'll take more money and never give it back." I accepted his challenge and gave more money, and they paid me back again. Returning to the banking official to report my success, I heard him say, "Well, maybe you can do it in one village, but it won't work in two villages." So I hurried out to try it in two villages, and it worked as well as before.

Soon, a race developed between the banking community and me; the bankers kept advancing larger numbers—5 villages, 20 villages, hundreds of villages—expecting that my stream of successes would eventually be broken. But I tried my approach in 5 villages, 20 villages, hundreds of villages, until I thought those bankers could not deny my results. Yet, they still would not accept my successes. The bankers had been trained to believe that poor people are not capable. Their minds were blind to the results they had seen. Luckily, my mind had not been trained that way.

Finally I thought, why am I trying to convince *them?* I am already convinced that poor people can take money and pay it back. So why don't I just set up a separate bank of my own? I wrote up a proposal and went to the government for permission. It took two years to convince them.

"Why should you have a bank for the poor people?" the government officials asked. "There are a lot of banks already under way that are having trouble. Why do you want to create more problems for us?"

"I don't want money from you," I clarified, "all I want is permission to set up a bank." At last, in 1983, Grameen Bank—a formal, independent financial institution—was opened. I opened it as an alternative to the current banking system, which I found to be biased against both the poor and women.

Bankers got angry with me, asking, "Why do you accuse us of being biased against women?"

I answered, "You just give me your list of all the borrowers in all the banks. I'll bet you that not even 1 percent of the borrowers are women. So what else would you call it?" Recognizing this bias, I wanted to make sure that women made up half of all Grameen's borrowers. But it was not easy trying to persuade the women in Bangladesh to join the bank. A man is not even allowed to address a woman in the villages.

But after trying several roundabout methods, we at Grameen Bank finally found a way to communicate with the women. Even then, the usual response was, "No, I don't need money. Give it to my husband." We kept telling the women that we understood their husbands could take it, but we wanted to give the money to them if they needed it for an idea. Yet they would say, "No, I don't have an idea." And that was repeated in village after village, by woman after woman. It took a lot of convincing before any woman could believe that she, herself, could take a loan and use it to earn income. After she believed in herself, we asked her to find four friends who, as a group, would borrow from the bank. All we needed was patience. Once we could persuade one woman, our work was half done. She went around convincing others. If one woman got a loan and started her business, all women were ready to talk.

Getting the women of Bangladesh to see themselves as family "bread-winners" represented an enormous leap—for us and for them. But still another hurdle remained: they had to wait for the day they would receive their loan. It was this waiting that prompted many women to doubt whether they should go through with their loan. They spent the night before sleepless, tossing and turning, telling themselves that they had already created problems for their families—being a woman—and that they did not want to create more problems by borrowing money they might not be able to pay back. But by morning, with the help of friends and much internal struggle, they decided to go through with their first loan, usually about $12 or $15.

Many women could not believe someone trusted them enough to loan them such an amount of money. As tears rolled down their cheeks, they promised themselves to work very hard and make sure they paid back every penny. And they did. Grameen Bank requires tiny weekly payments so in one year loans can be paid back with interest. By the time the loans are paid, the women are completely different people. They have explored themselves, found themselves. Others may have told them they were no good, but on the day a loan is paid off, the women feel as though they can take care of themselves and their families.

We noticed so many good things happening in families where the woman was the borrower instead of the man that we focused more and more on the women, not just 50 percent of our business. Today, Grameen Bank works in 36,000 villages in Bangladesh; it has 2.1 million borrowers, 94 percent of them women; and it has more than 12,000 employees. The bank completed its first billion dollars in loans three years ago, and we celebrated; a bank that started its journey by giving $27 in loans to 42 people and has come all the way to a billion dollars in loans is cause for celebration. We felt good that we had proven all those banking officials wrong. We are now approaching our second billion dollar mark at the end of this year.

The year after celebrating the one billion dollar mark, we gave $300 million in loans, followed by more than $400 million the following year, and another $400 million last year.

When we started, we had no experience running such an organization. I was just a teacher at the university; I had never run any corporation or business of any kind. It was the same for my employees; they had no experience. Most of them were my students who were picked up along the way. Recently we have found out that the combined total of loans made to the rural population by all the banks in Bangladesh does not add up to $400 million.

I went back to the people who are now my banking colleagues and admonished them: "You said poor people are not creditworthy. But for 20 years, they've been showing every day who is creditworthy and who isn't. It has been the rich people in Bangladesh who haven't paid back their loans because only the rich people were granted them. With Grameen Bank, it's the poor people who are paying back." Our recovery rate has remained more than 98 percent since we began. So my question now is, are the banks "peopleworthy"?

Researchers say that there must be some trick: I'm not reporting the right figures, I'm hiding things. But when they do their own reports they see the same numbers. They come in with hostility and leave as great admirers of the bank. Researchers now say that the income of all our borrowers is steadily increasing. The World Bank reports that one-third of our borrowers have clearly risen above the poverty line, another one-third are only months or a couple of years away from this achievement, and the remaining one-third are at different levels below that. I say, if you can run a bank, lend money, get your money back, cover all your costs, make a profit, and get people out of poverty, what else do you want?

One of the topics discussed at the 1994 United Nations Conference on Population and Development held in Cairo, Egypt, was Grameen Bank. *Business Week* ran a cover story on the Cairo conference and half of the article was devoted to our bank. The magazine reported that the adoption of family planning practices by Grameen families is twice as high as the national average. Though we are not a family planning organization, we feel it nonetheless makes sense that when people start making decisions about their lives, they also start making decisions about the size of their families. Sanitation for Grameen families is also much better than for non-Grameen families; likewise for nutrition and housing. The bank also provides housing loans of $300 each—more than 350,000 of them so far, with no difficulty getting back our money. With that $300, a family can build a house with a tin roof, concrete columns, and a sanitary latrine that, by Bangladesh standards, feels like a royal palace. Never in their lives had families thought they could own such a house.

Are poor people creditworthy? Does the world still wait for evidence? Does it care? I keep saying that poverty is not created by poor people. Poverty is created by the institutions we have built around us. We must go back to the drawing board to redesign those institutions so they do not discriminate against the poor as they do now. We have heard about apartheid, and felt terrible about it, but we do not seem to feel anything about the apartheid practiced by financial institutions. Why should some people be rejected by a bank simply because they are *thought* to be unworthy of credit? By the evidence, it is clear that the opposite is true.

It is the responsibility of all societies to ensure human dignity to every member of that society, but we have not done very well in that endeavor. We talk about human rights, but we do not link human rights with poverty. Poverty is the denial of human rights. And it is not just the denial of one human right; put together all the many ways our society denies human rights and that spells *poverty*.

Grameen-type programs are now being established in many countries. To my knowledge, 56 countries, including the United States, are involved in such programs. But the effort does not have the momentum it needs. There are 1.3 billion people earning a dollar or less per day on this planet, suffering extreme poverty. If we create institutions capable of providing the poor with credit, they will see the same success we have seen in Bangladesh through Grameen Bank. I see no reason why anyone in the world should be poor.

The Micro-Credit Summit held in Washington, D.C., on February 2–4, 1997, declared as its goal to provide 100 million of the world's poorest families with credit for self-employment by the year 2005, preferably through the women in those families. If we can achieve that, the stage will be set for a poverty-free world. My fondest dream is to someday see the next generations—our children and grandchildren—have to visit a museum to see what poverty was like. They will accuse us and our ancestors of letting poverty continue until the twenty-first century. But if we take action now, the next millennium will not have to suffer the same fate as the current one.

CHAPTER EIGHT

Meeting World Food Needs for Micronutrients

Nevin S. Scrimshaw

Introduction

Speakers at this symposium have emphasized the population growth anticipated in the decades ahead and the challenge that it represents for the food supply. Economists and most agriculturists view this challenge as supplying *enough* food. For developing countries, these analyses tend to relate this problem to cereal production. However, solving world food problems is a qualitative as well as a quantitative problem. Although the classical nutritional diseases of pellagra, scurvy, beriberi, and kwashiorkor have almost disappeared, widespread deficiencies of a number of critical micronutrients persist (WHO 1989). These deficiencies have both immediate and far-reaching consequences, especially if they occur in infancy or pregnancy. Saving individuals from starvation by supplying sufficient dietary energy is not enough. The goal must be nutritional security.

What Food Must Supply

There are two kinds of hunger: overt and hidden. First, food must supply adequate dietary energy and protein. If it does not, the result is clinically manifested as weight loss, reduced activity, and eventually wasting and death. Second, the body needs a variety of micronutrients—amino acids, minerals, vitamins, and essential fatty acids. A deficiency in *any* of these micronutrients can cause functional damage that is not necessarily clinically apparent. To meet world needs, food must supply sufficient amounts of macro- and micronutrients to prevent both overt and hidden hunger.

Dietary energy can come from any combination of food providing carbohydrate, fat, and protein. However, dietary protein needs cannot be met from cereals alone, and cassava as an energy source provides almost no protein. Without legumes to provide relatively good-quality protein that complements the protein of cereals, the poor are likely to be protein deficient unless they have access to food of animal origin.

The Green Revolution has been tremendously successful in increasing the yield and profitability of cereals. Government emphasis on production of cereals, their greater profitability for the farmer, and the lack of comparable research breakthroughs for legumes have jeopardized protein security due to the decreasing per capita availability and rising cost of legumes. Figure 8.1 illustrates the seriousness of this global trend (FAO 1992).

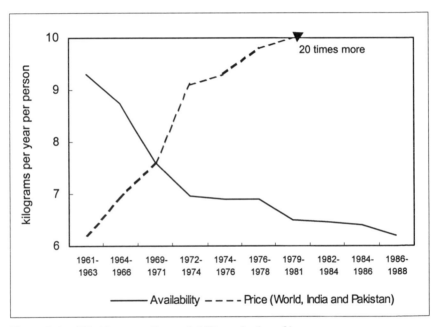

Figure 8.1. World per capita availability and price of legumes
SOURCE: FAO 1992 database (AGROSTAT/PC, Food Balance Sheets, FAO, Rome).
(Reprinted with permission)

Consequences of Specific Nutritional Deficiencies

Before discussing the policies and programs required to prevent protein and micronutrient deficiencies, I will review briefly the consequences of failing to supply sufficient dietary energy, usable protein, and the essential micronutrients. All of these deficiencies are common in developing-country diets today.

Dietary Energy

When dietary energy is less than required for *ad libitum* activity, there is an enforced decrease in physical activity that has social and economic consequences. For young children, decreased activity reduces the stimulus needed for normal brain development. If the energy deficit is greater than can be compensated for by reduced activity, then physical wasting occurs and may reach the extreme of death.

Protein Deficiency

In infancy and early childhood protein deficiency impairs not only physical growth but also cognitive development. At any age it compromises immunity and increases susceptibility to infectious diseases.

Iron Deficiency

In infancy the adverse effects of iron deficiency on learning and behavior are likely to be permanent (Lozoff, Jimenez, and Wolf 1991). At later ages cognitive performance is also affected but is reversible with iron supplements (Beard, Feagans, and Frobose 1995), although not if it is a continuation of deficiency in infancy (Pollitt et al. 1989). At any age iron deficiency can damage immune status, increase morbidity from infectious diseases, impair physical capacity, and even affect the ability to maintain body temperature (Scrimshaw 1990).

Iodine Deficiency

Iodine deficiency during pregnancy, when sufficiently severe, interferes with normal development of the fetal brain (Stanbury 1994), manifested as varying degrees of neurological impairment from deafness to significant permanent cognitive impairment (Boyages 1994). The most severe manifestation is the feebleminded cretinous dwarf.

Vitamin A Deficiency

Severe vitamin A deficiency is a historic cause of eye disease resulting in blindness and is still seen in severely disadvantaged infants and preschool children in some populations. However, of far greater concern is that even mild to moderate subclinical vitamin A deficiency is associated with increased mortality from infectious disease (Beaton et al. 1993).

Deficiency of Other Nutrients

The diet must also supply a variety of other nutrients, the lack of which causes serious health consequences (Scrimshaw and Young 1976). Folic acid deficiency in pregnancy contributes to neural tube defects in the newborn. Dietary vitamin C improves iron absorption; its deficiency leads to spongy, bleeding gums and other symptoms of scurvy. Thiamin and niacin deficiencies, although rare today, are associated with the classical deficiency diseases of beriberi and pellagra. In China, selenium deficiency leads to Caisson's disease. Borderline zinc deficiency impairs immune status in some populations.

Stunting

Almost any nutrient deficiency in infancy and childhood reduces growth (Waterlow and Schürch 1994), which in turn is an indicator of other potential developmental impairments. Moreover, there is a strong tendency for stunted children to become stunted adults who, as has been shown in many countries, have a reduced capacity for physical work (Scrimshaw 1987).

The major point to be recognized is that if *any* of these nutrients is sufficiently deficient in the diet, disability and death can ensue regardless of the diet's caloric content. There are currently several million deaths annually associated with iron and vitamin A deficiencies in developing countries, and lower IQ scores are found in populations with endemic iodine and iron deficiency. The classical deficiency diseases have reappeared in contemporary refugee populations. Moreover, the low body mass index of individuals in developing regions, particularly South Asia, reveals widespread chronic energy deficiency (James and Ralph 1992). The high prevalence of stunting in developing-country populations confirms the almost universal prevalence of some degree of malnutrition during early childhood.

Malnutrition in Pregnancy

The consequences of a diet that fails to provide adequate micro- as well as macronutrients are especially serious for pregnant women because they can seriously affect the fetus and the young child. I have already mentioned the lasting damage to the fetus from iodine, iron, and folic acid deficiency during pregnancy. A series of studies in England by D. J. P. Barker and collaborators, as well as 50 articles and two books (Barker 1992, 1994), identified infants with low birth weight in the years 1913 to 1943 and were able to determine either their cause of death or their current health.

Surprisingly, they found a significant increase in premature morbidity or deaths from chronic degenerative diseases, including hypertension, coronary heart disease, type-II diabetes, chronic bronchitis, obstructive lung disease, and thyrotoxicosis for individuals who were born small-for-date or were retarded in weight at one year of age. Similar findings have been reported in Sweden (Leon et al. 1996) and India (Stein et al. 1996). Related studies have been initiated in the United States and several other countries to replicate these observations.

Impact of Adult Diets on Chronic Diseases

In the last few decades there has been increasing evidence of the influence of adult diets on chronic diseases in later life, including some forms of cancer (Committee on Diet 1982). Some of this has been due to rising affluence that has led to more saturated fat and cholesterol in the diet and to more obesity. Diets rich in vegetables and fruits as well as cereals seem to be protective. It is believed that dietary fiber, dietary antioxidants such as vitamins E and C and carotenoids, selected other vitamins, and some of the multiple phytochemicals in fruits and vegetables are responsible for this protection (NRC 1989). Food supplies are not adequate for the health of either children or adults unless they include sufficient fruits and vegetables.

Nutritional Security

The focus of agricultural and policy efforts of international and bilateral agencies and organizations as well as national governments is commonly stated to be food security. But the concept of food security is complex. Adequate food supplies do not represent food security unless families in need are able to acquire them in some way, whether through purchase, barter, or charity. Moreover, "sufficient" food even at the family level is not enough unless its intrahousehold distribution meets the needs of every household member, particularly pregnant women and young children.

Clearly, food security requires not only adequate food production, but also distribution mechanisms that ensure that families actually acquire the food they need. Food security also needs educational and health interventions to help guarantee appropriate intrahousehold food distribution. It requires the stimulus of both production and effective demand for protective food as well as traditional staple food. If the objectives of programs for the prevention of hunger in underprivileged populations were expressed as nutritional security rather than food security, it would better signal the importance of quality as well as quantity of food. A balanced diet is required for sustained health, not only for individual but also for societal development. This transforms nutritional insecurity from a purely ethical matter to a major economic concern.

Actions Needed to Achieve Nutritional Security

The barriers to achieving nutritional security are economic, political, and social as well as agricultural. There are multiple and concurrent actions that must be taken to complement agricultural approaches to food security in order to achieve nutritional security.

Health measures are essential to nutritional security. Among the health interventions that are also nutritional interventions are immunization against common communicable diseases, improved environmental and personal hygiene to prevent enteric diseases including intestinal parasites, and control of systemic parasitic diseases, such as malaria and schistosomiasis. Promotion of breast-feeding and the timely and appropriate introduction of complementary feeding are also important. Health measures to reduce infant and preschool mortality must be complemented by the availability of effective family planning methods.

Social factors are hard to separate from political and economic ones, but food and health beliefs and practices associated with disadvantaged socioeconomic and ethnic groups often contribute to their nutritional insecurity. Nutrition education can be an important contribution to nutritional security.

Economic barriers interfere with the ability to acquire food by agricultural production, purchase, or barter. Measures to improve incomes or make food available at lower prices will have a favorable influence on nutritional security. Initiatives such as the Grameen Bank of Muhammad Yunus, the 1994 World Food Prize laureate, are extremely important. Economic considerations also limit the capacity of governments to support programs for agricultural extension, nutrition education, health interventions, and food distribution to the needy. However, apparent economic limitations are often really the result of bad public policy.

Political factors include corruption, greed, indifference, tyranny, ignorance, and bad judgment. They may lead to the dislocation and suffering from war and civil disturbances that afflict so many poor countries today. Examples of the food production consequences of bad political leadership are many.

When I visited Accra, Ghana, in 1977, food was scarce and black market food was expensive because the price controls imposed by the government to keep food cheap in the cities were below the cost of production. A year later, with a change in price policy, food was once again plentiful in city markets.

In 1972, India was desperately short of food and famine was averted only by massive external food assistance. Within a decade, India had amassed grain surpluses and large reserve stores. This success was due largely to the outstanding leadership of the 1987 World Food Prize laureate, M. S. Swaminathan, who adapted Green Revolution technology to India and helped bring about a fundamental change in agricultural policies.

In Mexico during the 1960s and 1970s, the Green Revolution technologies resulted in dramatic increases to the point of self-sufficiency in cereal production, as shown in Figure 8.2 (Wellhausen 1976). However, by the mid-decade the limits of its application were being reached. The Green Revolution technologies were not applicable to the rain-fed agriculture of the small dryland farmers who could not compete with the subsidized irrigation provided to large farmers. As a result, hundreds of thousands of peasants were forced to abandon their land and most of them moved to Mexico City. Mexico's food imports surged once again.

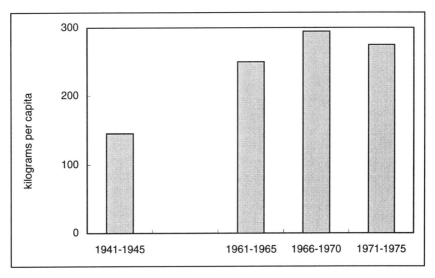

Figure 8.2.   Basic food crop production in Mexico.
SOURCE: Information from Wellhausen 1976.
NOTE: Since 1970 per capita production in Mexico has declined, as the population (now 62 million) has continued to grow at an annual rate of about 3.5 percent.

In 1980, Mexico began to implement the Sistema Alimentaria Mexicana (SAM), developed by a commission appointed by President López Portilla. The plan provided price guarantees; credit for seeds, fertilizer, and pesticides; and extension advice. As detailed by Austin and Estevez (1987), it is a striking example of political factors that can be either the cause of or the cure for food problems. Within one year migration to Mexico City had reversed, agricultural production increased, and food imports decreased. By the second year SAM had reached self-sufficiency targets for maize and beans. It was confidently predicted that the trend would make Mexico self-sufficient in food once again. In its third year, however, it was discarded by a new president and the gains were soon lost.

Agricultural factors that contribute to food insecurity and agricultural measures that improve food security have been identified by other participants in this symposium. However, there is often a subtle difference between the agriculturists' approach to food security and the nutritionists' concept of what is required. There are concurrent multiple actions that must be taken if agricultural approaches to food security are also to achieve nutritional security. Several of these actions have already been successfully pioneered by previous World Food Prize laureates and others participating in this symposium.

Solutions for Achieving Nutritional Security

Food-based solutions to improve protein availability include the following:

1. We must focus on increasing the supply and lowering the cost of legumes, the traditional complement to cereal proteins, through a combination of supportive government policies and research.

2. We must take full advantage of the remarkable breakthrough development of quality protein maize. When fed on an equal protein basis to either preschool children in Guatemala (Bressani, Alvarado, and Viteri 1969) or MIT students (Young et al. 1971), there was no difference in net nitrogen retention compared with milk. If consumed in sufficient quantity, quality protein maize can fully meet human protein requirements.

3. We must also improve the availability of a variety of animal protein sources. Verghese Kurien was awarded The World Food Prize in 1989 for "Operation Flood," an inspiring program that increased both milk supplies and the purchasing power of thousands of poor farmers in India. Other programs elsewhere in the world have increased the production of poultry by poor farmers, helped fishermen to improve their catches, improved livestock production, and improved the availability of animal protein for poor families.

There are several food-based solutions for supplying sufficient micronutrients.

1. Despite the greater difficulties in doing so, the intensity of agricultural research that led to the Green Revolution in cereal production must be extended to fruits and vegetables, to both their production and their postharvest conservation. The 1988 World Food Prize laureate, Robert Chandler, was recognized for

leading the dramatic improvement in rice cultivars at the
International Rice Research Institute (IRRI), and his subsequent
leadership of the Asian Vegetable Research and Development
Center in Taiwan was also important. The international agricultural research centers of the Consultative Group on International
Agricultural Research (CGIAR) and many national institutes
have diversified their programs to include multiple cropping and
a variety of noncereal crops. This is essential to future nutritional security from agricultural production. Continued improvements in both conventional and unconventional animal husbandry and fisheries must also be sought to try to keep pace with
rising population and demand.

2. The fortification of appropriate staples with selected micronutrients must be used as a biologically and economically effective
   way to prevent important deficiencies. In fact, iodine deficiency
   cannot be corrected by local food production in endemic areas;
   the addition of iodine to salt can completely eliminate the
   problem as it already has in more than 60 countries. When
   potassium iodate was successfully added to salt in Guatemala in
   1956, its cost was five cents per 100 pounds, not enough to
   justify an increase in the retail price.[1]

   The addition of iron to cereal flours can do much to prevent
   iron deficiency in a population. It is routine in most industrialized countries and is being adopted rapidly by developing countries as well. Studies in India and Canada have demonstrated
   that double fortification of salt with both iodine and iron holds
   promise for countries experiencing both of these problems.

   The addition of vitamins A and D to margarine is routine in
   most industrialized countries and adding vitamin A to cooking
   oils need not affect their cost to the consumer but will contribute
   significantly to vitamin A nutriture. Although congenital neural
   tube defects resulting in hideously deformed or nonviable off-
   spring are relatively uncommon, they can be virtually eliminated
   in the United States by adding folic acid to the standard
   enrichment mix for cereal flours that already includes thiamine,
   riboflavin, niacin, calcium, and iron.

   Although the classical diseases associated with thiamine and
   niacin deficiencies (beriberi and pellagra) are no longer a threat
   in the United States, low riboflavin intakes are being found in
   the elderly. In developing countries where dietary sources of
   riboflavin in the diet are often meager, the riboflavin contained
   in the standard enrichment mix can be important. It should be
   noted that the cost of the standard enrichment mix is so low that
   there is no difference between the retail prices of enriched and

nonenriched flour and bread in the United States, and there need not be in other countries.

3. A weekly supplement of 60 mg of iron must be provided to children and to women of childbearing age in populations in which anemia is a public health problem. A 60 mg daily or 120 mg weekly supplement of iron and folic acid should be given to all pregnant women in developing countries.

The Role of Education and Extension

A food-based approach to nutritional security requires nutrition education. There are two kinds of demand: a desire for a food, and the ability to acquire the food that is desired. The latter is commonly referred to as *effective* demand. It is obviously futile to expect agricultural success with foods that, however nutritious, will not be eaten. Desire for foods can be modified by education in schools, organizations, and the mass media and by direct extension efforts at the family level. An outstanding example is the promotion of the ivy gourd in Thailand, an effort that significantly increased consumption of this rich source of vitamin A. Agricultural research into less commonly used or new food varieties must progress together with an educational effort to promote the use of such foods.

Agricultural extension is needed not only to show farmers how to grow better staple crops but also to assist families with home gardens, fish ponds, and small animal production. Time and again I have seen striking differences in nutritional status between those villages with a strong tradition and knowledge of household gardens and those in which there are few or no gardens, even in the same country or region. Even in cities, urban gardens can contribute to food and nutrition security.

It is a serious mistake for any developing country to neglect the development of a strong extension system that reaches to the household level. It is necessary to emphasize approaches that have been successfully employed by the industrialized countries for at least 75 years, despite the temptation to focus on the potential of new research and biotechnology advances to solve food problems. We need both approaches, and ultimately research advances will require extension efforts at the farm level for their application.

Health Measures

It must be recognized that health interventions contribute to both food security and nutritional security. In addition to the true nutritional interventions mentioned above, I would be exceedingly remiss if I did not also emphasize the importance of access to a wide choice of family planning methodologies as an essential component of sustainable solutions to nutrition and health problems, as well as economic, social, and environmental ones. The high population growth rates of the poorest developing countries make food and nutrition security almost impossible to achieve without external assistance, and this is usually insufficient and uncertain. Developing countries, such as Thailand and Indonesia, that have made family planning methods accessible *and* introduced social, economic, and health measures that make their use advantageous have had striking success. The result has been not only a reduction in the rate of population growth but also economic and social development, along with a greater degree of food security. However, promotion of family planning without other social and health measures, as in Egypt and India, has had little success.

Economic Measures

While, thus far, I have stressed food production, fortification, nutrition education, agricultural extension, and health measures as essential elements of nutritional security, ultimately all of these depend on economic and political factors. I have given three examples—India, Ghana, and Mexico—to illustrate this point. I am sure that many more will be cited in the course of this symposium. Support for agricultural research and extension and for preventive and primary health care depends on government actions.

Too often in today's world, corrupt and despotic governments, indifferent to the welfare of their people, do not take the actions that would enhance food and nutrition security. Worse still, some promote ethnic conflict that destroys food security through the horrors and suffering from civil war or the mass exodus of refugees. Pragmatically, we must promote the measures that we know to be effective with those governments that are responsive. Fortunately, despite serious and tragic setbacks in several parts of the world, the number of countries that are willing to support actions that improve food and nutrition security is increasing steadily.

Improving the Status of Women

The status of women has an important influence on the nutritional security of the family. Education and freedom of choice for women should be strongly promoted along with efforts to lower the legal, economic, social, and political barriers to achieving equity.

Summary

Food production to meet effective demand is a readily obtainable goal in most countries under most circumstances. However, where need is not matched by the ability to acquire food, food availability is not enough and malnutrition and starvation will occur. Food security has often been measured and discussed primarily in terms of the quantity of food at the national, regional, community, and family level, but nutritional security will not be achieved unless qualitative needs are met and there is appropriate intrahousehold distribution of food.

The ultimate goal should be nutritional security. To achieve this countries need, in addition to agricultural research and productivity, a series of complementary measures. These include health interventions to reduce the burden of disease on nutritional status and food production, effective nutrition education to increase demand for a nutritionally desirable variety of food in the diet, economic measures that will benefit both food production and consumption, and most importantly, the necessary political decisions and support.

Every country can achieve food security and the broader goal of nutritional security, but only if it takes a comprehensive approach to the problem using the knowledge and experience that exist today and that will increase in the future.

Note

1. A one-time cost must be added to the ingredient cost for salt fortification with potassium iodate. This cost includes a mixer, which can be locally made, a motor to drive it, and a precision feeder. Depending on plant size, the investment required is only a few thousand dollars. For flour fortification, most mills already have mixing equipment and need only a precision feeder that costs a few hundred dollars.

References

Austin, J., and G. Estevez. 1987. *Food Policy in Mexico.* Ithaca: Cornell University Press.

Barker, D. J. P., ed. 1992. *Fetal and Infant Origins of Adult Diseases.* London: BMJ Publishing Group.

_____. 1994. *Mothers, Babies, and Disease in Later Life.* London: BMJ Publishing Group.

Beard, J. L., L. Feagans, and C. Frobose. 1995. Cognitive Dysfunction in Iron Deficient Adolescents (Abstract). *Federation of American Societies for Experimental Biology Journal* 9(4):A975.

Beaton, G. H., R. Martorell, K. A. L'Abbé, B. Edmonston, G. McCabe, A. C. Ross, and B. Harvey. 1993. *Effectiveness of Vitamin A Supplementation in the Control of Young Child Morbidity and Mortality in Developing Countries.* Toronto, Ontario: University of Toronto.

Boyages, S. C. 1994. The Damaged Brain of Iodine Deficiency: Evidence for a Continuum of Effect on the Population at Risk. In *The Damaged Brain of Iodine Deficiency,* edited by J. B. Stanbury. Elmsford, NY: Cognitive Communications Corp., 251–58.

Bressani, R., J. Alvarado, and F. Viteri. 1969. Evaluación en Niños de la Calidad de la Proteína del Maíz Opaco 2. *Archivos Latinoamericanos de Nutricion* 19:129–40.

Committee on Diet, Nutrition, and Cancer, Assembly of Life Sciences, National Research Council. 1982. *Diet, Nutrition and Cancer.* Washington, D.C.: National Academy Press.

Food and Agriculture Organization (FAO). 1992. *1992 Database.* AGROSTAT/PC, Food Balance Sheets. Rome: FAO.

James, W. P. T., and A. Ralph, eds. 1992. The Functional Significance of Low Bodymass Index. *European Journal of Clinical Nutrition* 48(S3).

Leon, D. A., I. Koupilova, H. O. Lithell, L. Berglund, R. Mohsen, D. Vågerö, U. Lithell, and P. M. McKeigue. 1996. Failure to Realise Growth Potential *In Utero* and Adult Obesity in Relation to Blood Pressure in 50-Year-Old Swedish Men. *British Medical Journal* 312:401–6.

Lozoff, B., E. Jimenez, and A. W. Wolf. 1991. Long Term Developmental Outcome of Infants with Iron Deficiency. *New England Journal of Medicne* 325:687–95.

National Research Council (NRC). 1989. *Diet and Health: Implications for Reducing Chronic Disease Risk.* Washington, D.C.: National Academy of Sciences Press.

Pollitt, E., P. Hathirat, N. J. Kotchabhakdi, L. Missell, and A. Valyasevi. 1989. Iron Deficiency and Educational Achievement in Thailand. *American Journal of Clinical Nutrition* 50(S3):687–97.

Scrimshaw, N. S. 1987. The Phenomenon of Famine. *Annual Review of Nutrition* 7:1–21.

_____. 1990. Functional Significance of Iron Deficiency. In *Functional Significance of Iron Deficiency,* edited by C. O. Enwonwu. Nashville, TN: Meharry Medical College, Center for Nutrition, 1–13.

Scrimshaw, N. S., and V. R. Young. 1976. The Requirements of Human Nutrition. *Scientific American* 235(3):50–64.

Stanbury, J. B., ed. 1994. *The Damaged Brain of Iodine Deficiency.* Elmsford, NY: Cognitive Communications Corp.

Stein, C. R., C. H. D. Fall, K. Kumaran, C. Osmond, V. Cox, and D. J. P. Barker. 1996. Fetal Growth and Coronary Heart Disease in South India. *Lancet* 348:1269–73.

Waterlow, J. C. , and B. Schürch, eds. 1994. Causes and Mechanisms of Linear Growth Retardation. Proceedings of an I/D/E/C/G workshop held in London, January 15–18, 1993. *European Journal of Clinical Nutrition* 48(S1).

Wellhausen, E. J. 1976. The Agriculture of Mexico. *Scientific American* 235(3):128–50.

World Health Organization (WHO). 1989. *Global Nutrition Status Update 1989*. Geneva: WHO.

Young, V. R., I. Ozalp, B. V. Cholakos, and N. S. Scrimshaw. 1971. Protein Value of Colombian Opaque-2 Corn for Young Adult Men. *Journal of Nutrition* 101:1475–82.

PART II

Pillars of Food Security

CHAPTER NINE

Agricultural Science and the Food Balance

- ◆ MODERATOR
- ◆ RESOURCE PAPER
- ◆ DISCUSSANTS

Moderator ◆ *Kenneth J. Frey*

Food is *the* most critical need for humankind. Provision for a sustained and adequate food supply for humankind remains a delicate balancing act. In recent times, however, this balancing act has moved from a local level to national and international levels of concern.

Throughout the nineteenth century and until the 1930s, additional food could be obtained by increasing the amount of land under cultivation. However, by the end of the 1950s, with the world's human population nearing 3 billion, little additional land was available for agricultural production. Enter science and technology. In 1960, annual world grain production was approximately 0.8 billion metric tons. In the mid-1990s, annual world grain production approaches 2.0 billion metric tons. But more significantly, 1960's production and 1990's production occur on virtually the same land area. What has been responsible for this 2.5-fold increase in grain production is the application of science and technology to production problems: science, to understand plant-environment interactions; technology, in the forms of better-adapted and more productive crop varieties, fertilizer availability and application, irrigation, and chemical pesticides. To put this increase in perspective, we have been using 1.5 billion hectares—an area nearly the size of South America—for crop production since 1950. Without science and technology, today another 2.5 billion hectares of land—an area the size of North America—would be needed to feed the world's human population.

In the 1960s, Dr. Norman Borlaug averred that the Green Revolution would buy about three decades of time for the world to bring human population growth under control. Well, here we are 30 years later with 6 billion persons (i.e., two times the 3 billion on earth when the Green Revolution was initiated), and the population is still increasing at 90 million per year. Science and agricultural technology came to the rescue when no new land area for crop production was available. But, can science and technology continue to provide increased food supplies to meet the needs of an annual human population growth of 90 million? As usually occurs with a question of this type, some experts are pessimists and some are optimists. I happen to be an optimist. I have to be an optimist when I consider the power of molecular biology, biotechnology, and electronics to change agriculture around the world—even in the remotest areas of this planet.

Molecular biology is opening the secrets of plant and animal development and growth. This knowledge will lead to a thorough understanding of how biology can be harnessed for agricultural use.

Biotechnology will permit the manipulation of germ plasm—yes, and genes—to give virtually new plants and animals that resist pests and diseases, tolerate stress conditions, provide more nutritious products, are more amenable to storage, and are higher yielding.

Electronics is already being used for instantaneous information transfer and communication even to the remotest areas of the world.

Yes, with the power of science and technology available today and tomorrow, we will be able to produce the food needed by the new human inhabitants each day, week, month, and year.

Resource Paper ♦ *Per Pinstrup-Andersen and Rajul Pandya-Lorch*

Agricultural Science and the Food Balance

At the threshold of the twenty-first century, about 800 million people— 20 percent of the developing world's population—are food insecure. They lack economic and physical access to the food required to lead healthy, productive lives. About 34 percent of the preschool children in developing countries, 185 million, are seriously underweight for their age. Failure to act to rid the world of such human misery is immoral and unethical and results in massive economic loss. Furthermore, the coexistence of widespread hunger and malnutrition in developing countries with extensive food supplies in industrialized countries is not conducive to a stable and just world and bears considerable potential to destabilize global economic, social, political, and environmental conditions.

Overview of Food Security Factors

It is of critical importance that enough food be produced sustainably to meet the food needs of every person on the planet and that all people have economic and physical access to the available food. Access to food is closely related to poverty and income growth: the poor usually do not have adequate means to gain access to food in the quantities needed for healthy, productive lives. More than 1.1 billion people in the developing countries—30 percent of the population—live in absolute poverty, with incomes of only one dollar a day or less per person. Every second person in South Asia and sub-Saharan Africa and every third person in the Middle East and North Africa is absolutely poor. While per capita incomes grew rapidly in East Asia during the 1980s and early 1990s, economic growth all but bypassed Africa, parts of Latin America, and the Middle East. Unless concerted action is taken, poverty is likely to remain entrenched in South Asia and Latin America and to increase considerably in sub-Saharan Africa.

The productivity of poor people must be increased, and their access to remunerative employment and productive resources improved. Investments in education, health care, clean water, sanitation, and housing, which are essential for human resource development, are far below required levels, particularly in rural areas of low-income developing countries. Underinvestment in the health and education of females is particularly severe.

Efforts must be made to lower fertility rates and slow population increases. Between now and 2020, it is likely that the world population will increase by about 40 percent to a total of nearly 8 billion; in other words, about 90 million people are likely to be added to the world's population every year for the next 25 years—the largest annual population increase in history. About 94 percent of the population increase is expected to occur in developing countries, most of it in the cities. Rapid urbanization could more than double the urban population in developing countries to 3.6 billion by 2020, by which time urban dwellers could outnumber rural dwellers. Meeting the increasing and changing food needs resulting from population growth, rising incomes, and changing lifestyles will be a fundamental challenge.

The extent to which food needs will be converted into effective market demand will depend on the purchasing power of the poor. The International Food Policy Research Institute (IFPRI) projects that between 1990 and 2020 per capita demand for food grains in developing countries as a group will increase by only 11 percent while demand for livestock products will increase 56 percent. In sub-Saharan Africa, however, the amount of food demanded per person in the market will show virtually no change, which is cause for serious concern as per capita food consumption is already low in that region.

Although the world is far from approaching the biophysical limits of food production, there are warning signs that growth in food production has begun to lag. For instance, food production increases did not keep pace with population growth in more than 50 developing countries in the 1980s and early 1990s. The growth rate of global grain production dropped from 3 percent in the 1970s to 0.7 percent in the 1985 to 1995 period. Growth rates in yields of rice and wheat have begun to stagnate in Asia, a major producer. Production from marine fisheries has peaked at 100 million tons and is now in decline. World food grain production is expected to grow on average by 1.5 percent per year between 1990 and 2020, while world livestock production is projected to grow by 1.9 percent a year. Aquaculture production, which doubled between 1984 and 1992, is projected to increase at a slower rate between 1990 and 2020, and marine fish catches in 2020 are likely to be no higher than current levels.

Yield increases will have to be the source of most food production increases as the amount of cultivated area is likely to decline in many developed countries and to increase only marginally in developing countries, except in sub-Saharan Africa and Latin America where some area expansion is still possible. However, these yield increases depend on continued research and effective dissemination of the results. Should public investment in agriculture, particularly in agricultural research, continue to decline, the aggregate food situation could worsen significantly and world food prices could rise.

Natural resources and agricultural inputs are critical determinants of food supply. Degradation of natural resources such as soils, forests, marine fisheries, and water undermines production capacity, while availability of and access to agricultural inputs such as water, fertilizer, pesticides, energy, research, and technology determine productivity and thereby production.

Since 1945, about 2 billion of the world's 8.7 billion hectares of agricultural land, permanent pastures, and forest and woodlands have been degraded. Overgrazing, deforestation, and inappropriate agricultural practices account for most of the degradation. To a large extent, these problems result from or are exacerbated by inadequate property rights, poverty, population pressure, inappropriate government policies, lack of access to markets and credit, and inappropriate technology. Crop productivity losses from degradation are significant and widespread. In the absence of efforts to protect nondegraded soils and to restore currently degraded soils, increasing population and persisting poverty will hasten soil degradation.

Small-scale, poor farmers clearing land for agriculture to meet food needs accounted for roughly two-thirds of the 15.4 million hectares of tropical forests worldwide that were converted to other uses every year during the 1980s. Such forest conversion, driven by food insecurity, is likely to continue, particularly in Africa, unless farmers have alternative ways of meeting food needs. And these food needs will accelerate with population growth in rural areas.

Fisheries are collapsing in some parts of the world, and international disputes over fish stocks are increasing. Following a period of rapid expansion of harvesting from the oceans, more than one-fourth of the 200 primary marine fisheries worldwide are overexploited, depleted, or recovering, while another two-fifths are fully exploited. Resource management has failed to coordinate fishers and restrain them from exploiting natural fisheries beyond sustainable limits, leading to increasing scarcity of and conflict over fish. Even where access is restrained, most fisheries have too many fishers with legitimate access. Future fish catches are likely to be, at best, no more than current levels; the challenge is to maintain the present levels of harvest from natural fisheries while sustainably increasing aquaculture production.

There are several major water-related challenges. New sources of water are increasingly expensive to exploit because of high construction costs for dams and reservoirs and because of concerns about environmental effects and displacement of people. Efficiency of water use in agriculture, industry, and urban areas is generally low. There are mounting pressures to degrade land and water resources through waterlogging, salinization, and groundwater mining. Water pollution from industrial effluents, poorly treated sewage, and runoff of agricultural chemicals is a growing problem. Unsafe water, compounded by inadequate or nonexistent sewage and sanitation services, is a major cause of disease and death, particularly among children, in developing countries. Inappropriate policies, distorted incentives, and massive subsidies provide water at little or no cost to rural and urban users, encouraging overuse and misuse of water. Water for irrigation, the largest use, is essentially unpriced. The overarching challenge is to treat water as the scarce resource that it is.

The use of mineral fertilizers will have to be substantially increased to meet food needs, although organic sources can and should make a larger contribution to supply plant nutrients. Depletion of soil nutrients is a critical constraint to food production in sub-Saharan Africa. While the negative environmental consequences of fertilizer use and production must be avoided, in most developing countries the problem is not excessive, but insufficient, fertilizer use. The major task is to promote a balanced and efficient use of plant nutrients from both organic and inorganic sources at farm and community levels to intensify agriculture in a sustainable manner.

Crop production losses from pests are significant; reducing these losses would contribute notably to improving food supplies. However, past levels of pesticide use cannot be sustained. Concerns are multiplying that pesticides compromise human health; contaminate soils and water and damage ecosystems; exterminate species; and lead to pesticide resistance, pest resurgence, and evolution of secondary pests. Moreover, evidence shows that overuse of pesticides leads to decreased food production. Environmentally sound alternatives must be developed and adopted. Available pest control means must be combined to achieve effective pest control with little or no negative environmental effects or health risks.

The efficient functioning of markets, especially agricultural input and output markets, supported by governments that have the capacity to perform their role, is of critical importance. In recent years, many countries have embarked upon market reforms to move away from state-controlled, or parastatal, organizations toward reliance on private firms operating in free markets. While clearly desirable, such reforms must be undertaken with care, taking into account the organizational structure of the affected markets. In many cases, inefficient parastatals are being replaced by oligopolistic or monopolistic private firms, with little or no improvement in performance. The ongoing and unprecedented transition from controlled to market economies and from patrimonial to open political systems has generated confusion about the appropriate role of government and weakened the capacity of governments to perform needed functions.

In many regions, especially sub-Saharan Africa, food marketing costs are extremely high. Lowering these costs through investment in improved transportation infrastructure and marketing facilities, which also increase competition, may be as important in lowering food prices to consumers as increasing agricultural productivity. Many countries have made considerable improvements in recent years, but investments in infrastructure, especially transport and communication, which are considered the leading elements, are far below needed levels. The increasing integration of developing countries into the global economy through international trade will benefit developing and developed countries considerably through expanded markets, job creation, and income generation. Many developing countries have gained enormously from increased participation in world markets as world trade has become increasingly liberalized. However, for a few countries—especially the low-income, food-importing countries—global trade liberalization coupled with agricultural subsidy policy reform in developed countries is a mixed blessing. These food-importing countries gain from increased access to developed-country markets, but they are less able to compete in those markets than other, better situated, developing countries. And they lose from increased food prices that may occur in the medium term with cutbacks in agricultural subsidies in the developed countries.

Without increased domestic resource mobilization—savings and investments—developing countries will not be able to accelerate the investments in economic growth and human resources that underpin future food security. For low-income countries, the vicious circle of poverty still exists: low income leads to low savings, low investment, low growth, continued poverty, and continued low savings. In sub-Saharan Africa, the share of the gross domestic product (GDP) devoted to investment has fallen in the past two decades from 20 percent to 16 percent, while the domestic savings rate has fallen from 18 percent to 15 percent. These rates are not high enough to raise economic growth rates significantly.

International assistance has a critical role to play in supporting developing countries as they implement the actions required to achieve food security and embark upon broad-based economic development. Private aid flows to developing countries have increased substantially since the late 1980s. Most of these flows, however, go to a small number of medium-income, semi-industrial countries in Latin America and Asia. Poorer countries, especially in sub-Saharan Africa, are left out and depend much more on public aid flows. However, official development assistance (ODA) to developing countries is slowing. Driven by cuts in foreign assistance in 17 countries, ODA from the Organization for Economic Cooperation and Development (OECD) countries has dropped significantly since 1993. Assistance from non-OECD countries has dropped even more significantly; the Arab Organization of Petroleum Exporting Countries (OPEC) provided 25 percent of world ODA in the early 1980s but contributed less than 2 percent in 1993. Assistance to agriculture in developing countries declined in real terms from $12 billion to $10 billion during the 1980s. Given observed trends in aid availability, developing countries are challenged to devise strategies for accomplishing their goals with less aid.

The Role of Agricultural Science

Existing technology and knowledge will not permit the necessary expansion of food production to meet future needs. Low-income developing countries are grossly underinvesting in agricultural research compared with industrialized countries, even though agriculture accounts for a much larger share of their employment and income. Their public sector expenditures on agricultural research are typically less than 0.5 percent of agricultural gross domestic product, compared with about 2 percent in higher income developing countries and 2 to 5 percent in industrialized countries. Developing countries have far fewer agricultural researchers relative to the economically active population engaged in agriculture or to the agricultural acreage. Growth in public sector expenditures for agricultural research in developing countries has slowed to 2.7 percent per year in the past decade, compared with 7.0 percent in the 1960s. Many developing countries are even reducing their support for agricultural research. This trend has been under way for several decades in Africa (agricultural research expenditures per scientist have fallen by about 2.6 percent per year since 1961) and is more recent in Latin America. Further reductions in public investment in agricultural research will have severe consequences for global food production by reducing yield growth.

Investment in agricultural research must be accelerated if developing countries are to ensure future food security for their citizens at reasonable prices and without irreversibly degrading the natural resource base. Accelerated investment in agricultural research is particularly important and urgent for low-income developing countries, partly because these countries will not achieve reasonable economic growth, poverty alleviation, and improvements in food security without productivity increases in agriculture, and partly because so little research is currently undertaken in these countries. As we mentioned, the negative correlation between investment in agricultural research and a country's income level is very strong. Poor countries, which depend the most on productivity increases in agriculture, grossly underinvest in agricultural research.

Sub-Saharan Africa, which desperately needs productivity increases in agriculture, has only 42 agricultural researchers per million economically active persons in agriculture and 7 researchers per million hectares of agricultural land. During the 1960s, the number of agricultural researchers in Africa grew at an annual rate of 9.1 percent (Pardey, Roseboom, and Beintema 1994). The growth rate decreased to 6.1 percent in the 1970s and 3.5 percent during the 1980s. The growth rate in African agricultural research expenditures dropped even more, from 6.6 percent during the 1960s to -0.3 percent during the 1980s. In 1961, Africa's agricultural research expenditures were 0.36 percent of the agricultural GDP. These expenditures increased to 0.58 percent in 1971 and to 0.84 percent in 1981, but during the 1980s they dropped to 0.55 percent. International agricultural research undertaken under the auspices of the Consultative Group on International Agricultural Research (CGIAR) has fared even worse in recent years.

Agricultural research has successfully developed yield-enhancing technology for the majority of crops grown in temperate zones and for several crops grown in the tropics. The dramatic impact of agricultural research and modern technology on wheat and rice yields in Asia and Latin America since the mid-1960s is well known. There have been less dramatic but significant yield gains from research and technological change in other crops, particularly maize. Investment in agricultural research results in large economic returns (Tables 9.1.1 and 9.1.2).

Table 9.1.1. Annual rates of return (RR) to agricultural research, by region

	Less than 10 percent RR	More than 50 percent RR
	(percent of programs analyzed)	
Africa	20	40
Asia	3	63
Latin America	7	46

SOURCE: Information from Evenson and Rosegrant 1993.

Table 9.1.2. Annual rates of return (RR) to agricultural research, by commodity

	Less than 10 percent RR	More than 50 percent RR
	(percent of programs analyzed)	
Wheat	7	43
Rice	0	68
Maize	0	31
Other commodity programs	12	53
Aggregate research programs	1	62

SOURCE: Information from Evenson and Rosegrant 1993.

Large yield gains now occurring for many crops at the experimental level offer great promise for future yield and production increases at the farm level. In addition to raising yield levels, research resulting in tolerance or resistance to adverse production factors such as pests and drought, research leading to biological and integrated pest control, and research to develop improved varieties and hybrids for agro-ecological zones with less than optimal production conditions, all reduce risks and uncertainty. These kinds of research enhance sustainability in production through better management of natural resources and reduced environmental risks.

Accelerated agricultural research aimed at more-favored areas will reduce pressures on fragile land in less-favored areas. Future research for the former must pay more attention to sustainability than in the past to avoid a continuation of extensive waterlogging, salinization, and other forms of land degradation. However, a continuation of past priority on more-favored agro-ecological zones is inappropriate and insufficient to alleviate poverty, improve food security, and appropriately manage natural resources. The relative allocation of research resources to more-favored areas with agricultural potential, and away from less-favored fragile land with poor rainfall and high risks of environmental degradation must be changed. A large share of the poor and food insecure reside in agro-ecological zones with high risk of environmental degradation. In low-income developing countries, poverty, rapid population growth, low agricultural productivity, and poorly defined ownership of and user rights to natural resources are the major causes of land degradation, deforestation, and inappropriate water use. Attempts by households to meet basic needs for survival in the short run take priority over longer-run sustainability.

The low priority given in the past to research on developing appropriate technology for less-favored agro-ecological zones is a major reason for the current rapid degradation of natural resources and high levels of population growth, poverty, and food insecurity. Much more research must be directed toward developing appropriate technology for these areas. Outmigration is not a feasible solution for these areas in the foreseeable future because of the large number of poor people who reside there and the lack of alternative opportunities elsewhere. Strengthening agriculture and related nonagricultural rural enterprises is an urgent need and must receive high priority.

Following on the tremendous success popularly referred to as the Green Revolution, international agricultural research centers under CGIAR have recognized the importance and urgency of research to ensure sustainable agricultural intensification through appropriate management of natural resources. Thus, natural resource management, germ plasm conservation, and germ plasm enhancement are high priorities in current and future research at these centers.

Declining investment in agricultural research for developing countries since the mid-1980s, both by developing-country governments and international foreign assistance agencies, is inappropriate and must be reversed. While privatization of agricultural research should be encouraged, much of the agricultural research needed to achieve food security, reduce poverty, and avoid environmental degradation in developing countries is of a "public goods" nature and will not be undertaken by the private sector. Fortunately, while private rates of return may be insufficient to justify private sector investment, expected high social rates of return justify public investment. The major share of this investment should be in the developing countries' own research institutions such as the National Agricultural Research Systems (NARS); there is an urgent need to strengthen these institutions to expand research and increase the probability of high payoffs.

The centers under CGIAR have a well-defined role to play in supporting the work of NARS, namely, to undertake research of a public goods nature with large international externalities and to strengthen the research capacity of NARS and networking among NARS, international centers, and research institutions in the industrialized nations. Research institutions in the industrialized nations have played an extremely important role by undertaking basic research required to support strategic, adaptive, and applied research by the international centers and NARS and by providing training for developing-country researchers. Collaboration among developed-country research institutions, CGIAR centers, and NARS in developing countries is widespread. Further strengthening is required to utilize the comparative advantages of each of the three groups for the ultimate benefit of the poor in developing countries.

Biotechnology research in national and international research systems should be expanded to support sustainable intensification of agriculture in developing countries. Effective partnerships among developing-country research systems, international research institutions, and private- and public-sector research institutions in industrialized countries should be forged to bring biotechnology to bear on the agricultural problems of developing countries. Incentives should be provided to the private sector to undertake biotechnology research focusing on the problems of developing-country farmers. Failure to expand agricultural research significantly in and for developing countries will make food security, poverty, and environmental goals elusive. Lack of foresight today will carry a very high cost for the future. As usual, the weak and powerless will carry the major burden, but just as we must all share the blame for inaction or inappropriate action, so will we all suffer the consequences.

References

Evenson, R. E., and Mark W. Rosegrant. 1993. Determinants of Productivity Growth in Asian Agriculture: Past and Future. Paper presented at the 1993 American Agricultural Economics Association International Pre-Conference on Post-Green Revolution Agricultural Development Strategies in the Third World: What Next? held in Orlando, Florida, August.

Pardey, P., J. Roseboom, and N. Beintema. 1994. *Agricultural Research in Africa: Three Decades of Development.* Briefing Paper No. 19. The Hague, Netherlands: International Service for National Agricultural Research.

Background References

International Food Policy Research Institute (IFPRI). 1995. *A 2020 Vision for Food, Agriculture, and the Environment: The Vision, Challenge, and Recommended Action.* Washington, D.C.: IFPRI.

Pardey, P., J. Roseboom, and J. R. Anderson, eds. 1991. *Agricultural Research Policy: International Quantitative Perspectives.* Cambridge: Cambridge University Press.

Pinstrup-Andersen, Per, and Rajul Pandya-Lorch. 1994. *Alleviating Poverty, Intensifying Agriculture, and Effectively Managing Natural Resources.* Food, Agriculture, and the Environment Discussion Paper No. 1. Washington, D.C.: International Food Policy Research Institute.

_____. 1995. *Poverty, Food Security, and the Environment.* 2020 Brief No. 29. Washington, D.C.: International Food Policy Research Institute.

Rosegrant, Mark W., Mercedita Agcaoili-Sombilla, and Nicostrato D. Perez. 1995. *Global Food Projections to 2020: Implications for Investment.* Food, Agriculture, and the Environment Discussion Paper No. 5. Washington, D.C.: International Food Policy Research Institute.

Scherr, Sara, and Satya Yadav. 1996. *Land Degradation in the Developing World: Implications for Food, Agriculture, and the Environment to 2020.* Food, Agriculture, and the Environment Discussion Paper No. 14. Washington, D.C.: International Food Policy Research Institute.

Williams, Meryl. 1996. *The Transition in the Contribution of Living Aquatic Resources to Food Security.* 2020 Brief No. 32. Washington, D.C.: International Food Policy Research Institute.

Discussant ♦ *John S. Niederhauser*

One of the major bottlenecks to agricultural progress in many places around the world during the past few years has been the fact that technology which has been developed does not reach the producer, i.e., the farmer. With respect to government policies, the use of foreign aid in many developing countries has been inefficient. In one case, for example, a large grant from a generous donor was given to enhance the production of wheat in a certain country. A considerable amount of the aid, however, went to administrative expenses, not only for the implementation of the project but also to cover so-called "administrative costs" at the local level. Thus, the supposed recipient of this aid, i.e., the actual producer of food in the recipient country, ended up receiving a very small fraction of the grant.

Another important issue is that agriculture has remained very responsive to an expanded private incentive. We recognize today that the individual farmer is faced daily with multiple decisions, and these decisions have never been successfully implemented in what I will call a "more centralized socialistic system of agriculture." The wonderful story of increasing agricultural production in China during the past decade is a good example of how a country has returned the private incentive to the farmer while still maintaining a centralized government.

Since we are talking about issues of nutrition, food, and hunger around the world, one crop of particular interest is the potato. During the past 45 years, the potato has increased greatly in importance in the developing world. In 1950, only 3 percent of the total world potato production was in the developing countries. By 1990, this number had increased dramatically to over 30 percent. It has been the work done directly with farmers that has made the potato so important to these developing countries.

The potato crop is considered to be highly productive partly because it has a relatively short season, ranging from 90 days to 100 days in many locations. The potato provides more calories per hectare than other food crops, as well as more protein and greater amounts of most vitamins, particularly vitamin C. We get more output per unit of water with the potato than with any of the other crops that are grown as basic food crops. The potato also can adapt very well to various climates and latitudes. The issue here is not to try to substitute for any crop. If grain is the basic food of a particular country, then the potato can fit well into rotations. This is what our experience has shown. Growing more potatoes does not necessarily mean using more land and, in most cases, it means using very little additional water.

Today the greatest single limitation to a continuing expansion of the importance of the potato in the developing world is resistance to the late blight disease. During the past 30 years, a significant amount of the increase in potato production has been on irrigated land. We know, however, that irrigated land and water are becoming more scarce, so if producers are going to grow more potatoes, this crop will have to be grown under rainfall. However, given the blight-susceptible commercial varieties of potato that are available, potato crops grown under rainfall are very exposed to serious losses due to late blight. The crops must be sprayed to control this disease. In many developing nations, it is either too expensive to spray with chemicals or these chemicals are simply not available. We are working on this problem more intensively today than we have for years, with the aim of providing potato producers with a blight-resistant potato. In Mexico today, over 40 percent of the national production is in blight-resistant varieties. The same is true in many countries in Latin America and Asia but, unfortunately, only a few in Africa.

However, one of the most important achievements would be to have these disease-resistant potatoes grown even more extensively in the industrialized countries as well. There has been a rather complacent attitude towards these varieties in most of the industrialized Northern Hemisphere countries because chemical sprays are readily available and, therefore, using disease-resistant potato varieties has not been considered a high priority. This has to change, and this change can be done most promptly and economically through international cooperation.

Finally, we need more international cooperation, particularly in the developing countries, whether it concerns one crop or a group of crops, disease problems, or the establishment of sustainable agriculture. We need the type of international cooperation that is illustrated by a wonderful program established in 1978 called PRECODEPA (Regional Cooperative Potato Program). In this program 11 participants, including Mexico and countries in Central America and the Caribbean, cooperate on production and marketing problems of mutual interest. For each project, one country is chosen as leader and is available to the other countries for technology, training, and materials. Thus, the 11 countries, with a very modest budget, share whatever technologies are developed by each leader. For example, Costa Rica is in charge of providing assistance for fighting bacterial disease, Mexico for enhancing seed production, Guatemala for storage, and so forth. This program has been financed by the government of Switzerland since 1978, and it is now in its eighteenth year. According to the Swiss government this has been the most cost-effective agricultural research and development program they have implemented anywhere in the world.

This kind of international cooperation is going to be more and more important in solving the food production problems facing developing countries. I am optimistic that there is a growing awareness among countries of the need to cooperate and make this type of cost-effective production program work. I hope that we have the discipline to respond to this challenge in a positive way. I hope also that during the next millennium we find not only that we can produce enough food to feed a growing world population, but also that we can develop the strategies for distributing the food and reducing the hunger that is a curse in the world today.

Discussant ♦ *Vernon W. Ruttan*
Meeting the Food Needs of the World

We are in the closing years of the twentieth century, completing one of the most remarkable transitions in the history of agriculture. Prior to this century, almost all of the increase in food production was obtained by bringing new land into production. By the end of the first decade of the twenty-first century, almost all of the increases in world food production must come from higher yields—from increased output per hectare and increased output per animal unit.

World Food Futures

Perspectives on world food futures have cycled rapidly over the last several decades. A 1989 study by the International Institute for Applied Systems Analysis (IIASA) advanced what came to be referred to as "the 2-4-6-8" scenario: a doubling of population, a quadrupling of agricultural production, a sextupling of energy production, and an octupling of the size of the global economy by 2050. Note that it is the growth of the global economy—particularly per capita income growth in the presently poor countries—that is the source of approximately one-half of the growth in food demand.

More recent research has criticized the IIASA projections as implying "whole nations of obese gluttons." It now seems apparent, on the basis of newer population, income growth, and consumption behavior projections that global food demand growth will be in the 2 to 3 percent range over the next 50 years (Table 9.2.1). Substantially higher growth rates will obtain at the beginning and lower growth rates toward the end. We need to be as concerned with the "income gap" as with the potential "food gap." *People who have money to buy food do not need to do so.*

The recent draw down in grain stocks and run-up in grain prices have, however, caused some observers to announce that the long-term decline in grain prices, which had made food available to consumers on increasingly favorable terms since the middle of the nineteenth century, has finally come to an end. In assessing these current predictions it should be recalled that almost identical predictions, triggered by similar events, were made by some of the same observers in the early and mid-1970s. My own sense is that the recent decline in per capita production of cereals and the run-up in cereal prices was largely policy induced.

Table 9.2.1. Projection of commodity demand for 2050

Year	Cereal	Bovine and ovine meat	Dairy	Other animal products	Protein feeds	Oil and fats	Sugar
2000	2082	86	642	27	104	96	286
		Medium population-Low income growth (MP-LG)					
2050	3262.4	175.5	855.9	130.5	128.3	158.6	324.7
		High population-High income growth (HP-HG)					
2050	4916.4	270.5	1347.3	184.5	205.3	288.6	523.8
		Commodity demand ratios					
MP-LG ratio 2050/2000	1.57	2.04	1.33	4.83	1.21	1.65	1.14
HP-HG ratio 2050/2000	2.36	3.15	2.10	6.83	1.94	3.01	1.83

Units: Cereals: Million tons   Bovine & ovine meat: Million tons carcass wt.
Dairy: Million tons wh. milk equiv.   Other animal products: Million tons protein equiv.
Protein feeds:Million tons protein equiv.   Oils: 10^4 tons oil
Sugar: 10^4 tons refined sugar equiv.

SOURCE: Kirit S. Parikh. 1994. Agricultural and Food System Scenarios for the 21st Century. In *Agriculture, Environment and Health: Sustainable Development in the 21st Century*, edited by Vernon W. Ruttan. Minneapolis, MN: University of Minnesota Press, 34. For projections of population and income on which the demand projections are based see pp. 30–31. (Reprinted with permission)

Constraints on Food Production

As the world's farmers attempt to respond to the demands that will be placed on them over the next half century, they will be confronted by a number of serious constraints.

1. Scientific and technical constraints. Gains in agricultural production will be achieved with much greater difficulty than in the immediate past. Biotechnology is not yet living up to its promise to provide an "encore to the Green Revolution." Agricultural research budgets have declined in many developed and developing countries. And "maintenance research"—the research required to prevent yields from declining as a result of land degradation and the co-evolution of pests and disease—is rising as a share of research effort.

2. Resource and environmental constraints. Intensification of industrial and agricultural production is imposing increasingly severe environmental constraints on agricultural production. These constraints range from the impact of fossil fuel consumption on global climate change to the loss of soil resources due to erosion, water logging, and salinization, as well as the resistance of weeds, insects, and pathogens to present methods of control.

3. Health constraints. A number of indicators suggest that health could emerge as a serious constraint on agricultural production in the early decades of the twenty-first century. These health problems include the resurgence of malaria and tuberculosis, the emergence of acquired immune deficiency syndrome (AIDS) and a number of other infectious diseases, the declining efficacy of available antibiotics, and the high cost of developing new drugs for the control of infectious diseases. Little progress has been made in the control of several important parasitic diseases. And we are only beginning to confront the environmental health effects of agricultural and industrial intensification.

If several of these health threats emerge simultaneously in specific geographic locations, it is not difficult to visualize scenarios in which the number of sick people in rural areas becomes a serious constraint on agricultural production.

Institutional Innovation

I am cautiously optimistic about the possibilities of responding to the demands that will be placed on agricultural producers over the next half century. My optimism is tempered, however, by the capacity of the global community to realize a number of important institutional innovations. First, our capacity to monitor (a) changes in the sources of productivity change in agriculture, (b) the sources (driving forces) and impact of environmental change, and (c) the sources and incidence of the emerging insults to health is inadequate and must be strengthened if we are to respond effectively.

Second, more effective bridges must be built, both in research and practice, among the agricultural, environmental, and health communities. At present these three tribes occupy separate and mutually hostile "island empires."

Discussant ◆ *Luc M. Maene*

As a panelist in this "Agricultural Science and the Food Balance" session, I am especially pleased to contribute some remarks, not only because it is a great pleasure to address such a distinguished audience, many of you well-known and good friends, but also because the industry I represent is a significant factor in the food production equation.

The International Fertilizer Industry Association represents approximately 500 companies and organizations in over 80 countries, all of which are involved in some way in the production, distribution, and use of mineral fertilizers. The critical importance of soil fertility in sustainable and productive farming systems worldwide is well known, and it is my role to explore how the availability of plant nutrients can be improved to enhance that productive potential.

I am grateful to Per Pinstrup-Andersen and his team at the International Food Policy Research Institute (IFPRI) for producing such excellent background material for our discussions. We are fortunate to have such objective, well-researched, scientific assessments. It is a concern to us that all too often information reaches the public domain from less reliable sources, leading to misunderstanding, misinterpretation, and overreaction. The declining trend in research funding at international and national levels must be reversed, and all sectors involved in the food chain, including the agricultural input industries, should play their part; the fertilizer industry is no exception.

Practicing agriculture in a sustainable way and producing enough food to meet current and future needs are not possible without the use of mineral fertilizers. Indeed, it is widely acknowledged that perhaps 40 to 50 percent of the global harvest already comes from the application of fertilizers. Nitrogen availability is clearly the single most important consideration. However, in many locations there is going to have to be much greater attention to the *mix* of nutrients, including phosphates and potash.

Such an overwhelming importance attached to fertilizers is, however, no reason for complacency by our industry. Many factors will help to raise world cereal yields in the coming years. Per Pinstrup-Andersen's resource paper states that "availability of and access to agricultural inputs such as water, fertilizer, pesticides, energy, research, and technology determine productivity and thereby production" (p. 154).

Detailed analysis of the future potential and performance of the fertilizer industry is therefore vital in any serious assessment of agricultural science and the food balance. But it is important to look at fertilizers in two distinct phases: production and consumption.

On the *production* side, the most important considerations are

- the environmental impact of the production process—*much has been done;*
- the quality of the products—*much is being done;* and
- the research and development of new products—*much remains to be done.*

On the *consumption* side, environmental impact and economics direct our efforts towards

- *balanced* use;
- *efficiency* of application; and
- research and development in *precision* agriculture techniques.

A suitable enabling economic environment governs the availability of fertilizers and can, in turn, affect the mix of nutrients which is applied.

While we recognize that in some areas of intensive agriculture the overuse of fertilizers has caused environmental problems, it is also true that vast areas of potentially productive land in many countries are grossly infertile, a result of decades of progressive soil nutrient "mining." Indeed, with over 40 percent of our Association's members in developing countries, the complexity of the issues facing our industry will be understandable.

I note that Pinstrup-Andersen's paper clearly states that "While the negative environmental consequences of fertilizer use and production must be avoided, in most developing countries the problem is not excessive, but insufficient, fertilizer use." His next sentence has become almost a mission statement for our Association: "The major task is to promote a balanced and efficient use of plant nutrients from both organic and inorganic sources at farm and community levels to intensify agriculture in a sustainable manner" (p. 156).

I wish to highlight two key words in this statement—balanced and efficient. We are talking about balance—in this session, the food balance. We are all working towards a common goal: balancing the food-population equation. In our industry we calculate supply/demand balances. It is crucial for investors and planners to make accurate predictions of trends in both food production and fertilizer supply. Serious imbalances help no one, as we have witnessed from the overreaction to the food and energy crises of the early 1970s, a response that has taken two decades to rectify, often with painful restructuring and rationalization.

We must also find a balance between the environmental constraints placed on the production of plant nutrients and the need to produce an agronomically and economically acceptable product for the farmer. I am pleased to state that our industry is in the forefront of efforts to search for constant improvement in environmental management techniques and standards. We are working closely with the United Nations Environment Programme (UNEP), for example, to establish widespread appreciation of the best practices in this respect.

With regard to efficiency, the fertilizer industry recognizes that it always must be on the lookout to improve the performance of its products. There is surely great room for improvement, not only in helping farmers manage plant nutrients more effectively, but also by devoting more research effort to developing products that can supply the crop's nutrient requirements more efficiently and improve the overall efficiency of transport and distribution.

To convey these new concepts throughout the industry and beyond, we have embraced the principle of Integrated Plant Nutrient Management as defined in Agenda 21, chapter 14: "The *integrated plant nutrition* approach aims at ensuring a sustainable supply of plant nutrients to increase future yields without harming the environment and soil productivity" (United Nations Conference on Environment and Development, Rio de Janeiro, 1992).

In detail, integrated plant nutrition involves the following factors.

1. Improve the efficiency and productivity of the farming system by upgrading capital resources of nutrients stored in soils, crop residues, and manures; improving the flow of nutrients according to the crop's needs; and limiting plant nutrient losses to optimize nutrient supply and use.

2. Design processes for the accumulation of plant nutrients in the crop-soil-livestock system through judicious use of fertilizers and other plant nutrient sources.

3. Increase the productive capacity of farmland by encouraging technical and economic development and by identifying alternative crop-specific management practices.

4. Enhance the decisionmaking capacity of farmers by promoting proper understanding of the mechanisms of plant nutrient uptake, flows, cycles, balance, and losses, using indicators to facilitate management.

Finally, I would like to quote from a distinguished guest at this symposium, Dr. Norman Borlaug, who made these remarks in July 1994 at the 15th ISSS World Congress of Soil Science in Acapulco, Mexico.

For those of us on the food production front, let us all remember that world peace will not be built on empty stomachs and human misery. Deny the small-scale, resource-poor farmers of the developing world access to modern factors of production—such as improved varieties, fertilizers, and crop protection chemicals—and the world will be doomed—not from poisoning, as some say, but from starvation and social and political chaos.

CHAPTER TEN

Population and the Environment

- ♦ MODERATOR
- ♦ RESOURCE PAPER
- ♦ DISCUSSANTS

Moderator ♦ *Samuel Enoch Stumpf*

The security of the world's food supply is always threatened by the dynamics of human populations. If it is a fact that the population of the world is growing at a far more rapid rate than is the quantity of food, then it is simply a question of time before there are more people than the world can feed, and it is predicted by some experts that the moment when this imbalance will occur is virtually upon us. By now we are familiar with the estimate that we are running out of arable land, pure water, and sources of energy at a time when we are adding each year about 80 million human beings to the world's population.

Whether we face unique problems depends on the accuracy of our information and predictions. Members of our panel have considered the forces at work affecting the size of our populations and the stability of the food supply. Ours is not the first era in history when catastrophe was expected. When the English economist Thomas Malthus formulated his famous theory in 1798, he was convinced that population would always "press against food." Malthus predicted that because the population increases geometrically while the food supply increases only arithmetically, in 225 years the population would be to the means of subsistence as 512 is to 10, and that in 300 years, it would be as 4,096 is to 13.

Almost 200 years have passed since Malthus presented his gloomy forecast, raising the question of whether or not he had sufficiently taken into account the capacity of human ingenuity to derive from nature a vastly increased basis for subsistence. Even if the ratio of people to food were as permanent in nature as are Newton's laws of physics, would not the possible tragic consequences to mankind be averted by breaking through the slow rate of growth in the food supply by means of technological advance? Moreover, was it inevitable that the only check on the growth of population would be the twin forces of "vice and misery," or wars, famine, and disease?

It is a striking feature of Malthus's work that he was so little conscious of the changes being effected by the "industrial revolution." For example, although the steam engine had been invented in 1776, over 20 years before he published his first *Essay on Population,* Malthus did not see in this device the possibility of substituting machine power for human energy, which would thereby radically alter the impact of his theory, which held, among other things, that there is a principle of diminishing returns at work reducing the output of the newest additions to the workforce. This view was still held by John Stuart Mill in the nineteenth century when he wrote that it is useless to say that all mouths that are brought into existence by population increase bring with them hands. The new mouths, Mill cautioned, require as much food as the old ones while the new hands do not produce as much. But mechanization has changed all that.

And there might be additional ways to control population, a point Malthus emphasized in a revised edition of his *Essay.* Although he did not favor mechanical means of birth control, and might not have been pleased to know that later on contraceptives were referred to by some as "Malthusian devices," Malthus did focus upon the need for some deliberate method of population control. The most acceptable way to control population, he argued, was through "moral restraint," which would delay marriage until prospective parents had some assurance they could support their children.

Whether our present worries about the food supply confirm Malthus's worst predictions or whether we stand on the threshold of major discoveries that would alter these gloomy predictions depends upon information not fully available at present. There are reasons for considerable optimism based upon the spectacular achievements of the Green Revolution; the novel modes of planting, fertilization, and irrigation; the many forms of mechanization; the genetic alteration of plants and animals; and the rapid advances in biochemistry. Moreover, the use of more sophisticated and effective techniques of birth control can provide a check upon the expanding population.

But these advances are not without some complexities. Already we are told that there is an upper limit to what we can expect from the Green Revolution. Not only can we expect a limit to the yield per acre, but the cost of achieving this yield is rising rapidly beyond the ability of some countries to afford it. Objectively, it costs more in energy to produce some foods than there is energy, measured in calories, contained in the food. But unlike Malthus, who could not even envision technological developments, we can aggressively pursue likely alternative solutions, as, for example, the possibility of harnessing solar energy. We can therefore afford to suspend final judgment of our present predicament on the reasonable assumption that the causes of our present crisis are temporary or transitory.

In one sense we do stand where Malthus stood in spite of our immense technology, for, as I have just pointed out, the developments in food technology might not continue at their recent rapid rate, nor might sufficient resources exist in the long run to provide the prerequisites for a universal Green Revolution among the countries of the world. To the extent that the supply of food has increased, the problem has been aggravated, as this increased food supply has caused the population to expand and its capacity to reproduce to increase. Even the provision of health services among previously deprived peoples has had the effect of nullifying some of the gains in the food supply by further increasing the size of the surviving population.

Underlying all discussions of the impact of a rising tide of population upon the limited food supply is the question of the moral responsibility of those who have toward those who suffer privation. Whatever the predictions concerning the world's food supply, whether optimistic or pessimistic, virtually every moral issue that surrounds this problem is provoked by some degree of scarcity. In his discussion, Professor Norman Myers focuses upon the many factors that cause a decrease in resources, such as soil erosion, water deficits, extreme weather conditions, and tropical deforestation. Added to these deficit-producing factors is the matter of poverty or widespread joblessness. Myers concludes that the overall problem is as much a poverty problem as it is a food problem. Among other things, it becomes the moral responsibility of the wealthier nations to provide food at prices the poorer peoples can afford.

Throughout his paper, Myers refers to the moral dimension of this global food problem. This should not surprise us as there is an intuitive recognition among human beings that they are bound together in such a special way that, although some are willing to enjoy affluence while others suffer severe privation, most people would be reluctant to make the pains of others the condition of their own enjoyment. William James once wondered what people would do if they were offered a world in which millions of people were permanently happy on the condition that one lost soul far away would lead a life of lonely torture. James suggested that a special and independent kind of emotion would make us immediately feel—even though an impulse arose within us to grasp at the happiness offered—how hideous the enjoyment of such happiness would be when deliberately accepted as the fruit of such a bargain.

Resource Paper ♦ *Norman Myers*

Population Dynamics and Food Security

The Challenge Ahead

Three decades from now we shall need to be feeding another 2.5 billion people, 42 percent more than today. The global total will then be 8.2 billion, and we should aim to provide them with at least 50 to 60 percent more food to ensure they are all adequately nourished and to cater to increased affluence (Brown et al. 1996; Harris 1996; Islam 1995; Mitchell and Ingco 1993). Some observers foresee a doubling of food demand (Abdulai and Hazell 1995; Crosson and Anderson 1992; ESDAR/The World Bank 1996; McCalla 1994). This challenge compares with that which faced the world's farmers in 1950 before the Green Revolution, described by many agronomic experts as "mission impossible" since it would far exceed any previous advance in such a short space of time. Fortunately, the farmers expanded grain production from 631 million tonnes (metric tons) in 1950 to 1,664 million tonnes in 1985, a 164 percent increase; and during those 35 years the world's population grew from 2.5 billion people to 4.9 billion, a 96 percent increase. The average annual rise in grain production was a whopping 4.7 percent, making it an agricultural accomplishment of unprecedented scale. Meantime, the average annual growth in the world's population was unprecedentedly rapid, but fortunately, at 2.6 percent, it was well below the growth rate for grain harvests. For all of 35 remarkable years, the plow kept ahead of the stork (Bongaarts 1994; Dyson 1995; Grigg 1993). Can it do so again?

The Recent Record

Today's challenge is at least as large in salient respects. If the world's farmers are to achieve an increase of at least 50 to 60 percent in food production by 2025, the average annual growth rate in the grain harvest must be 2.0 percent. However, it achieved a cumulative growth total of only 1.8 percent throughout the period 1985 to 1995, while population grew by 18 percent, resulting in a per person decline in grain consumption of 14 percent (Brown, Flavin, and Kane 1996). (See Table 10.1.1 and Figures 10.1.1a and 10.1.1b). Since rich people continued to receive all the grain they wanted, poorer people suffered a still greater shortfall. The harvest drop-off has been worst in recent years; the 1995 grain harvest of 1.69 billion tonnes was a full 5 percent below the 1990 record harvest of 1.78 billion tonnes (USDA 1995). (For a preliminary estimate of the 1996 harvest, see the final section of this chapter.) Even if the modest amount of cropland set aside through soil conservation and farm commodity programs in the United States and Europe had remained under the plow, the 1995 harvest would still have fallen far short of demand. For three consecutive years, 1993 through 1995, more grain was consumed than produced, and world carryover stocks declined to their lowest level in more than three decades (USDA 1996).

Table 10.1.1. World grain production, 1950 to 1996

Year	Grain total	Per capita
	(million tonnes)	(kilograms)
1950	631	247
1960	847	279
1970	1,096	296
1980	1,447	325
1981	1,499	331
1982	1,550	336
1983	1,486	317
1984	1,649	346
1985	1,664	343
1986	1,683	341
1987	1,612	321
1988	1,564	306
1989	1,685	324
1990	1,780	336
1991	1,696	315
1992	1,776	316
1993	1,703	307
1994	1,745	309
1995	1,694	295
1996 (preliminary)	1,820	315

SOURCE: Brown, Flavin, and Kane 1996. (Reprinted with permission)

Nor can there be much relief ahead if recent farming trends persist. Since 1981 there has been a 9 percent shrinkage in the world's grainland; since 1989 there has been a 16 percent decline in world fertilizer use; and since 1990 there has been almost no growth in irrigation water supplies (which is critical because irrigated land, comprising 17 percent of all crop-land, produces almost 40 percent of our food) (Brown, Flavin, and Kane 1996). On top of this, there has been severe or extreme soil erosion, desertification, and other forms of land degradation affecting 15 percent of damaged cropland; if this cropland were producing average harvests, it could feed 775 million people (Oldeman, Hakkeling, and Sombroek 1990).

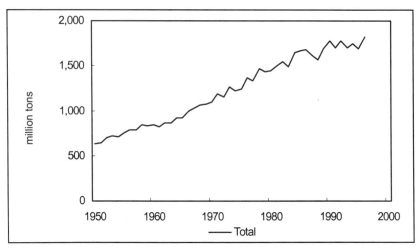

Figure 10.1.1a.   World grain production, 1950–96.
SOURCE: Brown et al. 1996 (Reprinted with permission); USDA 1996.

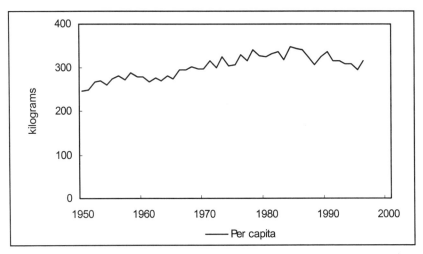

Figure 10.1.1b.   World grain production, per person, 1950–96.
SOURCE: Brown et al. 1996 (Reprinted with permission); USDA 1996.

All in all, grain shortages are likely to keep on mounting (Brown 1996; USDA 1995). Yet world population growth of 87 million people a year requires an annual expansion in grain production of 27 million tonnes (Brown and Kane 1994). If there were no population growth, these 27 million tonnes could be used to make up the diets of 700 million chronically malnourished or semistarving people, almost as many as the official count of all such people in the world today. The actual total could be higher, conceivably twice as high or even more, if one considers lack of micronutrients (Bouis 1995; McMichael 1993).

The Influence of Affluence

What counts in the population/food calculus is not only how many mouths there are to feed but how well they want to be fed. As people become more affluent, they move up the food chain, whereupon they consume more grain indirectly. As much as 36 percent of the world's grain is fed to livestock each year (over 50 percent in rich nations and 65 percent in the United States). This amount is enough to meet the basic grain needs of 3.1 billion people, or 67 percent of the developing world's total (Pimentel et al. 1995). One kilogram of feedlot-produced poultry or fish requires 2 kilograms of grain, 1 kilogram of pork requires 4 kilograms, and 1 kilogram of beef (the most preferred meat) requires 7 kilograms. Americans consume 800 kilograms of grain per person per year, one-third of it directly and the rest through meat consumption. Italians consume 400 kilograms—and they live longer than Americans even though they spend much less on health care. If the 1995 grain harvest of almost 1.7 billion tonnes were boosted to 2.0 billion tonnes, and if it were evenly distributed worldwide, it would support 2.5 billion people at the American level of consumption, 5 billion at the Italian level, or 10 billion at the Indian level of 200 kilograms per year. And if Americans were to cut their grain intake by just 16 percent (taking them back to their 1975 level), this would save 35 million tonnes of grain a year, enough to make up the diets of nearly 900 million people, or more than the total hungry today (Brown and Kane 1994).

This is where China's rising consumption of meat becomes significant for the world. It has jumped from 8 kilograms per person in 1977 to 32 kilograms in 1994. The soaring preference for meat has transformed China from a net grain exporter of 8 million tonnes a year to a net importer of 16 million tonnes during just the two years 1993 and 1994 (Brown 1996). The knock-on effects for the global grain economy are momentous. China's overnight emergence as a leading importer of grain, second only to Japan, is driving up world grain prices for all nations and people. The combination of increasing consumer demand and tightening markets during the period from mid-1994 to mid-1996 has caused grain export prices to double for two of the main staples, wheat and corn, while rice prices have risen by one-third (Brown 1996). This hits particularly hard at those poorer nations competing in global grain markets for the roughly 200 million tonnes of grain currently traded each year, that amount being some 11 percent of all grain grown. Sub-Saharan Africa (excluding South Africa) must import a full one-third of its grain, and do so with economies that collectively are no larger than Belgium's (Myers and Kent 1995).

The future prospect could well be worse still. If recent production and consumption trends persist until 2030, China alone with its explosive affluence may be seeking some 200 to 300 million tonnes of grain from overseas, or well over the total amount traded worldwide today. This would drive up grain prices to altogether unprecedented levels for all countries, and especially those developing countries looking for a further 190 million tonnes of grain (Brown 1995). Some analysts concur with this assessment of China's possible imports; for instance, Harris (1996) estimates 157 to 240 million tonnes. By contrast, others such as Pinstrup-Andersen and Garrett (1996) and Rosegrant (1996) believe that China's 2025 imports will not be nearly so large; nonetheless, they accept the premise that the country will become an ever-greater importer of grain, perhaps seeking 20 million tonnes as early as 2000 and 40 million tonnes by 2020. For a strongly dissident view, see Alexandratos (1996), Mei (1996), and Yansheng and Huang's paper in chapter 14. Worldwide, the demand for grain imports by 2025 could be anywhere between 350 million tonnes and 500 million tonnes (Brown and Kane 1994).

Nor is there likely to be much respite for the poorest nations in the form of food aid, even though it often comprises as much as 8 percent of all grain imports (Islam 1996). Because of rising grain prices, plus fiscal stringencies in the main donor nations, food aid has dropped from an all-time high of 15.2 million tonnes of grain in fiscal 1993 to 7.6 million tonnes in fiscal 1996 (Brown et al. 1996). The amount of food aid required in 1995 to meet the minimum calorie needs of the most malnourished people was 25 to 30 million tonnes—a total projected to rise to 55 million tonnes by 2010 (Missiaen, Shapouri, and Trostle 1995).

Tightening Patterns and Trends

These tightening patterns and trends apply to several other basic factors of agriculture. During the period 1950 to 1989, per capita cropland declined by around 50 percent, and during 1990 to 2010, it is expected to decline by an additional 21 percent. During 1950 to 1978, per capita irrigated land, which supplies almost 40 percent of our food from one-sixth of our cropland, expanded by 5 percent; however, during the period from 1979 to 1989 it contracted by 5 percent and from 1990 to 2010 it is expected to shrink by a further 12 percent. During 1950 to 1989 per capita production of fish more than doubled, whereas during 1990 to 2010 it is projected to drop by 10 percent (Postel 1994). All this will mean dearer food—and poor people will increasingly be in direct and tougher competition in the marketplace with rich people who are expanding in both numbers and per capita demands.

To cap it all, there has been a marked economic slowdown, hitting hardest at the poor. During the 1950s the global economy grew per capita by an annual average of more than 3 percent, but during the 1980s the rate slumped to little more than 1 percent, and from 1990 to 1995 it averaged only a little more than one-half of 1 percent (Brown, Flavin, and Kane 1996). In the case of the 1.3 billion poorest of the poor, those with cash incomes of less than one dollar a day and including 800 million malnourished people (almost every sixth person in developing countries), it is often the lack of money to buy food rather than outright shortage of food that is the main cause of their hunger. During the period 1980 to 1993, per capita gross national product (GNP) in 140 developing countries increased by an annual average of only 0.1 percent (in 38 of these countries, with collective populations of 587 million, it actually declined by 0.1 to 5.7 percent), whereas in rich nations it expanded by 2.2 percent (The World Bank 1995). In Mozambique it declined from $100 in 1990 to $80 in 1995, whereas in the United States it increased from $19,780 to $24,750 (WRI 1996). All this adds up to a profound watershed in consumption patterns and trends for rich and poor alike (Jackson and Marks 1995; Mayo and MacGillivray 1994).

At the same time the rich nations, with one-fifth of the world's population, use their affluence to consume more than one-half of the world's grain. This disproportionate consumption now contributes to reduced consumption on the part of developing nations. Many of the 3 billion people in nations with a per capita GNP of less than $725 have to spend more than 50 percent, and often as much as 70 percent, of their cash income on food (Serageldin 1993; WFP 1995). This means that even a marginal increase in food prices can tip them over the edge into hunger, if not starvation.

It is sometimes said by certain traditional economists that if there were a true food crisis, it would instantly reveal itself through rising food prices, whereupon farmers would be motivated to grow more food and the crisis would resolve itself through grace of the marketplace. Yet food prices remained more or less constant in real terms throughout most of the 1970s and 1980s, and even into the early 1990s—so it is assumed everyone was enjoying food security. This overlooks the fundamental fact that the people who suffer the worst food shortages, the poorest people, hardly possess the means to enter the marketplace at all, so they are denied the chance of registering their "dollar preferences" with respect to food. Thus arises the profound difference between "effective" demand and real demand for food, the first being only a part (and in some regions and communities, a small part) of the second. To a major extent, the overall problem is as much a poverty problem as a food problem.

At least 1.3 billion people have a cash income of less than one dollar per day, and another 1.7 billion people scarcely manage three dollars per day. These people amount to more than one-half of humankind, yet they account for less than one-tenth of the global economy. Their poverty causes them to be economically disenfranchised and nutritionally destitute. So it is strictly inaccurate in economic terms (let alone humanitarian terms) to assert that marketplace prices have much to do with the food purchases of those who most need food of any kind. Yet there are economic planners in the Food and Agriculture Organization (FAO) and the World Bank who still speak about food demand and supply as if all people enjoy the affluence to eat whatever they want. Anyone with a wallet can order a five-course dinner in Addis Ababa at the height of a famine.

Blinkered assertions about food prices are on a par with those about the supposed absence of major mortality from starvation—no "mass die-offs" despite supposedly alarmist warnings from doom-and-gloom environmentalists. During each of the past 25 years the total number of people dying from hunger or hunger-related diseases has averaged 10 million. This amounts to one-quarter of a billion people within a single generation. How many more would have to die to make the phenomenon qualify as a starvation disaster?

All this means that the challenge ahead is far from limited to producing enough food. It is a case of producing enough food at prices acceptable to everyone, especially the poor, and doing it while safeguarding the environmental basis of agriculture for today and tomorrow, a comprehensive approach known as food security (Barnett, Payne, and Steiner 1995; Berck and Bigman 1993; Falcon 1995; Pinstrup-Andersen 1994; Speth 1993; Swaminathan 1996a).

Decline of the Environmental Resource Base

The world's farmers will have to embark on their great enterprise with many of the environmental cards stacked against them. It is remarkable, then, that many of the leading analyses of our food prospects give almost no attention to this key factor. For some focus on these environmental constraints and how they are becoming ever more constraining, see Ehrlich, Ehrlich, and Daily (1993); Myers (1996a); Pimentel et al. (1997a); Pinstrup-Andersen (1994); Scherr and Yadav (1996); and Swaminathan (1996b).

Soil Erosion

During the past 20 years, some 500 billion tonnes of topsoil (roughly equivalent to all the topsoil in India's cropland) have eroded away, followed by another 25 billion tonnes each year (Brown et al. 1993). During the past 40 years, at least 4.3 million square kilometers of cropland have been abandoned because of soil loss, an expanse equivalent to 30 percent of today's cropland (Kendall and Pimentel 1994; Oldeman, Hakkeling, and Sombroek 1990). Without better soil conservation practices, between 1.4 million and 2.0 million square kilometers (the smaller expanse is equal to that of Alaska) will lose most of their good-quality soil over the next two decades, making the dust bowl look like a sandbox in comparison (Brown et al. 1993; Daily 1995; FAO 1993a; Pimentel et al. 1995).

Soil erosion is all the more serious today in light of the declining use of fertilizer, which has long served to offset loss of soil nutrients through erosion. If erosion is allowed to continue virtually unchecked, it could well cause a decline of 19 to 29 percent in food production from rain-fed cropland during the 25 years from 1985 to 2010 (Jarnagin and Smith 1993; Lal and Stewart 1990). In countries as disparate as Mexico, Costa Rica, Mali, and Malawi, soil loss causes annual declines in farm output worth 0.5 to 1.5 percent of GNP (Brown and Kane 1994), and a much higher proportion of the household incomes of the poorest people—precisely the ones who need to buy food to offset shortfalls from their eroded farmland. Yet upwards of half a billion impoverished people in developing countries find themselves obliged to farm hillsides where they cause extensive erosion (Boardman, Dearing, and Foster 1990; Pimentel 1993).

Desertification

Desertification, whether resulting from natural climatic change or human activities or both, threatens 45 million square kilometers, nearly one-third of Earth's land surface (UNEP 1995). It undermines the livelihoods of at least 900 million people in 100 countries, of whom 135 million are experiencing the rigors of severe desertification (Dregne and Chou 1992; Glantz 1994; for some dissident analysis, see Thomas and Middleton 1994). It currently eliminates 60,000 square kilometers of agricultural land each year and reduces another 200,000 square kilometers to a state of grossly depleted productivity. The cost of agricultural output lost is around $42 billion per year (Dregne and Chou 1992; Davies, Buchanan-Smith, and Lambert 1992).

One of the main causes of desertification is overgrazing by domestic livestock. Yet the Intergovernmental Panel on Climate Change has projected a 45 percent increase in meat and dairy output by 2025 to cater to population growth and dietary upgrading (Houghton, Jenkins, and Ephramus 1990). This, in the absence of unexpectedly productive breakthroughs in agro-technologies, implies a marked increase in grass-fed livestock.

Salinization and Waterlogging of Irrigated Lands

To reiterate a key factor, irrigated land produces almost 40 percent of our food from only one-sixth of our cropland. Equally important, this land contributed more than one-half of the increase in food production during the Green Revolution's heyday from the mid-1960s to the mid-1980s (Kendall and Pimentel 1994; The World Bank and UNDP 1990). Given the expected demand for additional food in the foreseeable future, plus the lack of further good-quality land to open up for agriculture, we shall need to look to existing cropland for ever-greater harvests, and especially to irrigated land as the most productive of all cropland. This irrigated land now covering some 2.6 million square kilometers (out of 14.4 million square kilometers of all cropland), or an aggregate expanse equivalent to almost the whole of India, may have to produce one-half or even two-thirds of our additional food in the future (Alexandratos 1995; Seckler 1996).

These prime food-producing areas expanded by an average of 2.8 percent per year during 1950 to 1980, but only by 1.2 percent since then (Brown, Flavin, and Kane 1996). Future expansion is likely to be no more than 0.3 percent per year, and possibly none at all (Raskin, Hansen, and Margolis 1995). In per capita terms, irrigated land has shrunk by 6 percent during 1978 to 1990, and is expected to contract by an additional 12 percent during 1991 to 2010 (Postel, Daily, and Ehrlich 1996). As many as 250,000 square kilometers of irrigated land, or almost one-tenth of the total, have been salinized enough to reduce crop yields, and another 15,000 to 25,000 square kilometers are so severely affected every year that many of them have to be abandoned (Kendall and Pimentel 1994; Seckler 1996). Because of salinization together with water logging, 450,000 square kilometers of irrigated land in developing countries need immediate and costly reclamation if they are not to be lost to agriculture (FAO 1993a).

Water Deficits and Droughts

Currently, 550 million people experience shortages of renewable water. In 88 developing countries with 40 percent of the world's population, the problem has become a serious constraint on development in general and on agriculture in particular (Harris 1996; Pimentel et al. 1997a). The total of water-short people is projected to reach approximately 3 billion by 2025 (a range of 2.8 billion to 3.3 billion), largely because of population growth; it could even grow to the order of 4.4 billion by 2050. The prospect in many regions could be made markedly worse by the climatic vagaries of global warming.

It is unlikely that demand will be met, if only because of practical upper limits of useable, renewable freshwater stocks. The principal areas at risk include (though are not confined to) parts of India and Pakistan, the Middle East, and much of Africa. More than 1 billion of the people who will be affected by 2025 are expected to be in Africa, comprising two-thirds of that continent's projected population (Engelman and LeRoy 1993; Falkenmark 1994; Gleick 1993; Postel 1992; Serageldin 1994). All of these regions, except for the Middle East, already suffer some of the most severe food shortages.

Tropical Deforestation

In many parts of the humid tropics, water shortages are aggravated by, if not primarily caused by, deforestation. Tropical forests are being destroyed at rates (usually 1 to 2 percent per year) that presage the elimi- nation of large expanses early in the next century. As the forests disappear, especially in upland catchments, so do many of their watershed services and hydrological functions, causing year-round water flows in downstream areas to give way to flood-and-drought regimes. As many as 40 percent of developing-world farmers depend upon regular flows of rivers and streams from healthy watersheds to irrigate their cropland (Myers 1992). In the Ganges Valley with its 500 million small-scale farmers in India and Bangladesh, the annual watershed costs of Himalayan deforestation were estimated as far back as the early 1980s at more than $1 billion in India alone (High Level Committee 1983). A similar decline of watershed ser- vices is becoming apparent in the densely populated valley lands of the Irrawaddy, Salween, Chao Phraya, and Mekong Rivers in Southeast Asia, a region where the biggest constraint on rice production is not lack of arable land or agronomic inputs but shortages of cropland water (Durning 1993).

In addition, there is evidence that deforestation sometimes results in reduced rainfall (Salati and Nobre 1992). In northwestern Peninsular Malaysia, the Penang and Kedah States have experienced disruption of rainfall regimes so severe that 20,000 hectares of paddy rice fields have been abandoned and another 72,000 hectares have registered a marked production drop-off in this rice bowl of the Peninsula (Chan 1986). Similar deforestation-associated changes in rainfall have been documented in parts of the Philippines, southwestern India, montane Tanzania, southwestern Ivory Coast, northwestern Costa Rica, and the Panama Canal Zone (Meher- Homji 1992).

Decline of Biodiversity

In contrast to environmental constraints, biodiversity decline exerts a mostly indirect and long-term impact. A good number of species and their populations supply genetic resources for new foods and improved forms of existing foods (Myers 1983; Oldfield 1989; Srivastava, Smith, and Forno 1996). We shall need all the genetic variability we can find to cope with both present pests and diseases and future problems such as the vagaries of climatic change. Thus, there is an unprecedented premium on crop types with resistance to too little or too much rainfall, increased ultraviolet (UV-B) radiation, and new pathogens. Yet we are losing approximately 50,000 species a year from a planetary complement of 10 million species (minimum reckoning), and probably a much higher proportion of species' populations. This extinction rate is accelerating rapidly: we may witness the demise of one-half of all species and most populations of surviving species by the middle of the next century (Ehrlich and Ehrlich 1991; Myers 1996b; Wilson 1992).

Still more importantly, all our agricultural crops are sustained through constant infusions of fresh germ plasm with its hereditary characteristics (FAO 1993b; Potter, Cahen, and Janczewski 1993). Certain of these infusions come from wild relatives of modern crops, others from land races and so-called primitive cultivars. Thanks to this regular "topping up" of the genetic or hereditary constitution of the United States' main crops, the Department of Agriculture estimates that germ plasm contributions lead to increases in productivity that average about 1 percent annually, with a farm-gate value that now tops $1 billion (U.S. NRC 1992).

Regrettably, wild gene pools are being depleted rapidly. Wheat, for instance, flourished across an expanse of more than 230 million hectares in 1995, with a rough average of 2 million stalks per hectare. This means that the total number of individuals exceeded 460 trillion, probably a record (Myers 1995). As a species, then, wheat is the opposite of endangered. But because of a protracted breeding trend toward genetic uniformity, the crop has lost the bulk of its populations and most of its genetic variability. In extensive sectors of wheat's original range where wild strains have all but disappeared, there is a virtual wipeout of endemic genetic diversity. Of Greece's native wheats, 95 percent have become extinct; in Turkey and extensive sectors of the Middle East, wild progenitors find sanctuary from grazing animals only in graveyards and castle ruins. As for wheat germ plasm collections, they were described more than a dozen years ago as "completely inadequate"—and that was without considering future broadscale problems such as acid rain, enhanced UV-B radiation, and global warming (Damania 1993).

We can tell a related story with respect to rice. In the early 1970s Asia's rice fields were hit by a grassy stunt virus that threatened to devastate rice production across more than 30 million hectares from India to Indonesia. Fortunately, a single gene from a wild rice offered resistance against the virus. Then in 1976 another virus, known as ragged stunt disease, emerged; again, the most potent source of resistance proved to be a wild rice. The economic returns from these wild rices would be more than enough to pay all the expenses of preserving the collection of rice germ plasm at the International Rice Research Institute (IRRI) in the Philippines. Several other diseases could impose widespread losses on Asia's rice crop, but at least 100 wild rices appear to harbor resistance (Chang 1986).

Landlessness

Landlessness—or rather, to be realistic, functional landlessness—can plainly reflect population pressures. Often it derives from environmental factors when land degradation has virtually eliminated the productive value of former farmland. When this happens there is less worthwhile land to go around in communities that may already find that population pressures make individual holdings unduly small. Landlessness usually stems too from a number of political, economic, social, and legal problems; it can often be deemed a poverty problem. It is unusually acute in Mexico, Colombia, Ecuador, Peru, Egypt, Nigeria, Ethiopia, Kenya, Tanzania, Pakistan, India, Bangladesh, Vietnam, Indonesia, China, and the Philippines. Their aggregate population today totals 3.2 billion (56 percent of humankind), and is projected to reach 4.6 billion (56 percent) by 2025.

The minimum amount of agricultural land necessary for sustainable food security, with a diversified diet similar to those of North America and Western Europe (i.e., including meat), is 0.5 hectares per person. This does not allow for any land degradation such as soil erosion, and it assumes adequate water supplies (Pimentel et al. 1995). Very few populous countries have more than an average of 0.25 hectares. It is realistic to suppose that the absolute minimum of arable land to support one person is a mere 0.07 hectares—and this assumes a largely vegetarian diet, no land degradation or water shortages, virtually no postharvest waste, and farmers who know precisely when and how to plant, fertilize, and irrigate (Engelman and LeRoy 1993). In India, the amount of arable land is already down to 0.2 hectares; in the Philippines, 0.13; in Vietnam, 0.10; in Bangladesh, 0.09; in China, 0.08; and in Egypt, 0.05. By 2025 the amount is expected to fall in India to 0.12 hectares; in the Philippines, 0.08; in China, 0.06; in Vietnam, 0.05; in Bangladesh, 0.05; and in Egypt, 0.03 (Engelman and LeRoy 1995a, b). For additional details, see Table 10.1.2.

Table 10.1.2.   Per capita cropland available in 1990 and projected in 2025

Country	1990	2025
	(hectares)	(hectares)
Mexico	0.29	0.18
Colombia	0.17	0.11
Ecuador	0.27	0.15
Peru	0.17	0.10
Egypt	0.05	0.03
Nigeria	0.34	0.14
Ethiopia	0.29	0.11
Kenya	0.10	0.04
Tanzania	0.13	0.05
Pakistan	0.17	0.07
India	0.20	0.12
Bangladesh	0.09	0.05
Indonesia	0.12	0.08
Vietnam	0.10	0.05
China	0.08	0.06
Philippines	0.13	0.08

NOTES: The minimum amount of arable land required to sustainably support one person is 0.07 hectares. This assumes a largely vegetarian diet, no farmland degradation or water shortages, virtually no postharvest waste, and farmers who know precisely when and how to plant, fertilize, and irrigate. The collective population of the above countries today is 3.2 billion (56 percent of the world's population), and is projected to grow by 2025 to 4.6 billion (a 56 percent increase).
SOURCE: Engelman and LeRoy 1995a. (Reprinted with permission)

By 2025 or shortly thereafter, certain of these countries, notably India, Bangladesh, Pakistan, Indonesia, Egypt, Ethiopia, Nigeria, and Mexico, may need to triple their grain imports (Brown 1996). Of course, to rely on foreign food is acceptable provided it is available at the right price. Japan is one of the most land-short countries of all, with only 0.06 hectares of arable land per person today, projected to fall to 0.04 by 2025. But Japan can afford to buy all the food it wants, and presumably will continue to do so in the future.

A related problem contributing to landlessness is the "jobs famine." The workforce in developing countries numbers about 2 billion people (UNDP 1996). Of these, at least 750 million are unemployed or grossly under-employed; their total exceeds the entire workforce of the developed countries (unemployment in countries of the Organization for Economic Cooperation and Development [OECD] amounts to approximately 30 million). By 2025, because of past population growth, the developing countries' workforce is projected to expand to more than 3 billion (The World Bank 1994). Note that this latter figure is hardly a projection subject to possible reduction through policy interventions—most of the future workers have already been born.

To supply employment for the new workers, let alone those without work today, developing countries will need to create an average of 40 million new jobs per year for the next three decades. The United States, with an economy one and a half times as large as the entire developing world's, often has difficulty generating an additional 2 million jobs each year. Many unemployed people in developing countries find they can gain a livelihood only by cultivating marginal land such as forests, dry savannahs, and hilly terrain, thus aggravating problems of deforestation, desertification, and soil erosion (Myers 1991; Pimentel 1991). Equally important, unemployed people are generally impoverished people, meaning they have little means to buy food even when it is available.

Agricultural Stress

The seven problems summarized here can be considered together as *agricultural stress.* Overall at least 2.9 million square kilometers of farmland, an area almost the size of India, have lost virtually all their productive capacity (Daily 1995; Tolba et al. 1992). In addition, 70,000 to 100,000 square kilometers of farmland are abandoned each year as a result of degradation, while another 200,000 square kilometers become essentially infertile (Barrow 1993; Pimentel 1993). In addition, there are problems of inadequate agro-technologies and extension services, as well as a lack of rural infrastructure such as road networks, marketing systems, and credit facilities. All these problems compound with population pressures and associated poverty. The worst degradation tends to be in countries with high population densities, notably China, the Philippines, Indonesia (Java), Vietnam, Thailand, Bangladesh, India, Pakistan, Ethiopia, Nigeria, and sectors of Andean South America. These countries and areas have an aggregate population today of 3.1 billion, projected to reach 4.3 billion by 2025. While some are advancing economically, several others are still trying to relieve widespread poverty.

As noted, we shall need to feed an extra 2.5 billion people by 2025. Yet during the two decades from 1994 to 2013, between 1.4 million and 2.0 million square kilometers of fertile land are expected to lose much of their agricultural value (FAO 1993b; Pimentel et al. 1995). If land degradation is not halted, it will more than negate all gains from the opening up of new agricultural land during the next two decades (at most, 50,000 square kilometers per year) (Kendall and Pimentel 1994). Indeed there are few areas at all to be opened up for agriculture. Cropland expanded at an average annual rate of 0.8 percent per year throughout the period 1950 to 1980, but since then the expanse has been contracting at an average annual rate of 0.5 percent. The amount of per capita arable land has declined since 1984 by an average of 1.9 percent per year, primarily because of population growth, urbanization and industrial spread into farming areas, and the sheer lack of new lands for agriculture (Engelman and LeRoy 1995a). At the same time, there is the agreed imperative of expanding grain production by 2.0 percent per year throughout the next three decades. While grain production increased by an average of 4.7 percent per year from 1950 to 1985, the growth rate dropped to an average of 0.3 percent per year from 1986 to 1995 (USDA 1996; Brown et al. 1996). It will be difficult to push it back up to 2.0 percent because of three factors: the shortage of good-quality cropland waiting to be opened up; the progressive loss of existing cropland to urbanization and industrialization; and a suite of intensifying environmental problems. It is true that the bulk of the grain harvest growth in recent decades has come not from cropland expansion but from yield increases, yield being output per unit of land (including the impact of both biological growth and cropping intensification). In turn, one-half of the yield increase from the mid-1950s to the mid-1980s came from irrigated land (The World Bank and UNDP 1990). As already mentioned, the expansion of irrigation has slowed because of lack of suitable additional land, deterioration of existing land, and competition for water supplies (Cassman and Harwood 1995; Postel 1992). By contrast, certain observers, such as Alexandratos (1995), expect that from one-half to two-thirds of increases in crop production between 1996 and 2010 will come from irrigated land. In addition, there are many other environmental problems overtaking cropland, whether irrigated or not.

In addition to these eight current environmental constraints, two others lie ahead. Like the rest, they are likely to exert progressively increasing impact and to be largely irreversible within the time horizon of interest to most people alive today.

Ozone-Layer Depletion and Increased UV-B

Depletion of the ozone layer allows additional UV-B radiation to reach Earth's surface, with an adverse impact on crop plants on land and phytoplankton-based food chains in the sea. Experiments show that enhanced UV-B lowers crop yields by anywhere from 5 percent for wheat to 90 percent for squash, with other sensitive crops in between. Soybean, for instance, a prime source of protein and second in value to corn in the United States, could lose one-quarter of its productivity. Other vulnerable crops include barley, peas, and cauliflower (Teramura 1986; Worrest and Caldwell 1986).

Phytoplankton is the most abundant type of ocean life both in weight and sheer numbers (1 liter of seawater can contain millions of phytoplankton plantlets), and no other life form appears to be as susceptible to UV-B radiation. As phytoplankton disappear, so will zooplankton (microscopic animals that feed on them), then fish, fish-eaters, and so on, to the end of the famished line (Worrest and Hader 1989). This does not bode well for marine fisheries, which supply more animal protein than any other source for more than 1 billion people in the developing world.

Extreme Weather Events and Global Warming

In recent years there has been one set after another of weather anomalies. In March 1993 the United States witnessed an east coast "storm of the century" followed a few months later by record flooding in the Midwest. At the end of the same year, Alaskans within 25 kilometers of the Arctic Circle enjoyed a temperature of 6°C, a heatwave by comparison with the usual -46°C, yet Michigan endured its lowest temperatures in 100 years, and Melbourne experienced its coldest Christmas in more than 100 years. In mid-1994 northern India had its highest temperatures in half a century, while several parts of the southern United States experienced a heat wave of 49°C, and the southeastern states recorded unprecedented rainfall. In January 1995 parts of California received as much rainfall in a single day as they typically receive in an entire year. Since early 1995 there have been more of these freak phenomena.

Indeed all of the 1980s and the first half of the 1990s have been characterized by what a leading British climatologist, Sir John Houghton (1994, 11), terms "the frequency and intensity of extremes of weather and climate." As experience has shown, these sharp weather shifts have the capacity to exert strong, adverse impacts on agriculture. Apart from the obvious impacts of floods or droughts, there is the impact of prolonged heat. In the late summer of 1995 in the U.S. grain belt, as well as in other grain basket regions of the European Union, the Ukraine, and Russia, prolonged heat led to withering of the crops, causing a 2 percent decline in the grain harvest, a total of 35 million tonnes (Brown 1996). This loss would have been enough to make up the diets of nearly 900 million malnourished people.

The recent extreme weather events suggest "something wrong in the climate works," and they are a possible portent of global warming, which may approach not gradually but in spurts. Global warming models predict more violent weather in many parts of the world as the atmosphere slowly and steadily warms (Bryant 1991; Downing et al. 1993; Pittock 1995). To quote the Chairman of the Intergovernmental Panel on Climate Change, Professor Bert Bolin (1994, 34), "Most of the damage due to climate change is going to be associated with extreme events, not by the smooth global increase of temperature that we call global warming." In the long run, global warming offers a broad scope for climatic vicissitudes: an event with a return period of 10 years today could well become a three-year event (Hennessy 1993).

Also significant is the prospect of global warming effects such as shifts in monsoon systems and the arrival of deep, persistent droughts, with all the implications these would have for agriculture. Monsoon patterns could be profoundly affected by a temperature rise of only 1°C to an extent that would dwarf the direct drought effects of such a temperature rise (Inter-governmental Panel 1990). The area most vulnerable to monsoon dislocations is the Indian subcontinent, projected to have 1.9 billion people by 2025—49 percent more than today. India relies upon the monsoon for 70 percent of its rainfall, hence its agriculture is critically dependent upon the stable functioning of the monsoon system (Rasmusson 1989). In broader terms, the entire Asia–Pacific region is markedly vulnerable to monsoon changes if only because it contains well over half the world's population, projected to become a still larger proportion by 2025 (Myers and Kent 1995).

Drought and its repercussions for agriculture are more of an uncertain issue since the climatic quirks are less well predicted through global cli-mate models at the regional level than is the case for monsoon patterns. As far as we can discern, droughts with only a 5 percent frequency today may increase to 50 percent by 2050 (Rind et al. 1990). Areas susceptible to drought include much of northern Mexico, northeastern Brazil, the Medi-terranean Basin, the Sahel, the southern quarter of Africa, and sectors of the middle and tropical latitudes of Asia (Magadza 1994; Schneider 1989; Williams and Balling 1994). In the Sahel, for instance, rainfall may de-crease even more than in the past three decades (Hulme and Kelly 1993). In China, more rain may fall in the north of the country and less in the fertile south where most of the population lives (Hulme, Wigley, and Jiang 1992; Schneider et al. 1990). Also likely to be affected by drought are parts of the United States, southern Canada, southern Europe, and Australia (Kaiser and Drennan 1995). These areas are important in this context since they produce much of the surplus food that sustains more than 80 developing countries today.

According to an innovative analysis involving drought among a host of other agricultural problems, a plausible global warming scenario for early next century indicates there could be a 10 percent reduction in the world grain harvest an average of three times a decade (Daily and Ehrlich 1990). The 1988 droughts in the United States, Canada, and China alone resulted in almost a 5 percent decline, and a mere 0.5°C increase in temperature could reduce India's wheat crop by 10 percent. (It is likely there will soon be a rise in temperature of 0.25°C per decade). Given the way the world's grain reserves dwindled to almost nothing as a result of droughts in the late 1980s (Brown et al. 1992, 1993), it is not unrealistic to suppose that each such grain harvest shortfall could result in huge numbers of starvation deaths—anywhere, according to computer model calculations, from 50 million to 400 million people (Daily and Ehrlich 1990).

Megascale famines are held at bay today in part through food shipments from the great grain belt of North America and from other food-exporting regions. In a globally warmed world, this grain belt could become "unbuckled" to an extent that there would be fewer such shipments as Americans find it harder to feed themselves, let alone other countries. The Great Plains, a region 2,100 kilometers long from north to south and 800 kilometers wide, already experiences weather extremes: severe droughts, heat waves, cold, winds, storms, floods, and tornadoes. The western portion between the 100th meridian and the Rocky Mountains verges on semidesert, with a drought about every 20 years. These phenomena could increase markedly through global warming.

This analysis has been paralleled by a more detailed and optimistic assessment (Rosenzweig and Parry 1994) that postulates global warming will reduce grain production, especially in the tropics (meaning primarily the developing world), by 10 to 15 percent by 2060. In conjunction with other factors, such as population growth and increased grain prices to reflect scarcity, this will increase the number of people at risk of hunger to somewhere between 640 million and one billion. Major grain shortfalls are expected to start by early next century, if not before.

This analysis is subject to a number of reservations. It assumes a significant and sustained increase in crop productivity, though this is by no means assured if only because of the "plateauing" of crop yields since 1985. It postulates a steady reduction of trade barriers, at least a moderate degree of economic growth, and only a moderate rate of population increase. More importantly, it assumes that increased atmospheric carbon dioxide will make a substantial contribution to the enhanced growth of crop plants (an outlook that is possible rather than probable). It supposes there will be no limitations on supplies of irrigation water, although as early as 2025 there are likely to be 3 billion people living in countries with pronounced water deficits. The analysis envisages no changes in the patterns of monsoon or tropical cyclone systems, or in the intensity of rainfall regimes. It discounts the impacts of shifts in the El Niño Southern Oscillation. It does not take account of changes in the occurrence and distribution of crop pests and diseases. All these factors are potentially important and possibly dominating constraints (Pittock, Whetton, and Wang 1994).

One area particularly vulnerable to climatic change in the form of droughts and other global warming phenomena would be Africa, where regions at special risk of drought include North Africa, the Sahel, the Horn of Africa, and southern Africa (Hulme et al. 1994; Magadza 1994). Their aggregate population today totals 334 million, projected to reach 619 million by 2025 (an 85 percent increase). The principal agricultural problem would be lack of soil moisture during the growing seasons. In sub-Saharan Africa there is now an annual food deficit of approximately 10 million tonnes, predicted to increase to 50 million tonnes by 2000 or shortly thereafter, and to 250 million tonnes by 2025 (Pinstrup-Andersen 1993). Even an optimistic scenario for sub-Saharan Africa (reduced population growth, enhanced soil and water conservation, and expanded irrigation) foresees that per capita grain production in 2025 would be well below today's sorely inadequate level (Kendall and Pimentel 1994). Hence, the region would be extremely vulnerable to even minor climatic disruptions.

An Overall Assessment of Environmental Constraints

These environmental constraints limit the outlook for agriculture as never before, precisely at a time when we need agriculture to become productive as never before to provide for population growth and its demographic dynamics. Certain constraints can be somewhat contained *provided* we recognize them in their full scope. Yet they are largely dismissed at one of the places where they should be most heeded, the FAO, which merely calls for "continued efforts to maintain and upgrade the world's food production capacity, e.g., investment in agricultural research, extension services, increased efficiency in water use, etc." (Alexandratos 1996, 11). Each of these three issues is headed in the wrong direction, as are the problems omitted altogether from the publication cited, notably soil erosion, desertification, deforestation, and other forms of land degradation, plus water shortages and emergent super-problems such as global warming. The FAO rationale is that our only guide to the future is the past—even though there is abundant evidence that the future will no longer be a simple extension of the past, and that we must expect a host of basic departures from former patterns and trends.

Future Priorities

In conclusion, we present several points that deserve priority attention from political leaders, policymakers, and strategy planners. Discussions by Falcon (1995), McCalla (1994), Pimentel and Pimentel (1996), Pinstrup-Andersen (1994), Rosegrant, Agcaoili-Sombilla, and Perez (1995), and Swaminathan (1995) will provide parallel assessments.

Demographic Determinants

Because the population explosion is still in its most explosive phase (the present annual increase of 87 million people will not decline much for at least another decade), the world's political leaders and agency officials should be increasing their efforts to generate more food. Yet the agricultural cause seems to have fallen off political agendas, as shown by the decline in funding for agricultural research, even though this research has represented some of the finest advances in science, at exceptionally low cost, and with an exceptionally high rate of return. Yet because of population pressures, together with constraints from the environmental resource base, there is now a tighter grain supply situation than since the mid-1970s (Pimentel et al. 1997b; Ruttan 1996). There is also an end to the grain buyer's market that has persisted for most of the last half century (Brown 1996). All this occurs precisely as we enter the main population bottleneck of the next two decades. By contrast, if population growth were to be zero, food production would need to increase by only one-third as much as is anticipated (Bongaarts 1994).

Humankind Half Urbanized

By around 2000, one-half of humankind will live in cities and other urban areas. We will have shifted from the pattern that has predominated since we came out of our caves, namely a largely rural community, to a primarily urban one. For the first time there will be as many urban consumers as rural producers of food. And by 2025, the number of urban dwellers in developing countries will have soared from 1.6 billion in 1990 to 4 billion. This rapid urbanization together with rising incomes will cause people to shift their preferences from low quality staple grains to higher quality grains such as rice and wheat, and to meat and dairy products, thereby increasing the demand for grain-based products (Mitchell and Ingco 1993).

All this not only alters the fundamental niche of agriculture in the human enterprise but, contrary to what one might intuitively suppose, it places a greater premium than ever on building rural infrastructure. This would be partly for the benefit of, and thus partly paid for by, urban communities. Infrastructure here includes marketing systems, credit facilities, road networks, irrigation systems, and agro-technology, as well as larger scope investments such as rural health and agriculture-oriented education.

Unstable Prices

For poor people who spend as much as 50 to 70 percent of their cash income on food, a 20 percent rise in grain prices can cut their food consumption by 10 percent (WFP 1995). During the past quarter century, grain prices have risen by as much as 30 to 35 percent in a single year, and fallen by 27 to 32 percent in another year (Islam 1996). But in the last two years world prices for two of three main grain crops have doubled, causing the most destabilizing changes in more than two decades. There may be further severe price increases if the United States and/or the European Union decide to impose export embargoes on feed grains such as corn and barley, allied to similar restraints by Vietnam and other grain-exporting countries.

Environmental Underpinnings of Agriculture

The environmental resource base that underpins agriculture—soils, water, and climate—is more vital than has often been recognized in the past, and it will become more significant in the future, if only because of soaring food demands. At the same time, these environmental underpinnings are being depleted rapidly. Thus, there is a policy premium on greatly increased safeguards for the environment. This is already emphasized in certain councils of power, but largely ignored in many others (Barnett, Payne, and Steiner 1995; Cassman and Harwood 1995; Ehrlich 1995; Pinstrup-Andersen 1994; Scherr and Yadav 1996).

Trade Liberalization

Much depends on the relaxation of trade restrictions on the part of developed countries, notably the United States and members of the European Union. If these countries were to reduce today's trade restrictions by 50 percent and phase out some of their domestic subsidies and export subsidies in the agricultural arena (Myers 1997), they would effectively transfer to developing countries export revenues on the order of $50 billion per year, a sum almost as large as all development aid (The World Bank 1993). This would enable developing countries to compete better in international grain markets.

The Role of the Poor

Poverty precludes some 2 to 3 billion people from participating in the marketplace for critical food supplies such as staple grains. This means that prices reflect effective demand rather than total need for grain, and the gap between the two is large and widening quickly. It is the poor who most need access to affordable food. For a fine overview see Pinstrup-Andersen, Pandya-Lorch, and Hazell (1994). This dilemma arises at a time when there is less food aid available. We must expect a steady (or perhaps speedy) increase in the number of people in absolute poverty and enduring chronic malnutrition. In addition, we must be ready for extreme contingencies such as famines and starvation on a scale not yet witnessed, notably in southern Asia and sub-Saharan Africa.

In this context, note the tie-in with developing world unemployment, a linkage not often noticed in the food arena. Without adequate jobs or any jobs at all, at least 750 million people today lack much of the purchasing power they need to buy food; the total could swell to a much larger number by 2025. Conversely, if there could be a policy push for supplying employment on a sufficient scale, this would not only enable poor people to buy food, it would also send a marketplace message to farmers that they would find purchasers if they grew more food for commercial outlets.

The Role of Women

Women bear and nourish children, so they have special nutritional needs. Yet women of every age in the developing world have disproportionately higher rates of malnutrition than men. In addition, women are overrepresented among poor and illiterate people (Cohen and Reeves 1995; Swaminathan 1995). Because of their efforts to feed their families, women are often more closely involved than men with agricultural resources such as soil and water. In parts of sub-Saharan Africa, women are responsible for 80 percent of the production, processing, and marketing of food (Adepoju and Opong 1994). Similarly, in the fuelwood sector with its many linkages to agriculture, women traditionally do most of the wood gathering, and often take the lead in tree planting (Jacobson 1993).

Environmental Refugees

In the early 1980s a new phenomenon emerged: environmental refugees. These are people obliged to abandon their traditional homeland for primarily environmental reasons such as soil erosion, water shortages, and desertification, as well as population pressures, extreme poverty, "marginalization," and other socioeconomic factors. Today they total at least 25 million people, many more than all the established forms of refugees—political, ethnic, and religious—combined. Their total could double within another 10 years, whereupon supplying them with relief food would require 10 million tonnes of grain. As global warming starts to cause unprecedented droughts and coastal flooding, their number could readily soar to 200 million or higher (Myers and Kent 1995), placing a far greater strain on relief food supplies.

Innovative Options

In certain instances, one way to meet food needs lies not with planting more grain but more trees. Windbreaks made of trees can enhance soil moisture by 15 to 25 percent and increase crop productivity by 5 to 10 percent (Newcombe 1984). Trees can rehabilitate denuded upland catchments and stabilize river flows; 40 percent of farmers in the developing world depend upon upstream watersheds for their water supplies. As a very preliminary and exploratory estimate, the amount of extra grain provided thereby can average 5 to 10 percent (Myers 1992). Water supplies of sufficient quantity and quality for domestic household use can help to reduce disease, since 90 percent of disease cases in the developing world are related to contaminated water. When a person is sick with diarrhea, malaria, or another major disease due to inadequate water supply, some 5 to 20 percent of the person's food intake is needed simply to offset the disease (Pimentel et al. 1997b). In these various ways, planting trees rather than grain crops can sometimes help provide 15 to 45 percent of food requirements. Yet this aspect of the food challenge is scarcely heeded.

Investment Constraints

Finally, let us consider a constraint that is more problematic than several of the items already discussed, while offering more scope for a highly productive solution. It comprises the severe constraints on agricultural investment in general and on research investment in particular. In addition, there are restrictions on extension services and other ancillary support, together with larger scope factors such as the maldistribution of farmland and the political neglect of agriculture (especially smallholder and subsistence agriculture) in many developing countries (Swaminathan 1994). We will look at the first two constraints because they are indicative of the investment context for agricultural advancement.

In order for general investment to meet food goals by 2010, we need to think in terms of $105 billion annually for primary agricultural production, plus $43 billion for postproduction facilities and $37 billion for public support services and infrastructure. The total of $185 billion is 30 percent above the annual average in recent years (Rosegrant, Agcaoili-Sombilla, and Perez 1995). Foreign aid for agriculture declined from $13.4 billion in 1988 to less than $10 billion in 1993, when it constituted only 14 percent of all foreign aid, down from 20 percent in 1980 (Von Braun et al. 1993).

As for research funding, government investments in agricultural R & D worldwide totaled $17 billion in 1990, of which $8.8 billion was in developing countries. The Consultative Group on International Agricultural Research (CGIAR) spent $319 million in 1992 to support its network of International Agricultural Research Centres. However, that amount dropped to $245 million by 1994 even though the network needed $270 million merely to maintain its activities at former levels, let alone to expand them at a time when research breakthroughs are more urgent than at any time since the 1960s (Greenland et al. 1994). In light of the returns on investment, which can be as high as 20 or even 40 percent per year, the CGIAR budget is absurdly small (Anderson, Pardey, and Roseboom 1994; Tribe 1994). There is all the greater urgency in bolstering research funding since we are aiming for an annual 2 percent increase in food production and there is often a time lag of 10 to 20 years before breakthrough research leads to major harvest increases in farmers' fields (McCalla 1994). Yet the United States has recently slashed its international research funding by more than 50 percent (see Per Pinstrup-Andersen and Rajul Pandya-Lorch's paper in chapter 9).

As a measure of needs and opportunities, consider the rice and wheat outlook in Asia, which is a large part of the entire food challenge worldwide. Some 1.28 million square kilometers are planted with rice and 620,000 square kilometers with wheat; three-quarters of both amounts are irrigated. Neither of the irrigated expanses is expected to increase; indeed, they may decrease somewhat. If demand for rice were to rise by 1.75 percent per year (it is actually projected to rise by 1.8 percent per year), then paddy rice production would have to expand by a total of 760 million tonnes by 2020 to achieve self-sufficiency. This translates into an increase in annual per hectare output from just under 5 tonnes to almost 8 tonnes, or a 61 percent increase. A similar rise in the productivity of irrigated wheat will be required to avoid an upsurge in the projected wheat deficit.

Both the scale and the rate of these increases are way beyond the frontier of today's agro-technologies. Indeed, they approach the limits of theoretical yield potential as we understand it. In any case, farmers rarely achieve more than 80 percent of yield potential even when they deploy optimal management techniques (Cassman and Harwood 1995). In short, there is already a massive premium on research, and funding should be boosted to higher levels immediately.

The 1996 Grain Harvest

The priorities listed here are all the more urgent in 1997 as we face a food crisis only slightly smaller than any we have tackled since we started on the challenge of feeding fast-growing numbers of people almost 50 years ago. While the 1996 grain harvest of 1.82 billion tonnes (a preliminary estimate) is the largest ever and exceeds the 1995 harvest of 1.69 billion tonnes by 7.4 percent, it is still only 4.3 percent higher than the average for the 1990s, which is 1.74 billion tonnes. It is a mere 40 million tonnes, or 2.2 percent, higher than the next highest grain harvest in 1990, which was 1.78 billion tonnes. All this means that the late 1996 worldwide per capita grain supply, 315 kilograms, is 6.8 percent higher than in 1995 but 9.0 percent lower than the all-time high of 346 kilograms in 1984.

Fortunately there are likely to be few immediate large-scale restrictions on trade. The United States, which accounts for about one-half of all grain exports and had considered early in 1996 imposing an export embargo for corn, will be inclined to keep an eye on grain export revenues worth around $40 billion a year, which are needed to offset oil imports of $60 billion a year. The European Union may prohibit barley exports, and Vietnam has temporarily suspended rice exports. Otherwise, grain trading is expected to continue unfettered, albeit at rather higher prices than we have been accustomed to for at least two decades, and with an end to the 50-year-long buyers' market.

The year 1996 concludes with 87 million more people than in 1995. To reiterate a key point of population dynamics in relation to food security: 27 million of the 40 million tonnes increase in the 1996 grain harvest over 1995 will be taken simply to feed extra mouths. Let us note, however, a positive prospect. It concerns the region where the population growth rate remains highest, where per capita food output is lowest and declining, and where the specter of starvation is unprecedentedly stark: sub-Saharan Africa. If the region were able to reduce its population growth rate by one-half by 2025, and at the same time to double its annual increase in crop productivity, it would become self-sufficient in food. That would be an achievement indeed.

Let us remind ourselves, moreover, that we have exceeded best expectations on various occasions in the past. Who in 1950 would have anticipated the unprecedented accomplishments of the Green Revolution? Or, who in 1985 would have expected Zimbabwe to buck the sub-Saharan Africa trend and switch from being a food importer to a food exporter? In a broader political perspective, who would have bet in 1989 that by 2000 we would have gotten rid of the Berlin Wall, Communism, the Soviet Union, and the

Cold War, and that peace would be breaking out in South Africa and the Middle East? Can we now break down the Berlin Walls of ignorance and "ignore-ance" in the minds of political leaders and senior planners with regard to agriculture? Nothing less will do if we are to measure up to the ever greater challenges of feeding humankind.

Acknowledgments

I am glad to acknowledge the many helpful comments on an early draft of this paper from Lester Brown, Gretchen Daily, Anne Ehrlich, Hal Kane, David Pimentel, and Emma Small. It is a special pleasure to recognize the extensive research and analyses, notably the many incisive insights, supplied by my research associate, Jennifer Kent.

References

Abdulai, A., and P. Hazell. 1995. The Role of Agriculture in Sustainable Economic Development in Africa. *Journal of Sustainable Agriculture* 7:101–19.

Adepoju, A., and C. Opong, eds. 1994. *Gender, Work and Population in Sub-Saharan Africa*. London: James Curry Publishers.

Alexandratos, N., ed. 1995. *World Agriculture: Towards 2010*. Rome: Food and Agriculture Organization.

_____. 1996. China's Projected Cereals Deficits in a World Context. *Agricultural Economics* 15(1):1–16.

Anderson, J. R., P. G. Pardey, and J. Roseboom. 1994. Sustaining Growth in Agriculture: A Quantitative Review of Agricultural Research Investments. *Agricultural Economics* 10(2):107–23.

Barnett, V., R. Payne, and R. Steiner, eds. 1995. *Agricultural Sustainability in Economic, Environmental and Statistical Terms*. London: John Wiley and Sons.

Barrow, C. J. 1993. *Land Degradation: Development and Breakdown of Terrestrial Environments*. Cambridge, U.K.: Cambridge University Press.

Berck, P., and D. Bigman. 1993. *Food Security and Food Inventories in Developing Countries*. Wallingford, U.K.: CAB International.

Boardman, J., J. A. Dearing, and I. D. L. Foster, eds. 1990. *Soil Erosion on Agricultural Land*. New York: John Wiley and Sons.

Bolin, B. 1994. *Report on Behalf of the Intergovernmental Panel on Climate Change to the Intergovernmental Negotiating Committee on the Framework Convention on Climate Change*. Geneva, Switzerland: World Health Organization.

Bongaarts, J. 1994. Can the Growing Human Population Feed Itself?
Scientific American 270:36–42.

Bouis, H. E. 1995. Plant Breeding for Nutrition. *Journal of the Federation
of American Scientists* 48(4):1, 8–16.

Brown, L. R. 1995. *Who Will Feed China?* New York: W. W. Norton.

_____. 1996. The Acceleration of History. In *State of the World 1996,*
edited by L.R. Brown and ten others. New York: W. W. Norton.

Brown, L. R., C. Flavin, and H. Kane. 1996. *Vital Signs 1996: The Trends
That Are Shaping Our Future.* New York: W. W. Norton.

Brown, L. R., and H. Kane. 1994. *Full House: Reassessing the Earth's
Population Carrying Capacity.* New York: W. W. Norton.

Brown, L. R., and twelve others. 1992. *State of the World 1992.* New York:
W. W. Norton.

Brown, L. R., and eleven others. 1993. *State of the World 1993.* New York:
W. W. Norton.

Brown, L. R., and ten others. 1996. *State of the World 1996.* New York: W.
W. Norton.

Bryant, E. A. 1991. *Natural Hazards.* Cambridge, U.K.: Cambridge University Press.

Cassman, K. G., and R. R. Harwood. 1995. The Nature of Agricultural
Systems: Food Security and Environmental Balance. *Food Policy*
20:439–54.

Chan, N. W. 1986. Drought Trends in Northwestern Peninsular Malaysia: Is
Less Rain Falling? *Wallaceana* 44:8–9.

Chang, T. T. 1986. *Unique Sources of Genetic Variability in Rice Germplasm.*
Los Banos, Philippines: International Rice Research Institute.

Cohen, M. J., and D. Reeves. 1995. *Causes of Hunger.* Washington, D.C.:
International Food Policy Research Institute.

Crosson, P., and J. R. Anderson. 1992. *Resources and the Global Food
Prospects: Supply and Demand for Cereals to 2030.* Washington, D.C.:
The World Bank.

Daily, G. C. 1995. Restoring Value to the World's Degraded Lands. *Science*
269:350–54.

Daily, G. C., and P. R. Erlich. 1990. An Exploratory Model of the Impact of
Rapid Climate Change on the World Food Situation. *Proceedings of the
Royal Society of London.* Series B 241:232–44.

Damania, A. B., ed. 1993. *Biodiversity and Wheat Improvement.* Chichester,
U.K.: John Wiley and Sons.

Davies, S., M. Buchanan-Smith, and R. Lambert, eds. 1992. *Resources,
Environment and Population: Present Knowledge, Future Options.* New
York: Oxford University Press.

Downing, T., E. J. A. Lohman, W. J. Maunder, A. A. Olsthoorn, R. Tol, and H.
Tiedemann. 1993. *Socio-Economic and Policy Aspects of Changes in the
Incidence and Intensity of Extreme Weather Events.* Oxford, U.K.:
Environmental Change Unit, University of Oxford.

Dregne, H. E., and M. T. Chou. 1992. *Global Desertification: Dimensions and
Costs.* Lubbock, Texas: Texas Tech University.

Durning, A. T. 1993. *Saving the Forests: What Will It Take?* Washington, D.C.: Worldwatch Institute.

Dyson, T. 1995. *Population and Food: Global Trends and Future Prospects.* London: Routledge.

Ehrlich, A. H. 1995. Implications of Population Pressure on Agriculture and Ecosystems. *Advances in Botanical Research* 21:79–104.

Ehrlich, P. R., and A. H. Ehrlich. 1991. *Healing the Planet.* Menlo Park, CA: Addison Wesley.

Ehrlich, P. R., A. H. Ehrlich, and G. C. Daily. 1993. Food Security, Population, and Environment. *Population and Development Review* 19(1): 1–32.

Engelman, R., and P. LeRoy. 1993. *Sustaining Water: Population and the Future of Renewable Water Supplies.* Washington, D.C.: Population Action International.

_____. 1995a. *Conserving Land: Population and Sustainable Food Production.* Washington, D.C.: Population Action International.

_____. 1995b. *Sustaining Water: An Update.* Washington, D.C.: Population Action International.

Environmentally Sustainable Development Agriculture Research and Extension (ESDAR)/The World Bank. 1996. *Role of Research in Global Food Security and Agricultural Development.* Washington, D.C.: ESDAR/The World Bank.

Falcon, W. 1995. *Food Policy Analysis, 1975–95: Reflections by a Practitioner.* Washington, D.C.: International Food Policy Research Institute.

Falkenmark, M. 1994. Landscape as Life-Support Provider: Water-related Limitations. In *Population: The Complex Reality,* edited by Sir F. Graham-Smith. London: The Royal Society, 103–16.

Food and Agriculture Organization (FAO). 1993a. *Soil Loss Accelerating Worldwide.* Rome: FAO.

_____. 1993b. *Forest Resources Assessment for the Tropical World.* Rome: FAO.

Glantz, M. H., ed. 1994. *Drought Follows the Plough.* Cambridge, U. K.: Cambridge University Press.

Gleick, P. H., ed. 1993. *Water in Crisis: A Guide to the World's Fresh Water Resources.* Oxford, U.K.: Oxford University Press.

Greenland, D. J., G. D. Bowen, H. Eswaran, R. Rhodes, and C. Valentin. 1994. *Soil, Water and Nutrient Management Research: A New Agenda.* Bangkok, Thailand: International Board for Soil Research and Management.

Grigg, D. 1993. *The World Food Problem* (2nd ed.). Oxford, U.K.: Blackwell.

Harris, J. M. 1996. World Agricultural Futures: Regional Sustainability and Ecological Limits. *Ecological Economics* 17:95–115.

Hennessy, K. 1993. Extreme Temperatures and Rainfall. In *Climate Impact Assessment Methods for Asia and the Pacific,* edited by A. J. Jakeman and A. B. Pittock. Canberra, Australia: Australian International Development Assistance Bureau.

High Level Committee on Floods, Government of India. 1983. *Report on the Emergent Crisis.* New Delhi: High Level Committee on Floods, Government of India.

Houghton, J. 1994. *Global Warming: The Complete Briefing.* Elgin, IL: Lion Publishing.

Houghton, J. T., G. J. Jenkins, and J. J. Ephramus, eds. 1990. *Climate Change: The IPCC Scientific Assessment* (Final Report of Working Group I). Cambridge, U. K.: Cambridge University Press.

Hulme, M., D. Conway, P. M. Kelly, S. Subak, and T. E. Downing. 1994. *Climate and Africa: An Assessment of African Policy Options and Responses.* Norwich, U.K.: Climatic Research Unit, University of East Anglia.

Hulme, M., and M. Kelly. 1993. Exploring the Links Between Desertification and Climate Change. *Environment* 35(6):5–11, 39–45.

Hulme, M., T. Wigley, and T. Jiang. 1992. *Climate Change Due to the Greenhouse Effect and its Implications for China.* Gland, Switzerland: World Wide Fund for Nature International.

Intergovernmental Panel on Climate Change. 1990. *The Potential Impacts of Climate Change: Impacts on Agriculture and Forestry.* Geneva, Switzerland: World Meteorological Organization; Nairobi, Kenya: United Nations Environment Programme.

Islam, N. 1996. *Implementing the Uruguay Round: Increased Food Price Stability by 2020?* Washington, D.C.: International Food Policy Research Institute.

_____, ed. 1995. *Population and Food in the Early Twenty-First Century.* Washington, D.C.: International Food Policy Research Institute.

Jackson, T., and M. Marks. 1995. *Measuring Sustainable Economic Welfare: A Pilot Index.* London: New Economics Foundation.

Jacobson, J. L. 1993. Closing the Gender Gap in Development. In *State of the World 1993,* edited by L.R. Brown and eleven others. New York: W. W. Norton, 61–79.

Jarnagin, S. K., and M. A. Smith. 1993. *Soil Erosion and Effects on Crop Productivity: Project 2050.* Washington, D.C.: Food and Agriculture Sector, World Resources Institute.

Kaiser, H. N., and D. E. Drennan, eds. 1995. *Agricultural Dimensions of Global Climate Change.* Washington, D.C.: St. Lucie Press.

Kendall, H. W., and D. Pimentel. 1994. Constraints on the Expansion of the Global Food Supply. *Ambio* 23(3):198–205.

Lal, R., and B. A. Stewart. 1990. *Soil Degradation.* New York: Springer-Verlag.

Magadza, C. H. D. 1994. Climate Change: Some Likely Multiple Impacts in Southern Africa. *Food Policy* 19:165–91.

Mayo, E., and A. MacGillivray. 1994. *Growing Pains.* London: New Economics Foundation.

McCalla, A. F. 1994. *Agriculture and Food Needs to 2025: Why We Should Be Concerned.* Washington, D.C.: Consultative Group on International Agricultural Research, The World Bank.

McMichael, A. J. 1993. *Planetary Overload: Global Environmental Change and the Health of the Human Species.* Cambridge, U. K.: Cambridge University Press.

Meher-Homji, V. M. 1992. Probable Impact of Deforestation on Hydrological Processes. In *Tropical Forests and Climate,* edited by N. Myers. Dordrecht, The Netherlands: Kluwer Academic Publishers.

Mei, F. 1996. Food Development in China in the Early 21st Century. *Food and Nutrition in China* 2:27–9.

Missiaen, M., S. Shapouri, and R. Trostle. 1995. *Food Aid Needs and Availabilities: Projections for 2005.* Washington, D.C.: United States. Department of Agriculture, Economic Research Service.

Mitchell, D., and M. Ingco. 1993. *The World Food Outlook.* Washington, D.C.: The World Bank.

Myers, N. 1983. *A Wealth of Wild Species.* Boulder, CO: Westview Press.

_____. 1991. *Population, Resources and the Environment: The Critical Challenges.* London: Banson Books.

_____. 1992. *The Primary Source: Tropical Forests and Our Future.* New York: W. W. Norton.

_____. 1995. Biodiversity and the Precautionary Principle. *Ambio* 22(2–3) 74–9.

_____. 1996a. *Ultimate Security: The Environmental Basis of Political Stability.* Washington, D.C.: Island Press.

_____. 1996b. The Biodiversity Crisis and the Future of Evolution. *The Environmentalist* 16:37–47.

_____. 1997. *Perverse Subsidies.* Chicago: The MacArthur Foundation.

Myers, N., and J. Kent. 1995. *Environmental Exodus: An Emergent Crisis in the Global Arena.* Washington, D.C.: Climate Institute.

Newcome, K. 1984. *An Economic Justification for Rural Afforestation: The Case of Ethiopia.* Washington, D.C.: The World Bank.

Oldeman, L. R., R. T. A. Hakkeling, and W. G. Sombroek. 1990. *World Map of the Status of Human-Induced Soil Degradation.* Wageningen, The Netherlands: International Soil Reference and Information Centre; Nairobi, Kenya: United Nations Environment Programme.

Oldfield, M. L. 1989. *The Value of Conserving Genetic Resources.* Sunderland, MA: Sinauer Associates.

Pimentel, D. 1991. Diversification of Biological Control Strategies in Agriculture. *Crop Protection* 10:243–53.

_____, ed. 1993. *World Soil Erosion and Conservation.* Cambridge, U. K.: Cambridge University Press.

Pimentel, D., C. Harvey, P. Resosudarmo, K. Sinclair, D. Kurz, M. McNair, S. Crist, L. Shpritz, L. Fitton, R. Saffouri, and R. Blair. 1995. Environmental and Economic Costs of Soil Erosion and Conservation Benefits. *Science* 267:1117–23.

Pimentel, D., J. Houser, E. Preiss, O. White, H. Fang, L. Mesnick, T. Barsky, S. Tariche, J. Schreck, and S. Alpert. 1997a. Water Resources: Agriculture, The Environment and Society. *BioScience* 47(2):97–106.

Pimentel, D., X. Huang, A. Cordova, and M. Pimentel. 1997b. Impact of
Population Growth on Food Supplies and Environment. *Population and
Development Review* (in press).

Pimentel D., and M. Pimentel. 1996. *Food, Energy and Society*. Niwot, CO:
University Press of Colorado.

Pinstrup-Andersen, P. 1993. *Socioeconomic and Policy Considerations for
Sustainable Agricultural Development*. Washington, D.C.: International
Food Policy Research Institute.

_____. 1994. *World Food Trends and Future Food Security*. Washing-
ton, D.C.: International Food Policy Research Institute.

Pinstrup-Andersen, P., and J. L. Garrett. 1996. *Rising Food Prices and
Falling Grain Stocks: Short-Run Blips or New Trends?* Washington,
D.C.: International Food Policy Research Institute.

Pinstrup-Andersen, P., R. Pandya-Lorch, and P. Hazell. 1994. *Hunger, Food
Security, and Role of Agriculture in Developing Countries*. Washington,
D.C.: International Food Policy Research Institute.

Pittock, A. B. 1995. Climate Change and World Food Supply. *Environment*
37(9):25–30.

Pittock, A. B., P. Whetton, and Y. Wang. 1994. Climate and Food Supply.
Nature 371:25.

Postel, S. 1992. *Last Oasis: Facing Water Scarcity*. New York: W. W. Norton.

_____. 1994. Carrying Capacity: Earth's Bottom Line. In *State of the
World 1994*, edited by L. R. Brown and eleven others. New York: W. W.
Norton, 3–21.

Postel, S. L., G. C. Daily, and P. R. Ehrlich. 1996. Human Appropriation of
Renewable Fresh Water. *Science* 271:785–88.

Potter, C. S., J. I. Cahen, and D. Janczewski, eds. 1993. *Perspectives on
Biodiversity: Case Studies of Genetic Resource Conservation and Development*. Washington, D.C.: American Association for the Advancement
of Science.

Raskin, P., E. Hansen, and R. Margolis. 1995. *Water and Sustainability: A
Global Outlook*. Stockholm: Stockholm Environment Institute.

Rasmusson, E. M. 1989. Potential Shifts of Monsoon Shifts Patterns
Associated with Climate Warming. In *Coping with Climate Change*,
edited by J. C. Topping. Washington, D.C.: Climate Institute.

Rind, D., R. Goldberg, J. Hansen, C. Rosenzweig, and R. Reudy. 1990.
Potential Evapotranspiration and the Likelihood of Future Drought.
Journal of Geophysical Research 95:9983–10,004.

Rosegrant, M. W. 1996. *Food Prices and Food Security: Where Are We
Headed?* Washington, D.C.: International Food Policy Research
Institute.

Rosegrant, M. W., M. Agcaoili-Sombilla, and N. D. Perez. 1995. *Global Food
Projections to 2020: Implications for Investment*. Washington, D.C.:
International Food Policy Research Institute.

Rosenzweig, C., and M. L. Parry. 1994. Potential Impact of Climate Change
on World Food Supply. *Nature* 367:133–38.

Ruttan, V. W. 1996. Population Growth, Environmental Change and
   Technical Innovation: Implications for Sustainable Growth in Agricul-
   tural Production. In *The Impact of Population Growth on Well-Being in
   Developing Countries,* edited by D. A. Ahlburg et al. Berlin: Springer-
   Verlag.
Salati, E., and C. A. Nobre. 1992. Possible Climatic Impacts of Tropical
   Deforestation. In *Tropical Forests and Climate,* edited by N. Myers.
   Dordrecht, The Netherlands: Kluwer Academic Publishers.
Scherr, S. J., and S. N. Yadav. 1996. *Land Degradation in the Developing
   World: Implications for Food, Agriculture, and the Environment to 2020.*
   Washington, D.C.: International Food Policy Research Institute.
Schneider, R., J. McKenna, C. Dejou, J. Butler, and R. Barrows. 1990.
   *Brazil: An Economic Analysis of Environmental Problems in the
   Amazon.* Washington, D.C.: The World Bank.
Schneider, S. H. 1989. *Global Warming: Are We Entering the Greenhouse
   Century?* San Francisco: Sierra Club Books.
Seckler, D. 1996. New Era of Water Resources Management: From "Dry" to
   "Wet" Water Savings. *Issues in Agriculture 8.* Washington, D.C.:
   Consultative Group on International Agricultural Research, The World
   Bank.
Serageldin, I. 1993. *Development Partners: Aid and Cooperation in the
   1990s.* Washington, D.C.: The World Bank.
_____. 1994. *Water Supply, Sanitation and Environmental
   Sustainability.* Washington, D.C.: The World Bank.
Speth, J. G. 1993. *Towards Sustainable Food Security.* Washington, D.C.:
   Consultative Group on International Agricultural Research.
Srivastava, J., N. J. H. Smith, and D. Forno. 1996. *Biodiversity and Agricul-
   ture: Implications for Conservation and Development.* Washington, D.C.:
   The World Bank.
Swaminathan, M. S. 1994. Sustainable Agriculture. *Environment* 36(3):3–4.
_____. 1995. Population, Environment and Food Security. *Issues in
   Agriculture 7.* Washington, D.C.: Consultative Group on International
   Agricultural Research.
_____. 1996a. *Sustainable Agriculture: Towards Food Security.* Delhi:
   Konark Publishers Ltd.
_____. 1996b. *Sustainable Agriculture: Towards an Evergreen Revolu-
   tion.* Delhi: Konark Publishers Ltd.
Teramura, A. H. 1986. Overview of Our Current State of Knowledge of UV-B
   Effects on Plants. In *Effects of Changes in Stratospheric Ozone and
   Global Climate.* Vol 1: Overview, edited by J. G. Titus. Washington,
   D.C.: U.S. Environmental Protection Agency, 165–73.
Thomas, D. S. G., and N. J. Middleton. 1994. *Desertification: Exploding the
   Myth.* Chichester, U.K.: John Wiley and Sons.
Tolba, M. K., O. A. El-Kholy, E. El-Hinnawi, M. W. Holdgate, D. F.
   McMichael, and R. E. Munn. 1992. *The World Environment 1972–1992:
   Two Decades of Challenge.* London: Chapman and Hall.

Tribe, D. 1994. *Feeding and Greening the World: The Role of International Agricultural Research.* Wallingford, U.K.: CAB International.

United Nations Development Programme (UNDP). 1996. *Human Development Report 1996.* New York: Oxford University Press.

United Nations Environment Programme (UNEP). 1995. *United Nations Convention to Combat Desertification.* Nairobi, Kenya: United Nations Environment Programme.

United States Department of Agriculture (USDA). 1995. *Grain: World Markets and Trade.* Washington, D.C.: USDA, Foreign Agricultural Service (September 1995).

_____. 1996. *Grain: World Markets and Trade.* Washington, D.C.: USDA, Foreign Agricultural Service (February 1996).

United States National Research Council (U.S. NRC). 1992. *Managing Global Genetic Resources: The U.S. National Plant Germplasm System.* Washington, D.C.: National Academy Press.

Von Braun, J., R. F. Hopkins, D. Puetz, and R. Pandya-Lorch. 1993. *Aid to Agriculture: Reversing the Decline.* Washington, D.C.: International Food Policy Research Institute.

Williams, M. A. J., and R. J. Balling. 1994. *Interactions of Desertification and Climate.* Geneva, Switzerland: World Meteorological Organization; Nairobi, Kenya: United Nations Environment Programme.

Wilson, E. O. 1992. *The Diversity of Life.* New York: W. W. Norton.

World Bank. 1993. *World Development Report 1993: Investing in Health.* New York: Oxford University Press.

_____. 1994. *World Development Report 1994.* New York: Oxford University Press.

_____. 1995. *World Development Report 1995.* New York: Oxford University Press.

World Bank and United Nations Development Programme (UNDP). 1990. *Irrigation and Drainage Research.* Washington, D.C.: The World Bank; New York: United Nations Development Programme.

World Food Programme (WFP). 1995. *WFP Mission Statement.* Rome: WFP.

World Resources Institute (WRI). 1996. *World Resources 1996.* New York: Oxford University Press.

Worrest, R. C., and M. M. Caldwell, eds. 1986. *Stratospheric Ozone Reduction, Solar Ultraviolet Radiation and Plant Life.* New York: Springer-Verlag.

Worrest, R. C., and D. P. Hader. 1989. The Effects of Stratospheric Ozone Depletion in Marine Organisms. *Environmental Conservation* 16: 261–63.

Discussant ◆ *Robert F. Chandler, Jr.*
Looking Ahead at the Global Food/Population Balance

The old adage "hindsight is better than foresight" is indeed applicable to my remarks. Predictions of what will happen to the food/population balance during the next half-century vary greatly among "authorities." Some, encouraged by the rapid strides during the past 30 years in upping the yield potential of the major food crops, predict that this increase will continue and that there will be no problem in meeting the food needs of 10 billion people 60 or 70 years from now.

Others, noting the slowed advances and per capita declines in food production during the past decade, feel that the big jump in yield potential has already occurred and that before a stable population double its present size is reached, the less-developed nations will be in worse condition than they are today.

Neither the optimists nor the pessimists deny the fact that the world is still in the midst of a population increase unprecedented in human history, and that by the time the population becomes stabilized, human numbers may well reach 10 billion, or even more.

In his comprehensive resource paper, Norman Myers gave an excellent presentation of the achievements of the past, of the current situation, and of the challenges that lie ahead. My brief remarks will supplement his discussion with examples of optimistic and pessimistic views of various authorities, along with my own opinion concerning the future food/population balance with particular reference to the food grains.

A recent example of an extremely optimistic view appeared in the 1996 summer issue of *Daedalus,* the prestigious journal of the American Academy of Arts and Sciences. The issue is devoted to a set of essays on the general theme of "The Liberation of the Environment." In general, the authors express the opinion that food, energy, and water resources will be adequate to meet the needs of a world population of 10 billion. In support of their stand they cite the increasing yield potential of the major food crops and the decreasing carbon-hydrogen ratio of sources of energy (from wood, to coal, to natural gas, and eventually to hydrogen). They state that water for irrigation will be adequate due to greatly increased use efficiency (the lining of irrigation ditches, water harvesting, the use of trickle and drip irrigation, and the like).

In support of the potential for increased food production, Paul Waggoner, among others, cites yields of maize by contest farmers in the United States of 16 to 21 metric tons per hectare (t/ha) while average yields in the nation are only 7 t/ha, which is several times the average yield in Africa. The conclusion is that worldwide increases in yield on farmers' fields of the basic food crops, especially food grains, will increase to the extent that considerably less land area than is now being cultivated will be needed to feed 10 billion people in 2050. Waggoner used as examples of yield increases not only maize yields in the United States but also wheat yields in Ireland and France, which are well above those in America. He devoted little attention to the difficulties of achieving such harvest increases in Third World countries plagued by low educational levels, inadequate infrastructure, high population growth rates, and widespread poverty, to name but a few.

I was surprised to find that several writers in that issue of *Daedalus* considered the views of the Club of Rome and those of the authors of the 1972 publication *The Limits to Growth* to be obsolete now and that current thought is that continued economic growth is essential and that a stable population is undesirable because it would cause economic stagnation. A more pessimistic and, in my opinion, more realistic view of the food/population balance is expressed in the publications of the World Watch Institute, headed by Lester Brown. In the annual issues of *State of the World*, Brown and his colleagues warn of impending crises of food and water shortages caused by the ever-increasing world population.

These views are well presented in Dr. Myers's paper and do not need further elaboration. Nevertheless, let me list the important points held by the World Watch Institute group.

1. World food grain production has leveled off since 1990, a bumper crop year.
2. Per capita grain production has been decreasing since 1985.
3. World carryover grain stocks in 1996 were the lowest in over 20 years.
4. World fish harvests have reached a maximum and further increases will have to come from aquaculture.
5. In many drier areas of the world, aquifers are being depleted faster than they can be recharged, and water tables are dropping.
6. Per capita land area devoted to grain production has been decreasing since 1955.
7. Lester Brown's book *Full House* and his and Hal Kane's *Who Will Feed China?* maintain that the carrying capacity of many regions of this finite planet is being exceeded already and that by the end of the next half-century there will not be enough food produced to feed the world.

I, as a member of the panel discussing "Population Dynamics and Food Security," am expected to express my views on this subject, and they are as follows.

I am not greatly concerned about energy supplies for the future as I am confident that through science and technology we will continue to reduce the use of carbon and increase the use of hydrogen. Furthermore, wind turbines and photo-voltaic cells as renewable sources of electric power are expanding rapidly.

I realize there is much room for improvement in the efficiency of water use in both rural and urban areas, and that governments need to change policies for the use of water. However, I am not at all confident that such moves will result in meeting the demand for water by an ever-expanding world population.

Approximately 65 percent of the world's freshwater is used for crop irrigation, leaving 25 percent for industries and 10 percent for households and municipalities. As urbanization continues, there will be less water for agriculture. One cannot help but be troubled by falling water tables and depleted aquifers not only in California, the High Plains, and the southwestern region of the United States, but in such far-flung regions as the Valley of Mexico, India, North China, and sub-Saharan Africa.

In my view, the greatest threat to the well-being of mankind is overpopulation—more so than global warming. Neither the optimists nor the pessimists can deny the fact that the land area on this planet is a constant and that our species cannot continue to increase its numbers indefinitely. I disagree with those who maintain that a stabilized population will result in economic stagnation. Millions of people in this world are living in extreme poverty, many of them earning no more than the equivalent of a dollar a day. Raising the incomes of these people will make it possible for them to purchase more goods and services, which will promote economic growth even though the population remains stable.

The yield potential and world production of wheat and rice have more than doubled during the past three decades. However, as shown by Myers, recent advances in yield and production of cereal grains have slowed, indicating that further increases will be more difficult to attain. Let me cite a few examples involving rice, a crop with which I am the most familiar.

China, the world's largest rice producer, now has an average yield of rough rice of approximately 6 t/ha and, undoubtedly, yields will continue to rise. However, its land area for crop production is shrinking by some 300,000 hectares annually.

Although China's rate of population increase is near 1 percent, it must be realized that, even at such a low level, over 12 million people are being added to China's population yearly. China is now a major importer of food grains, and the amount needed from outside sources will continue to grow.

India, the second largest producer of rice and the nation with the largest area devoted to the crop, has attained a national average yield of approximately 3 t/ha and is currently not only self-sufficient in rice but has a surplus in storage. Assuming a current population of 960 million and an annual increase rate of 1.8 percent, India will add 17.2 million people to its population in the next 12 months. The land area devoted to rice cultivation has remained essentially unchanged since 1977.

Because 55 percent of India's rice land is rainfed and further opportunities for expanding the irrigated area are limited, I feel that it will be difficult for the nation by 2020 to attain an average yield of rough rice of more than 3.6 t/ha (the mean yield of its *irrigated* rice land today). Assuming that the area devoted to rice remains at its current level of 42 million hectares, the country would produce a total of 151 million tons of rice, as compared with the current harvest of about 115 million tons. If these estimates should prove to be correct, India would have 36 million more tons a year to feed its people.

Conservatively, India will add no fewer than 384 million people to its population within the next 24 years, which would bring its total to 1.34 billion by 2020. If its current annual per capita consumption of rice of about 102 kilograms (rough rice equivalent) continues, India would require an annual rice production of 137 million metric tons by 2020. From these estimates, it would appear that India may continue to grow enough rice for its people during the next quarter-century. However, by extrapolating these figures to the half-century mark, it is clear that the country would be importing huge quantities of rice. A similar story could be told for wheat, India's other principal food grain.

Ironically, India, now with a surplus of food grains, has the largest number of undernourished and poverty-stricken people of any nation in the world. This demonstrates the great need for off-farm employment.

The situation in Bangladesh is not far different from that in India. Bangladesh has recently increased its rice production by growing more of the crop under irrigation in the dry season. However, its growing population and fixed land area will soon make Bangladesh a major rice importer.

Turning to Africa, per capita food grain production has been declining for the past two decades, a trend that will continue until crop yields are raised and the rate of population increase declines.

However, there are those who hold great hope for Africa. At a recent symposium in Johannesburg, South Africa, sponsored by the International Food Policy Research Institute (IFPRI), Peter Hazell expressed the belief that Africa, through the expansion of area planted to crops and with greater use of fertilizer and irrigation, could increase crop production by at least 4 percent annually. Hazell pointed out that India, with only 13 percent of the land area of Africa, feeds more people than now exist in all of that continent.

In my view, Africa suffers not only from a rapidly growing population—increasing faster than development—but also from both national and individual poverty; from a lack of industry to provide off-farm employment opportunities; from poor roads; and from low literacy rates and inadequate numbers of schools and qualified teachers—not to mention severe tribal conflicts.

To illustrate the population situation in Africa and Asia, I constructed Table 10.2.1, which shows the ranges in annual population growth rates for principal countries in the two continents. It is significant that only two countries in Africa (Zimbabwe and Gabon) have reduced their annual population growth rates to less than 2.0 percent and 16 nations have an annual rate ranging between 3.0 and 3.6 percent.

Table 10.2.1.   Current annual population growth rates of selected countries in Asia and Africa

Asia	
From 0.3 to 1.0 percent	Japan, South Korea, China, Taiwan
From 1.1 to 2.0 percent	Singapore, Sri Lanka, Thailand, Indonesia, India, Myanmar, North Korea, Vietnam
From 2.1 to 3.0 percent	Malaysia, Bangladesh, Bhutan, Philippines, Afghanistan, Laos, Cambodia, Pakistan
Africa	
From 1.5 to 1.9 percent	Gabon, Zimbabwe
From 2.0 to 2.5 percent	Egypt, Chad, Burundi, Congo, Uganda, Lethoso, Guinea, Tanzania
From 2.6 to 2.9 percent	South Africa, Sierra Leone, Zambia, Rwanda, Angola, Cameroon, Mozambique
From 3.0 to 3.6 percent	Sudan, Burkino Faso, Kenya, Senegal, Ethiopia, Ivory Coast, Ghana, Nigeria, Liberia, Zaire (Democratic Republic of the Congo), Madagascar, Somalia, Mali, Namibia, Niger, Togo

SOURCE: Information from World Almanac 1996.

Among Asian countries, Japan, South Korea, China, and Taiwan have reduced their growth rates to 1.0 percent or lower, and eight countries are below 2.0 percent. Furthermore, no countries have rates higher than 3.0

percent. The highest (between 2.1 and 3.0 percent) include Malaysia, Bangladesh, Bhutan, Philippines, Nepal, Afghanistan, Laos, Cambodia, and Pakistan. We must keep in mind that a 2.5 percent annual increase means that the population will double in about 30 years.

To conclude, I believe that although further increases in the yield and production of food grains will occur, the greatest advances have already taken place. We must not expect again to double the yield potential of wheat and rice. The carrying capacity of our planet is indeed limited. It is imperative that the human population be stabilized.

Heads of state in the less-developed countries must be convinced that the future well-being of their people depends greatly on reducing population growth rates. To achieve this, a national will for smaller families must be created. The more important ingredients of a successful family planning program include compulsory education (at least through grades six to eight, for females as well as males), the relief of poverty by creating more off-farm employment opportunities, monetary incentives for small families, and more health clinics with free advice and materials for birth control.

I cannot help but add that even here in the United States, we are faced with serious problems of overcrowding in inner sections of our major cities. Recent reports show that many school buildings in New York City and Los Angeles, for example, have been unable to accommodate the large numbers of pupils entering school. Moreover, there is an acute scarcity of qualified teachers. Among the industrialized nations, the United States has the highest population increase rate, partially due to the large numbers of both legal and illegal aliens that are being absorbed.

Back to the less-developed world, however, where over 90 percent of the population increases are likely to occur, I predict that unless the fertility rate in those countries is greatly reduced within the next three decades, the world will experience catastrophic increases in poverty levels, malnutrition, disease epidemics, and widespread famine in years of unfavorable weather—all conditions leading to outbreaks of civil strife.

As I said at the beginning, hindsight is far better than foresight. Naturally, I hope that my assessment is wrong and that conditions will be far better than I predict. Nevertheless, recent downward trends in *per capita* essential goods and services cause me to believe that unless major changes in total development take place, including decided reductions in population growth rates, we shall continue to witness severe destruction of the environment and widespread poverty.

Discussant ♦ *Helen A. Guthrie*

Dr. Myers eloquently and graphically painted a picture of the overwhelming challenge that lies ahead as we try to balance the advances in agricultural productivity—e.g., calories and protein yield per hectare of staple cereal grains such as rice, wheat, and corn—against the agricultural stresses—soil erosion, desertification, deforestation, salinization of the water, and landlessness—that act to decrease our agricultural capacity. These stresses also compromise our ability to provide qualitative food or nutrition security for a population continuing to grow at an exponential rate.

As a nutritionist looking at the issue of food/nutrition security, I sense an equal and compelling urgency to consider qualitative as well as quantitative aspects of the challenge—a need to direct our focus not only to maintain and enhance our current agricultural resources but to intertwine with it a focus on the health, productivity, and well-being of the population across the life span. This quality of component would be identified if we talked of nutritional as well as food security. An adequate intake of many of the micronutrients is essential before the growth-promoting potential of food can be realized. Even the most casual observer in an area of food insufficiency recognizes the signs of hunger—distended bellies, wasted limbs, lethargic and gaunt faces, prematurely aged features, stunted growth, and two-, three-, and five-year old siblings indistinguishable on the basis of size—all manifestations of the lack of food. The casual observer, however, may not be aware of the insidious nature of "hidden hunger" with symptoms that are not as obvious. These symptoms have devastating consequences on the quality of life and place an overwhelming drain on "human capital." An insufficiency of many micronutrients—trace minerals, vitamins, amino acids and fatty acids—is a major cause of morbidity and mortality for 2 billion people, one-third of the population, with women and children most commonly affected.

Specifically, I am referring to the common symptoms of vitamin A deficiency: night blindness, reduced resistance to infection, and eventual blindness. These symptoms can result from the limited intake of fruits and vegetables or from a marginal intake of fat, which interferes with the ability to absorb the fat-soluble carotene they provide. It is estimated that nearly 14 million preschool children have the clinical eye disease, xerophthalmia. Two hundred thousand to 250 thousand of these children become blind each year, and of these, two-thirds die within months of going blind. This reflects a concurrent breakdown in the children's immune defenses and their ability to combat infections such as measles, which are seldom if ever fatal for a well-nourished child.

Hidden hunger is also manifested in nutritional anemia, a condition affecting over 2 billion people, particularly women of child-bearing age and children. Among the compromised functional outcomes of iron deficiency are reduced work capacity, impaired learning ability, increased susceptibility to infection, and increased risk of death associated with pregnancy and childbirth. Although anemia may be the result of multiple causes—intestinal parisitism, malaria, deficiencies of vitamin B-12—over one-half of the cases are attributable to iron deficiency. Iodine deficiency is prevalent in many parts of the world, affecting as many as 1.6 billion people. Associated manifestations include reduced cognitive ability, mental, and neurological functions, which in the extreme appear as cretinism; increased rates of stillbirths; spontaneous abortions; and infant deaths. An additional 20 million people suffer from mental retardation, which is preventable if adequate iodine is available early in life.

Although iodine, iron, and vitamin A are the most extensively studied of the nutrients considered responsible for hidden hunger, there is growing evidence that inadequate intake of zinc is associated with impaired immune function and sexual development in males and that selenium deficiency is associated with cardiopathy. Keshen-Becks disease of crippling chondropdystrophy may be of equal concern. Another concern is that as the capability for greater yields of cereal crops increases, land that was used previously for legumes and vegetable crops will be diverted to cereals, reducing the diversity of the food supply and, hence, its nutritional quality.

All of these micronutrient deficiencies are recognized as major causes of human suffering on a global scale. Failure to focus on solutions to these nutritional shortfalls on either a global or a regional basis is to sentence a large number of survivors of protein/calorie malnutrition to a life of reduced functional quality that is at best unacceptable and unconscionable.

What can and must we do? Finding and implementing solutions to all of these food-related problems is as daunting an undertaking and as great, if not greater, a challenge as providing a subsistence level of grain for the expanding world population. Finding the means of distributing food in a timely manner demands innovative and greater efforts.

Public health approaches—such as regular injections or the use of supplements as part of the health care of children, the enrichment of a dietary staple such as salt, MSG, or the local staple cereal, or fluoridation of the water supply—techniques that have been useful in developed countries with central processing of the food supply and a central water supply do not lend themselves to the living situation of most of the world's population. Instead, solutions must look to regionally appropriate modification in the basic food supply and its distribution.

Plant geneticists and agriculturists are important allies of the nutritionist seeking a sustainable food-based solution. They have the capability of genetically manipulating the characteristics of our food supply to enhance the ability of a plant to extract essential trace elements, such as zinc and selenium, from the soil and deposit them in the endosperm or other edible portion of the plant, to lower the content of substances such as phytic acid that reduce the bioavailability of many nutrients, to increase yield, to shorten growing periods, and to strengthen disease resistance.

In the case of iodine, the conventional methods of iodizing salt or providing regular injection of lipid oil or other iodine-containing compounds may give way to a more local promising technique of "dripping" an iodide into irrigation water to enhance the iodine content of the water, the soil it irrigates, the plants that grow on the soil, the animals that eat the plants, and the humans that eat both the plants and animals.

National policies have the potential of helping the food security cause by providing incentives for the production and distribution of a diversity of food crops. Consumers also respond to incentives for behavioral modifications.

As we include the nutritional benefits of food in agricultural polices, we should not forget the concurrent needs for water and fuel to prepare the food—both of which have a high energy cost that will be significant in terms of additional food needs to meet the calorie costs of procuring food and water. The value of breast milk as an economic resource deserves more recognition, as does the potential of oilseeds—a renewable source of calories.

The ultimate solution to the challenge of providing food security and maximizing health is going to require collaboration among agricultural, nutrition, and health professionals, policymakers, communicators, and social scientists. Each group has a unique contribution to make to the complex behavioral and scientific issues involved. The objective of agricultural production should be to produce healthy people in addition to producing sufficient quantities of food to maintain current levels for an expanding population.

Discussant ♦ *Roelof Rabbinge*
World Food Production, Food Security, and Sustainable Land Use

Introduction

Food production and food security have been major issues for humankind throughout its existence. Food collection and production was the most time-consuming activity in all societies and only in this century has the decrease in employment in agriculture accelerated.

Currently, only 4 to 5 percent of the professional population is directly involved in agriculture in the industrialized world. In the developing world that figure still varies between 50 and 70 percent, and a decrease in that number is likely within the next decades. With only very few people involved in agriculture, the food production and food security for all may be guaranteed. Worries about that guarantee have been uttered frequently in the past.

Two centuries ago the limits to growth were proclaimed by English economist Thomas Malthus and world food security was, in his opinion, in danger. One century ago similar warnings were on the top of the policy agenda, and food production was again considered to be an important issue that would make the future very dark due to structural shortage.

Now at the turn of the century food production and food security for the next century are again on the top of the policy agenda. Worries about the future are widespread, and recently the number of conferences, symposia, workshops, and summits on food production and food security has mushroomed.

Not that food production has decreased during the last century. On the contrary, this age has seen a five-fold increase in the number of people on the globe and a six-fold increase in food production—an unprecedented achievement of humankind. In this century, land productivity in the industrialized world increased five- or six-fold; labor productivity, 200- to 300-fold; and although external inputs such as water and nutrients increased considerably, their efficiency also increased. Yields of 8 tons of wheat per hectare are no longer an exception. The amounts of water, labor, oil, nitrogen, and phosphorous per kilograms (kg) of wheat, rice, potato, or any other product that is produced have decreased. That increase in efficiency has been due to a synergistic effect among various inputs.

Productivity increase has been widespread and has shown discontinuities (Green Revolutions) during this age in the industrialized world after World War II and in many places in Asia in the late 1960s and early 1970s. Before those revolutions, productivity increase took place with 5- to 10-kg grain equivalents per hectare per year; after those revolutions, productivity changed to a rate of 80- to 150-kg grain equivalents per hectare per year. These Green Revolutions have not yet taken place in Africa, but they are urgently needed in the coming decade.

So at this moment, if you consider these developments during the last century you can, in fact, only be hopeful, because of what happened in the last age. Why should it not happen in the next few decades again? The present situation, however, is complicated, as indicated by Norman Myers in his opening discussion, because many important phenomena occur at the same time. Those phenomena include the leveling off of grain production, which we have seen during the last few decades; the decrease of grain stores, which has occurred during the last few months; the decrease in aquifers; the lowering of water tables; and the decrease in fish harvest. The leveling of grain production is, however, due to the effect of active policies in Europe and in the United States to promote set-aside. The per capita grain production is also decreasing as a result of the increasing use of the marginal areas instead of further promotion of the well-endowed lands.

We see indeed that fish harvest is decreasing and we see that fish harvests are not produced by aquaculture in fish ponds, but as much as 80 percent of the fish production is still caught in an old-fashioned way in the sea or in fresh water. If the method would change, then the productivity of fish would increase considerably. There are, of course, undeniable environmental side effects that are due to an unsustainability spiral in which much of agriculture—Africa, for example—finds itself at this moment.

One example of this unsustainability spiral is the decreasing soil fertility in Africa, which occurs as a result of poverty. If we consider the estimated shares of the various activities that cause decreased quality of land, and that may cause many difficulties for the continuity of production in agricultural land, we find the following: 35 percent of the unsustainability is due to overgrazing; 29 percent is due to deforestation; 7 percent is due to overexploitation; 28 percent is due to mismanagement on a world scale; and only 1 percent is due to overuse of inputs. In the Western world, we tend to overestimate, in fact, the consequences of the too-intensive agriculture systems. On a world scale, that is of minor importance.

Much more important, of course, is the unsustainability due to poverty. When you are not bringing enough external imports into a system, that creates a decrease in the production and results in an overuse of that land.

Two authors, Lester R. Brown of the Worldwatch Institute and Mark W. Rosegrant of the International Food Policy Research Institute, take the past and consider trends and use them to look ahead in time; they come to very different conclusions. Their conditions are being characterized as pessimistic views or optimistic views. Both authors are technically right as they look at the same statistics and the same tables, but they have different interpretations and predictions.

When you consider the productivity rise, and mention the efficiency increase, and realize what has been achieved during the last century, then you can have an optimistic view and say "Oh, that is going to repeat in the near future."

If you consider the negative side effects of many of the changes then you see the system's unsustainability spiral, which takes place at several places. If you see that there is a problem in the use of land in many places, then you can become a pessimist.

It is for this reason that predictive studies that only consider trends and extrapolate them are of limited value to investigate the long-term possibilities and future perspectives. For that reason "explorative" studies are needed.

Explorative Food Production Studies

An example of an explorative study of food production is given by the Netherlands Scientific Council for Government Policy. That study investigated the future possibilities of food production, food security, and sustainable land use.

For the purposes of the study a number of assumptions were made with respect to sustainability. In the first place, sustainability presupposes the aim of closing all cycles as effectively as possible. Agriculture makes use of nature's productive capacity, tapping outputs from the system in the form of products. If agriculture is to be maintained over a lengthier period, inputs need to be added to the system in order to compensate for the tapped-off outputs. By definition, inputs can never be 100 percent converted into outputs; this implies that part of the inputs will be lost to the environment as leakages. These leakages can never be sealed off completely, so that the substance cycles in agriculture can never fully be closed.

Various strategies may be pursued in order to obtain an optimal result. Efforts may be made to close cycles as far as possible at the regional level in order to bring the losses to the environment under local control. Another strategy is based on closing the cycles at the global level to maximize the efficiency of the system and thus minimize the overall losses. Within each of these two approaches the agricultural system can be organized in many different ways.

Two different systems of agriculture are considered to demonstrate the extreme estimates for the production potential without violating the principles of sustainable production. These are Globally Oriented Agriculture (GOA) and Locally Oriented Agriculture (LOA).

The first system, GOA, seeks to achieve sustainability by aiming at the maximum efficiency of agriculture at the global scale. It is based on the notion that the environment is best served by the *lowest possible loss of inputs per unit of output*. This then makes it possible for comparatively high local leakages to the environment to be accepted with a view to reducing the overall burden on the environment. By making use of efficiently produced fertilizer and transporting it to places where these nutrients can be converted as efficiently as possible into agricultural products, an attempt is made to limit the total losses as far as possible.

With a view to guaranteeing sustainability, LOA aims as far as possible at closing regional or local cycles. This is based on the underlying premise that the quality of the environment is best served by the *lowest possible loss of inputs per hectare*. This principle results in the deployment of techniques that avoid the use of external, alien substances such as fertilizers and pesticides wherever possible. Efficiency is therefore defined at a totally different level of scale.

Both globally and locally oriented agriculture aim at maximum efficiency within their own limiting conditions, with a view to the sustainable functioning of the entire system. Under the GOA system the output is ultimately limited by the available agricultural land and the local availability of water. Under the LOA system the output is limited not only by the local availability of land and water but also by the amount of nitrogen that can be fixed from the atmosphere by natural means. In addition, other physical conditions, such as the quality of the soil, play a role in determining the production potential. The computations have been made on the assumption that other aspects, such as energy, minerals, investments, and labor, do not impose constraints. Any demand for energy or investment can therefore be met in both the GOA and the LOA system. In relation to the present situation this represents a substantial expansion of production.

In many parts of the world, however, there are distinct limitations attributable to the lack of resources and manpower. The necessary quantities of fertilizer or energy are, for example, by no means procurable universally. Furthermore, the necessary infrastructure is lacking in many places. Even if money were available under development projects to buy fertilizer, it remains questionable whether the fertilizer could be applied in the right place in the right way. The assumptions on which this model study is based clearly indicate, therefore, that the calculations can provide insight into the maximum potentialities of both agricultural systems, but these potentialities tell us little if anything about probable developments in the various regions. For this, far more information is required on the range of obstacles impeding agricultural development in various places.

The availability of water does, however, constitute a possible obstruction in calculating the production potentials of both systems. The maximum possibilities have therefore been made dependent on physical factors. This has been calculated by examining how much water is available for irrigation purposes in each catchment area in the 19 regions shown in Figure 10.3.2. For each region, the demand that can be derived from possible population developments on the basis of United Nations (UN) scenarios has been combined with the production possibilities. The basic calculations of these potentials have been made on a 1°x1° grid basis. The comparison between demand and supply indicates whether the demand for food can be satisfied in each of the 19 regions, while at a world scale it is possible to establish whether agricultural production is able to feed the growing population. The differences between the regions are indicative of the need for the transport of food from surplus areas to shortage areas.

Opinions on sustainable food production differ not only in relation to the potential agricultural techniques but also as to the package of food which the average world citizen could consume in the future.

The choice between a *Western* diet or a *Moderate* diet is prompted by differing estimates of the environmental consequences. The choice in favor of a Moderate diet may be based on the view that in the long term the world population cannot be fed at the present level of Western consumption, as this would impose an undue strain on the environment. In the case of a Western diet, by contrast, the environmental risks are deemed acceptable. It may be noted that neither of these two diets is extreme; the Moderate diet is substantially higher than the present world average, whereas the Western diet is lower than the present level of consumption in, for example, the United States. The Western diet contains a comparatively high proportion of meat and is equal to the present level of average European consumption. This requires a primary production of around 4.2 kg of grain equivalents per person per day. (The use of grain equivalents enables various agricultural products, for example, wheat, rice, millet, and maize to be brought under a common denominator.) The Moderate diet requires around 2.4 kg of grain equivalents per person per day. The difference is due to the conversion of cereals into meat. In some countries, including the

Netherlands, a high degree of efficiency has been achieved in converting animal fodder into meat. This applies especially to intensive livestock farming. The global average is approximately 8 kg of grain per kilo of meat. A diet containing more meat therefore leads to a substantial increase in the necessary volume of grain equivalents.

The calculations have drawn on two estimates for the growth of the world population based on figures in the 1992 UN publication *Long-range World Population Projections (1950–2150)*. The low estimate produces a figure of 7.7 billion people in 2040 and the high estimate a figure of 11.2 billion. A decision in favor of either one of these variants will obviously have a major impact on the results of the calculations.

Action Perspectives

Four action perspectives are described that differ in terms of food supply with respect to the combinations of production techniques and diets. Table 10.3.1 indicates how the Utilizing, Saving, Managing, and Preserving action perspectives relate to the normative differences in insight.

Table 10.3.1. Action perspectives for sustainable development of the world food supply

	Luxury package	Moderate package
Globally oriented agriculture	Utilizing	Saving
Locally oriented agriculture	Managing	Preserving

SOURCE: Netherlands Scientific Council for Government Policy. 1995. *Sustained Risks: a Lasting Phenomenon.* 44:60.

Utilizing

The Utilizing perspective aims at the provision of a Western diet on a worldwide basis as quickly as possible. It is assumed that this level of consumption is consistent with the ambitions in large parts of the world. Potential environmental problems are regarded as not insuperable. In addition, there is marked confidence in technological solutions to environmental problems. In particular, increasing agricultural output on good soils can result in the highly efficient utilization of physical inputs such as fertilizers and pesticides, to the benefit of the environment. Per unit of product, this agricultural technique requires a minimum level of physical inputs. Furthermore, comparatively little land is taken up at maximum levels of production. The social risks associated with the introduction of a production-oriented agricultural system that is required to meet a sharply increasing demand for food are regarded as acceptable under this vision. Relevant know-how also is increasingly exploited by food producers throughout the world.

Saving

The Saving perspective considers that major environmental risks are attached to feeding a rapidly rising world population. Locally oriented agriculture would, however, involve an excessive change in relation to the present forms of agriculture, for which reason it is sought to minimize the risks for the environment by limiting the demand for food. This will involve a substantial reduction in the pressure exerted by the agricultural system on the environment. The aim is for a Moderate diet—without much meat—for each world citizen now and in the future. This situation must be realized by the redistribution of the food produced. Residual environmental problems that could arise under the GOA system are regarded as soluble. The system can be finely tuned to the point that alien substances such as fertilizers and pesticides need not be released in large quantities into the environment.

Managing

The Managing perspective departs from the aim of a Moderate diet on account of the associated social risks. This must not, however, be at the expense of subsequent generations. The risks to the environment of a GOA system are therefore regarded as excessive. The environment faces threats not so much from the losses per unit product as from the local losses to the various environmental compartments. Water, soil, and air must be or remain of high quality, and energy and resources must be used sparingly. The comparatively high uptake of land that may be expected under a LOA system is regarded as less of a problem, as are the necessary adjustments in the structure of production.

Preserving

Under the Preserving perspective the risks to the environment are regarded as so grave that the demand for food needs to be limited and local substance cycles optimized by the development of modified agricultural systems. The introduction of alien substances and the long-range transportation of potentially harmful substances (e.g., fertilizers) are considered to pose an undue risk to the environment. The social risks of "adjusting" the demand to a Moderate diet are regarded as acceptable. It must be possible at the global level to hold down the trend toward rising levels of consumption of animal protein. In the rich countries, the consumption of meat will therefore need to fall sharply to around 40 grams a week. The reduction in demand combined with careful chain-management at the local level would guarantee a sustainable world food supply. Here, too, the emphasis is on an equitable distribution of the by-no-means overabundant supply of food.

Translation of the Action Perspectives into Scenarios

The potential grain yields per hectare, corrected for storage and transport losses, are between 4 tonnes (metric tons) under LOA and 10 tonnes under GOA. In the tropics, two to three crops can be cultivated per year given sufficient irrigation water. An estimate of the suitable area is required in order to determine the potential yield. The land must be capable of supporting sustained farming over a number of years, so that vulnerable lands have been left out of account. By way of illustration, 128 million hectares of land are currently used for agriculture in the European Union, whereas on the basis of the soil properties, this study reaches the conclusion that 80 million hectares may be deemed suitable for agriculture in the longer term.

The available land in each of the regions may or may not be irrigated. Clearly, this depends on the amount of water available for irrigation purposes in the region in question. It will also be clear that the various production levels of LOA and GOA result in different water requirements. Taking everything into consideration, the distribution is as shown in Figure 10.3.1. In all cases over 8 billion hectares (approximately 70 percent of the total area) is unsuitable for agriculture. In the scenarios based on LOA, approximately 20 percent of the area is irrigated. The GOA scenarios result in the irrigation of approximately 14 percent of the total area. This distribution turns out to be comparatively insensitive to the various world

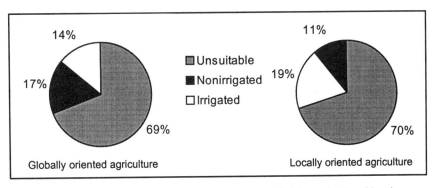

Figure 10.3.1.   Breakdown into suitable and unsuitable agricultural land
SOURCE: Netherlands Scientific Council for Government Policy. 1995. *Sustained Risks: a Lasting Phenomenon.* 44:61–2.

population growth variants. Although the demand for water for household and industrial purposes rises in line with the population, this increase has little if any impact on the total amount of water available for agriculture. The results of the calculations based on the number of people in the region, the preferred level of consumption, and the agricultural system in question are shown in Figure 10.3.2 for the high population growth variant and in Figure 10.3.3 for the low population growth variant.

From Figures 10.3.2 and 10.3.3 observe that self-sufficiency is realizable at the global level in all four scenarios. Even in the low population growth variant, however, there are shortages in a number of regions for three of the four scenarios. Only in the Saving scenario under low population growth can self-sufficiency be achieved in each region. This implies that in all other scenarios, certain regions suffer from shortages and that interregional trade is required in order to meet food needs.

A self-sufficiency index does not, of course, tell us much about the absolute quantities, for which reason the results of the scenarios are shown in somewhat different form in Tables 10.3.2 and 10.3.3. In these tables the maximum production per region is compared with the regional demand, which is equal to a self-sufficiency index of 1.1. A safety margin of 10 percent has therefore been built into all regions. Table 10.3.2 indicates the consequences of this under high population growth. In this case the Managing scenario turns out to be unattainable. The combination of LOA and the wish to provide a Western diet therefore proves out of reach given a high rate of population increase. At the world level there remains a shortfall of approximately 1.5 billion tonnes of grain equivalents.

The other scenarios show a surplus. The imposed 110 percent self-sufficiency can therefore be achieved. In all three of these scenarios, however, there remain regions with shortfalls, especially Asia (East, Southeast, South, and to a lesser extent West). These regions will need to be supplemented from other regions with a food surplus. It is not possible, however, to specify a criterion to indicate which other regions will become exporters with a view to redressing the global balance. What the scenarios can do is to indicate the total shortage in the regions with a deficit. This total shortage provides an initial indication of the trade flows that would be required in order to meet the demand in all regions.

The biggest trade flow is required under the Utilizing scenario and amounts to approximately 5.5 billion tonnes. This is followed by the Preserving scenario, with approximately 4 billion tonnes, and finally the Saving scenario with approximately 1 billion tonnes. The figures also reveal that the impact of a change in the diet is greater than that of the production technique applied. Different population growth figures also have a major

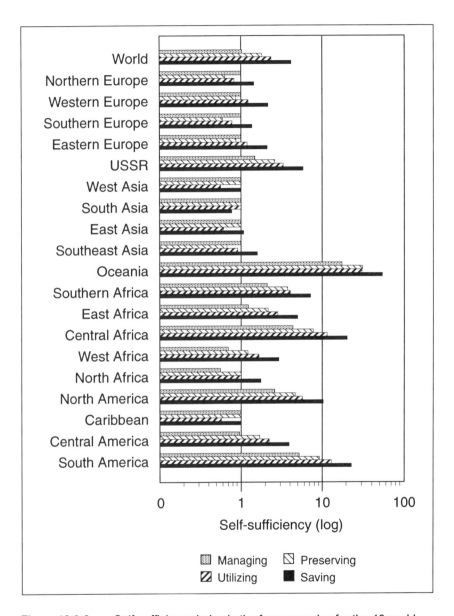

Figure 10.3.2.   Self-sufficiency index in the four scenarios for the 19 world
regions given high population growth, 2040

SOURCE: Netherlands Scientific Council for Government Policy. 1995. *Sustained Risks: a Lasting Phenomenon.* 44:63.

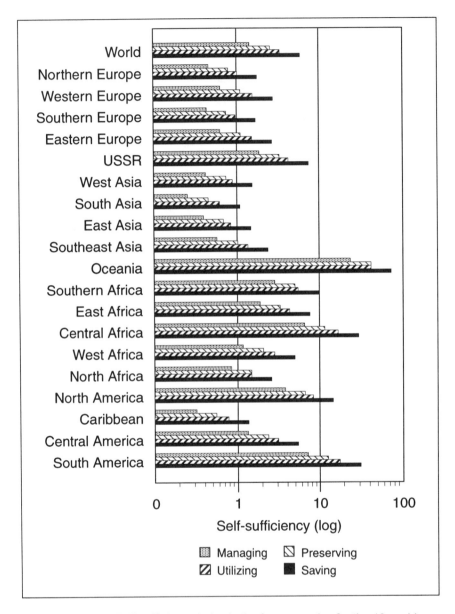

Figure 10.3.3.   Self-sufficiency index in the four scenarios for the 19 world
regions given low population growth, 2040

SOURCE: Netherlands Scientific Council for Government Policy. 1995. *Sustained Risks: a Lasting Phenomenon.* 44:64.

Table 10.3.2. Regional production balance in the four scenarios given a self-sufficiency index of 1.1 and high population growth (in 10^6 tonnes)

Country	Preserving (LOA/moderate) Production	Demand	Balance	Saving (GOA/moderate) Production	Demand	Balance
South America	5,353	630	4,724	13,173	630	12,543
Central America	420	268	152	976	268	709
Caribbean	23	62	-30	57	62	-5
North America	1,612	378	1,235	3,539	378	3,161
North Africa	359	398	-39	637	398	239
West Africa	847	758	90	2,049	758	1,291
Central Africa	1,944	271	1,672	4,966	271	4,695
East Africa	1,585	799	786	3,645	799	2,845
Southern Africa	399	116	282	768	116	652
Oceania	1,184	42	1,142	2,069	42	2,026
Southeast Asia	583	955	-372	1,386	955	431
East Asia	912	1,993	-1,082	1,958	1,993	-36
South Asia	775	2,744	-1,969	1,897	2,744	-847
West Asia	163	379	-216	340	379	-39
USSR	939	398	541	2,110	398	1,712
Eastern Europe	100	128	-28	245	128	117
Southern Europe	82	153	-71	188	153	36
Western Europe	128	163	-35	319	163	155
Northern Europe	52	91	-38	118	91	27
World	17,461	10,725	6,735	40,438	10,725	29,713

Country	Managing (LOA/Western) Production	Demand	Balance	Utilizing (GOA/Western) Production	Demand	Balance
South America	5,353	1,109	4,244	13,173	1,109	12,063
Central America	420	471	-52	976	471	505
Caribbean	23	109	-86	57	109	-52
North America	1,612	665	947	3,539	665	2,874
North Africa	359	700	-342	637	700	-64
West Africa	847	1,335	-487	2,049	1,335	714
Central Africa	1,944	478	1,466	4,966	478	4,488
East Africa	1,585	1,408	177	3,645	1,408	2,236
Southern Africa	399	205	194	768	205	563
Oceania	1,184	74	1,110	2,069	74	1,994
Southeast Asia	583	1,682	-1,099	1,386	1,682	-296
East Asia	912	3,511	-2,600	1,958	3,511	-1,554
South Asia	775	4,834	-4,059	1,897	4,834	-2,937
West Asia	163	667	-505	340	667	-328
USSR	939	701	238	2,110	701	1,409
Eastern Europe	100	226	-126	245	226	19
Southern Europe	82	269	-188	188	269	-81
Western Europe	128	288	-159	319	288	31
Northern Europe	52	160	-107	118	160	-42
World	17,461	18,894	-1,433	40,438	18,894	21,544

SOURCE: Netherlands Scientific Council for Government Policy. 1995. *Sustained Risks: a Lasting Phenomenon.* 44:66–67.

Table 10.3.3.   Regional production balance in the four scenarios given a self-sufficiency index of 1.1 and low population growth (in 10^6 tonnes)

Country	Preserving (LOA/moderate)			Saving (GOA/moderate)		
	Production	Demand	Balance	Production	Demand	Balance
South America	5,353	457	4,896	13,173	457	12,716
Central America	420	192	228	976	192	784
Caribbean	23	45	-22	57	45	11
North America	1,612	260	1,352	3,539	260	3,279
North Africa	359	263	96	637	263	374
West Africa	847	443	404	2,049	443	1,606
Central Africa	1,944	181	1,763	4,966	181	4,785
East Africa	1,585	510	1,075	3,645	510	3,135
Southern Africa	399	84	314	768	84	684
Oceania	1,184	31	1,154	2,069	31	2,038
Southeast Asia	583	625	-42	1,386	625	761
East Asia	912	1,428	-516	1,958	1,428	530
South Asia	775	1,866	-1,091	1,897	1,866	31
West Asia	163	236	-74	340	236	103
USSR	939	307	632	2,110	307	1,803
Eastern Europe	100	99	2	245	99	146
Southern Europe	82	120	-38	188	120	68
Western Europe	128	125	4	319	125	194
Northern Europe	52	71	-19	118	71	47
World	17,461	7,343	10,118	40,438	7,343	33,095

Country	Managing (LOA/Western)			Utilizing (GOA/Western)		
	Production	Demand	Balance	Production	Demand	Balance
South America	5,353	805	4,548	13,173	805	12,367
Central America	420	338	81	976	338	638
Caribbean	23	80	-57	57	80	-23
North America	1,612	459	1,154	3,539	459	3,080
North Africa	359	463	-104	637	463	174
West Africa	847	780	67	2,049	780	1,268
Central Africa	1,944	318	1,626	4,966	318	4,648
East Africa	1,585	899	687	3,645	899	2,746
Southern Africa	399	148	250	768	148	620
Oceania	1,184	54	1,131	2,969	54	2,015
Southeast Asia	583	1,101	-518	1,386	1,101	285
East Asia	912	2,515	-1,603	1,958	2,515	-557
South Asia	775	3,287	-2,512	1,897	3,287	-1,390
West Asia	163	417	-254	340	417	-77
USSR	939	541	398	2,110	541	1,569
Eastern Europe	100	174	-74	245	174	71
Southern Europe	82	211	-130	180	211	-23
Western Europe	128	220	-92	319	220	99
Northern Europe	52	125	-73	118	125	-7
World	17,461	12,935	4,526	40,438	12,935	27,503

SOURCE: Netherlands Scientific Council for Government Policy. 1995. *Sustained Risks: a Lasting Phenomenon.* 44:66–67.

impact on the necessary trade flows. Given low population growth (Table 10.3.3) all four scenarios prove attainable. The necessary transport flows amount in this case to 5.5 billion tonnes (Managing), 2 billion tonnes (Utilizing and Preserving), and zero (Saving).

Evaluation

Prior Conditions for Safeguarding World Food Production

Enough food can be produced to feed the entire world in almost any of the scenarios. Depending on the level of consumption selected, the agricultural system in question, and the availability of water, between 11 billion (Managing scenario) and 44 billion (Saving scenario) people can be fed worldwide. A sustainable food supply does not, therefore, run up against the limits of a physical environmental utilization space for the world as a whole. The extent to which the world population can be fed depends rather on political and socioeconomic factors.

An important demand made in many countries and regions is the ability to feed one's own population. Various economic blocs, such as the European Union (EU), the North American Free Trade Agreement (NAFTA), and the former COMECON attach major importance to food security, thereby underlining the strategic importance of food. The analysis in this report does not permit statements to be made at the individual country level, although the possibilities for self-sufficiency at the level of large regions can be established.

The results indicate that sufficient food can always be produced in South America, North America, Central Africa, and Oceania to meet the demand, irrespective of the preferred diet. In East and South Asia, however, this is only the case given a Moderate diet and a GOA system. Problems can arise in various regions. In a limited number of regions (North and South America and Europe) one can in fact afford the luxury of a Western diet combined with LOA, but this is an exception. For the rest of the world the distribution of food is a possibility. This presupposes an economic climate conducive to international trade, adequate purchasing power in the deficit regions, and a high degree of international solidarity. In terms of the present world community, these are extremely exacting conditions.

In the regions where more food is produced than is required for self-sufficiency, it is in principle possible to increase production in order to offset the shortages in other regions. The regions in which such extra production would need to take place would depend on the optimal level of production in each region. The desire to minimize the transportation of agricultural commodities throughout the world might, for example, lead to a choice to locate the additional production as close as possible to the areas of shortage. On the other hand, the optimization requirement might mean that the additional production was located in those areas where the highest yields could be obtained with the least amount of irrigation.

In all cases the scenarios outlined above will involve enormous changes in the agricultural system in comparison with the present structure. These adjustments will require across-the-board cooperation by all concerned. In both a system aimed more at self-sufficiency and one based on international trade, considerable demands will be made on international cooperation and solidarity. Just how likely this is to succeed can be assessed very differently.

Both the agricultural systems that have been elaborated are based on optimal management methods. This will require a great deal of know-how, insight, and "green" fingers—which will also need to be combined with "green" brains. The entire know-how innovation system will need to be geared to this end. In practice, this is asking a great deal. It implies, for example, that farmers will be well educated and that modern technologies will be available worldwide. This will require an enormous transfer of know-how and technology. A huge effort will be required to achieve this situation within a time frame of 50 years. As noted earlier, the results mainly indicate the potential, not the most probable development. If it proves impossible in large parts of the world to meet the prior conditions for optimal agriculture, this will create additional problems. The aim of a Western diet in a situation where regional food needs can scarcely be met will then be an illusion. There will probably be numerous physical and organizational obstacles towards bringing about the optimal developments.

Environmental Consequences of the Scenarios

The GOA system assumes the availability of the necessary external inputs. The requirement of best technical means assumes that the production techniques applied will have only limited negative effects on the environment. The quality of the soil will, for example, need to be maintained. In the present situation this is not the case. It will also be difficult to limit nitrogen losses under LOA, meaning there will be an impact on the environment. Locally, the leakages per hectare will be greater under GOA than under LOA. However, because the production under LOA is lower than under GOA, the leakages per unit product will be higher. Both LOA and the desire to provide a Western diet will necessitate a higher volume of interregional food trade. In the case of LOA this will in fact be at variance with the underlying premises, such as that of closing substance cycles.

The differences between nonirrigated and irrigated production are dramatic. In river basins there will often be no lack of water to realize maximum production. In Europe, for example, a good deal of water is available in a comparatively small area. Similarly, there is sufficient water for food production in Iran even though the availability of water in the catchment area is lower, because the agricultural area is extremely limited. In South America, which has a large catchment area, water is by contrast a limiting factor for food production because the area available for agriculture is so enormous.

The application of irrigation is based on the most efficient techniques. Even more would be possible if household and industrial wastewater were to be used for irrigation. All this makes heavy demands on the available technical know-how, on the institutional and social structure, and on the political stability in the region in question. There is the potential for conflict among the various categories of water users, not only among households, industry, agriculture, fishery, and the like but also among individual countries and regions. Access to water is already giving rise to conflicts, which can only be exacerbated as population pressures mount.

Apart from increasing yields, irrigation can also have negative effects on the environment, such as a growing nutrient load, salination, diseases, drying up of soils elsewhere, and soil erosion. The quality of the water available for irrigation is also important. High salt concentrations and other forms of pollution are harmful to agricultural crops and will lower the potential yield.

Relationship with Other Goals

The availability and suitability of land constitute vital factors for food production. Together with the availability of water, these factors determine the potential yield of a region. The availability of land depends in part on the agricultural system chosen. A production-oriented agricultural system will require less land than biological agriculture.

Few standard figures are available on the suitability of land on a worldwide basis. Suitability is at present determined on the basis of relatively simple criteria. Soils that cannot be farmed with modern, mechanized agricultural techniques are designated as unsuitable. Processes such as acidification, nutrient exhaustion, deforestation, and overexploitation are left out of account in determining suitability. These processes can, however, substantially damage both the quality of the soil and its suitability for agriculture.

A large area of land is required for food production. A good deal of land remains available for agriculture, for example in South and North America and in Central and North Africa. The area considered suitable for agriculture, however, includes the present tropical rainforests, including those in Central and Southern Africa. Realization of the agricultural potential in these regions will often involve large-scale deforestation, with an increased risk of soil erosion.

Clearly, other claims on land can come into sharp conflict with the demands made by food supply. The overall picture is, however, that 70 percent or more of the total area is not deployed for agricultural purposes in all the scenarios. This area is therefore available for other claims, e.g., nature conservation. These macro figures do, however, disguise the locally strong competition between the various forms of land use. In general it is fair to say that the less land demanded for agriculture the more opportunities there are for the realization of other goals. This means that both the aim of a Moderate diet and the development of a GOA system can contribute towards the solution of land-use conflicts.

In the Saving action perspective, confidence in the resilience of the environment does not extend across-the-board. On account of the enormous growth in the scale of human activities, the continuity of those activities is even regarded as under threat in the long term. A cut in living standards is therefore required, which is where policy comes to bear. The possibilities for applying technology must not be overestimated.

Under the Managing action perspective, the risks to the ecological system are avoided as far possible. This is, however, subject to the condition that the rise in living standards is largely left undisturbed. Under this perspective, the social risks of rigorous intervention are regarded as so great as to call into question the legitimacy of such intervention. Although the Managing perspective does provide for some moderation of consumption, the solutions are primarily sought in the technological sphere.

The Preserving action perspective exhibits little confidence in the resilience of the environment, for which reason adjustments are required to economic and other social activities that impose a burden on the environment. Measures can be brought to bear both in the field of consumer behavior and with respect to the production system. Ultimately, the necessary social willingness is deemed to exist under this perspective.

Epilogue

The four options described above demonstrate the possibilities to feed the world and to maintain sustainable land use. They illustrate that there are sufficient natural resources to feed many more people than there are at this moment in the world. They require, however, in all cases a tremendous effort of humankind. That is the mission and task of agriculturists, researchers, and farmers. That task is not impossible; it is a rewarding task but it demands a combined effort of many people.

Agricultural research is urgently needed to raise productivity, improve the efficiency of water and nutrient use, and promote the wise use of control measures for pests, diseases, and weeds. This is why research in agriculture should be stimulated and the tendency on a world scale for reduced investments in agriculture should be stopped.

The so-called "double Green Revolution" is urgently needed. There are ample possibilities for a double Green Revolution—a discontinuity in productivity rise per hectare and a better use of natural resources. That Green Revolution (a discontinuity in productivity rise per hectare per year) could be followed by white, yellow, and blue revolutions. Discontinuities in water use efficiency, nitrogen use efficiency, and phosphorous use efficiency, respectively, are possible. Various production ecological studies have demonstrated the biological possibilities to increase water use efficiency with at least a factor 5, nitrogen use efficiency with a factor 3, and phosphorous efficiency with a factor 5. To achieve such discontinuities, combinations of technologies, inputs, and social infrastructural changes will be needed. The world community has the possibility to supply or support such combinations and efforts. Why not use them?

CHAPTER ELEVEN

Food Security and Poverty

♦ MODERATOR
♦ RESOURCE PAPER
♦ DISCUSSANTS

Moderator ♦ *Donald Winkelmann*

Introduction

Our challenge is to talk about "Food Security and Poverty." What are the links among income distribution, economic growth, and food security? My task is to set the scene for my four colleagues: Irma Adelman, Verghese Kurien, Muhammad Yunus, and Robert Herdt.

Before turning to the relationship between poverty and food security, it is fitting to recognize the considerable improvements that have taken place in the food balance during the recent past. At times, there is a sense that things have not gone well, that there is little cause for satisfaction. But what about the track record? Data from the Food and Agriculture Organization of the United Nations (FAO) show that the global food supply has increased dramatically, with world per capita supplies more than 15 percent higher today than 25 years ago. In developing countries today, more than 90 percent of the people live in countries with more than 2,100 calories per capita per day, whereas in 1970, only about 35 percent of the people lived in such circumstances. Meanwhile, chronic undernutrition declined from 35 percent of developing-country population to less than 20 percent today. World Food Prize (WFP) laureates have played a notable role in achieving these dramatic improvements. Still, there is much to be done, as some 800 million people are still undernourished.

237

Terminology

Poverty is defined by low incomes. It has dramatic implications for health, longevity, and self-esteem. Most of those persons with low incomes reside in the 40 or so poorest developing countries. Although it is difficult to measure income in ways that permit comparisons across countries, the World Bank has framed a strategy for doing so, and has gone on to set a poverty line that is comparable from country to country (Chen, Datt, and Ravallion 1994). The strategy, which features the concept of purchasing power parity, is couched in 1985 dollars, and the poverty line is set at one dollar per day per capita. Applying that standard, it is estimated that there are approximately 1.3 billion people below the poverty line, with some 70 percent of them in Asia and 16 percent in sub-Saharan Africa.

Food security is a recent concept and its definition has not yet been fully settled. In one definition, and in its most basic form, food security is economic and physical access at all times to the food required for a healthy and productive life. The concept is notably difficult to measure empirically. Measurement usually includes the intake of calories. It is noteworthy that FAO data show a strong correlation between calorie availability and income at the country level.

Whatever its definition and measurement, *food security* has two dimensions: the *availability* of food and *access* to the food available. Globally, there is no current shortage of foodstuffs. Most experts who project availability for the future are confident that the global population's food needs will continue to be satisfied, given reasonable assumptions about growth rates in the important variables affecting availability—especially the rate of population growth and the amount of investment in research and infrastructure (Alexandratos 1995).

Perceptions

Although predictions about availability vary among observers, virtually all of them are deeply concerned about access to food for the more than 1 billion poorest people in developing countries. Food security is, overwhelmingly, a problem of the poor. There are short-run solutions to food security—e.g., food aid, subsidies, or charity—that do not involve income directly, but in any long-run scenario, promoting food security requires coping with poverty. And, in most of the world's poorer countries, coping with poverty starts with agriculture. In these countries, 60 to 80 percent of the work force is in agriculture; 40 to 50 percent of the average family's budget is spent on foodstuffs; and it is estimated that 75 percent of those below the poverty line are living in rural areas, heavily dependent on agriculture. One of life's ironies, then, is that most of those people with a food security problem depend for their livelihood on the production of food.

For the poor in rural areas, agriculture plays a major role in both the supply and, perhaps more importantly, the demand for food. The supply side is clear, as these people provide the resources—the land, labor, physical capital, and human capital—that are combined to produce foodstuffs. On the demand side, the returns generated by these resources are the principal source of income for the poor. Restrictions on resource holdings limit income streams, and low incomes limit access to food. The rural poor benefit from developments that increase their income streams, e.g., from higher food prices or from lower costs of production. However, because many of the poor in rural areas are themselves buyers of food, lower costs will be preferred to higher prices.

The remaining 25 percent of those below the poverty line live in urban areas. Urban poor benefit directly from lower prices for foodstuffs, as such prices give them greater access to food. The interests of rural and urban poor coincide through increased productivity in agriculture, because productivity changes lead to both lower food prices and higher incomes for those who hold productive resources.

The Way Forward

With poverty and food security inextricably linked, with much of the poverty in rural areas, and with virtually all of the poor—whether rural or urban—devoting a major portion of their incomes to procuring food, agriculture must play a primary role in alleviating the poverty that limits food security. To make that happen, productivity must be increased—whether through new technologies, policies, information, or infrastructure. The resulting reduction in production costs will increase incomes in rural areas and, simultaneously, will lower the price of food in urban areas. Moreover, the resulting higher real incomes will lead to widening rounds of spending and investment that will contribute to broad-based economic growth.

Agriculture and its dependent rural environs are, then, both at the heart of the poverty/food security problem and the principal vehicle for its solution. The imagination and innovation recognized by the WFP will be major ingredients in achieving the productivity increases required to reduce poverty and increase food security. Such increases will promote higher real incomes for the poor; broad-based economic growth, along with expanded opportunities outside of agriculture; and food security through greater access to foodstuffs.

On to the Questions at Hand

Now, what happens to income distribution as growth proceeds? Kuznets (1955) said that distributions were first made less, and then more, equal as growth progressed. Professor Adelman was among the first to look at the issue empirically. In her presentation, she develops a point of view on the theme. She notes that much depends on the representativeness of data, and on the extent to which the initial changes in productivity were themselves pro-poor. Our commentators add to our understanding from their experiences in facilitating productivity increases among the poor. What we should focus on are the factors that influence the relationship between income growth and its distribution. To what extent does growth lead to reductions of poverty and hence to food security? (See Deininger and Squire [1996] and Datt and Ravallion [1996] for recent discussions of this theme.)

References

Alexandratos, N., ed. 1995. *World Agriculture: Towards 2010, An FAO Study*. New York: John Wiley and Sons.
Chen, S., G. Datt, and M. Ravallion. 1994. Is Poverty Increasing in the Developing World? *Review of Income and Wealth*. Series 40.
Datt, G., and M. Ravallion. 1996. Why Have Some Indian States Done Better than Others at Reducing Rural Poverty? The World Bank's Policy Research Department, Policy Research Working Paper #1594. Washington, D.C: The World Bank.
Deininger, K., and L. Squire. 1996. A New Data Set Measuring Income Inequality. *The World Bank Economic Review*. 10:565–91.
Kuznets, S. 1955. Economic Growth and Income Equality. *American Economic Review*. 45:1–28.

Resource Paper ♦ *Irma Adelman*
Food Security and Poverty

The relationship between food security and poverty is complex. Some poor people, such as subsistence farmers, have a large measure of food security despite their poverty. And some relatively better off households, such as urban workers, suffer from a large degree of food insecurity because they are totally dependent on the vagaries of the market for their food supply. How households are affected by fluctuations in the market depends on whether they are net buyers, occasional participants in the food-staple market, or net sellers of food. It also depends on whether they are both producers and consumers of food; on how varied, how substitutable, how accessible, and how correlated their income sources are; on how vulnerable the technologies they use are to weather fluctuations, pests, and other environmental changes; and on whether the labor payments include some income in kind.[1]

Earlier Policies

The initial design of the economic development process during the 1950s and 1960s exacerbated food security problems. In the vast majority of developing countries, the economic development process neglected agriculture and treated the agricultural sector as a cash cow providing food, labor, and foreign exchange for industrialization.[2] It used a low agricultural terms-of-trade policy to subsidize urban workers and consumers at the expense of the poorer farmers, with negative incentives for domestic food production. It encouraged concentration on cash crops and monoculture at the expense of more diversified food agriculture. This process increased the market dependency and vulnerability of farmers. It stimulated rapid rural-urban migration, keeping urban wages low, creating an urban unemployment problem, and reducing the growth rate of the grain supply. It relied on cheap PL480 imports to meet urban needs and foster international competitiveness through exports based on cheap labor. And it banked on "trickle-down economics" to filter the benefits of accelerated industrialization to the poor.

In the early 1970s, development economists realized that their earlier prescriptions for development, although effective for increasing economic growth, were mostly counterproductive when viewed from the perspective of poverty reduction and food security increase. Urban unemployment problems and inattention to agriculture and poverty came to haunt us. The serious urban unemployment problems were first identified in various world-employment missions organized by the International Labor Organization (ILO) under the leadership of Louis Emmerij. The grain crisis of 1973 eliminated the supply of cheap, local-currency PL480 grain. At the same time, the oil crisis diminished the availability of foreign exchange for grain imports. And development economists saw that not only was trickle-down not working rapidly enough to reduce poverty, but also that poverty rates were actually rising in the early stages of development.

Income Distribution

In 1955, Simon Kuznets proposed the hypothesis that income distribution first becomes more unequal and then more equal as development proceeds. However, Kuznets was an optimist because he compared the distribution of income in three developing countries—Thailand, the Philippines, and India, for which he happened to have data—with the distribution of income in developed industrialized countries. He found that the distribution of income in industrialized countries was, on the average, more even than the distribution of income in the developing countries. And, therefore, he hypothesized that, with development, the share of income accruing to the poor would improve. However, he did not take into account differences in institutions and in the distribution of wealth and access represented in the comparison between industrialized countries and the three developing countries. The result was his hypothesis that income distribution improvement comes automatically with development.

In 1971, the United States Agency for International Development (AID) commissioned a report (Adelman and Morris 1973) that verified the first part of this hypothesis. The authors found that the upturn in the share of income accruing to the poor is not at all automatic; it depends on the nature of the policies that are adopted. In particular, the important considerations are how the agricultural sector is treated relative to the industrial sector, what is the nature of the educational policies, and what policies reduce the social economic dualism between agriculture and rural population on the one hand and urban population on the other. Thus, you can have countries like Brazil that are at a relatively high level of development for developing countries with a very poor distribution of income because of the nature of their elitist educational policies and the strong, industrial, large-scale, heavy industry bias that they have adopted. And you can have countries at a lower level of development and income, like Sri Lanka, that have a better distribution of income because they have adopted very different strategies in trying to raise their incomes. In sum, the authors found that for countries at the Newly Industrialized Country (NIC)-stage of development, whether or not there is a reestablishment of greater equality depends on the nature of the policies that are adopted. As a result of the finding that development can increase poverty, at least in the early stages, it became clear that a radical change in attitudes towards poverty, agriculture, and food security was called for.[3]

A Revised Perspective

In general, we can talk about a "j-shaped curve" rather than a "u-shaped curve" (Adelman, Morris, and Robinson 1976). The j-curve (one with little improvement in poverty) can be transformed into a u-curve (one with improvement in income distribution) by the adoption of pro-poor and pro-agricultural policies both for development strategy and for access to productive resources such as education and credit. It should not surprise us that, given the nature of the bifurcation between policy choices, the examination of the relationship between income distribution and development continues to be a matter of contention. Appropriate sampling is extremely important to studies of income distribution because if you omit some observations at the low end, and include a fair number of observations at the high end, what you will find is something like a j-curve. If, on the other hand, you omit some observations at the high end, then you will miss the possibility for an upturn. What recent studies seem to indicate is that in any case the bottom of the curve, whether it is a u-shaped curve or a j-shaped curve, is very flat. Thus, it takes a long time for development to actually improve the share of income accruing to the poorest members of its population.

The other difficulty with these kinds of studies, especially when applied to the 1980s, is that if you relate the rates of growth of per capita income to rates of change in income accruing to the poor, or to changes in the share of income accruing to the poor, you are catching countries at a very different stage of the adjustment problem. Because countries have adjusted at very different rates, some are still in the throes of an adjustment crisis with very low income, very low rates of economic growth, and so on. Others passed that stage five years ago, and are starting to experience an upturn. So if most of the countries that have adjusted successfully to the oil price crisis are in the Far East—that is, in East Asia and South Asia—whereas most countries that have not adjusted are concentrated in Latin America, and most countries that are still experiencing serious declines are in Africa, what you capture is a multidimensional difference between East Asia and South Asia on the one hand, and the rest of the developing countries on the other. This difference occurs when you are relating rates of growth of per capita gross national product (GNP) to rates of growth of the poor or shares of income accruing to the poor. What this does is create false optimism because it suggests that if you improve the rate of growth you will solve the poverty problem. But, the solutions to the poverty problem seem to be much more difficult, and require special policy attention.

It was realized that not only economic growth, but the right kind of economic growth, was required to alleviate poverty and increase food security. This entailed, at a minimum, labor-intensive growth, attention to agriculture, reductions in economic and social dualism among sectors and regions, and promotion of mass education (Adelman and Morris 1973). There was a call for explicitly redistributive policies.[4] Robert McNamara, while he was the head of the World Bank, learned that lesson very well and propagated it to developing countries, stressing agricultural development, labor-intensive export-oriented industrialization, and education as instruments of transferring access to resources and earning power to the poorer members of society.

The Food Security Problem

Let us now turn to the food security problem. Food security is not only a matter of mean levels of consumption of calories and nutrients, it is also a question, perhaps more seriously, of variance in the consumption of calories and nutrients. In that regard the shocks to the economy have different kinds of impacts on various groups in the economy. The important distinction is between those who are both food producers and consumers, i.e., the rural population, and those who are only food consumers, i.e., the urban population—more specifically, whether a group is a net buyer, a net seller, or on the average self-sufficient in food.

The food security problem can be treated in various ways. The first approaches to food security were supply-oriented, and focused on production. Policymakers sought technological improvements whose dissemination would improve crop yields. It was assumed that if the price of staples declined this would benefit the poorer members of society, in whose budgets grains made up the largest share of consumption. However, this approach neglected the role of the poor and near-poor as producers as well as consumers of grain. It created the "first generation" income disasters for farmers and accelerated rural-urban migration. It also increased the dispersion of income in the rural sector as larger farmers, who, with some delay, were able to increase their grain supply by more than the price decline, benefited from the technological innovations while smallholders and tenants, who could not, lost out. Once the larger farmers increased the amount of land devoted to miracle grains, the technological innovations in grain production did benefit the poorest of the poor—the landless rural labor—in the "second generation" response, because the new technologies were labor-intensive (Mellor 1976).

Food security policy requires attention to how supply-oriented measures affect income and demand, especially among the poor. The relatively low price elasticity for food says that a small increase in quantity, if left to the market, can vastly decrease the price of the grain or whatever is being produced and consumed. Thus, supply-oriented measures, which are not accompanied by demand-generating measures, can in fact, and frequently do, work as a food security disaster for the country as a whole. Such measures have tended to benefit urban dwellers, namely the capitalists, with the cheap supply of labor because, since food prices were low, they could pay low wages, and therefore could generate exportables at competitive prices.

But the impact on the rural sector has been mixed. Large supplies may affect the poor adversely. If the poor are net sellers of food, high supply conditions may reduce their incomes unless they are moved to increase their marketed surplus by a larger percentage than the fall in market price. If the poor are net buyers of food, low supply conditions reduce their real incomes unless demand and wages for their services increase by more than necessary to compensate. The extreme cases of low supply and catastrophic grain failures due to widespread drought, flooding, or predator infestation cause famine because the demand for labor in the affected villages dries up at the same time as grain prices rise (Sen 1981). Part of the result has been that, contrary to current advice, most countries have shifted to dual price policies by which rural producers are paid higher than market price, and urban consumers are given access to food at below equilibrium market price. The price to producers is kept more stable so that they can make decisions on allocation of their productive resources among crops and other food sources.

To sort out how various approaches that have been advocated to achieve food security at the national level affect the food security of the population, Professor Berck and I (Adelman and Berck 1991) have simulated these approaches in the framework of a stochastic, systemwide, interdependent simultaneous model of an actual economy, South Korea. The model was of a computable general equilibrium variety. It disaggregated households by rural-urban and by their degree of access to major assets: land and labor in the rural economy, and labor and capital in the urban economy. Consumers were assumed to maximize utility and producers to maximize net income. The model was price-endogenous: it calculated the impacts of various measures upon the prices of grains and other consumables and exportables as well as upon urban and rural wages from the simultaneous effects of each food security measure on demand and supply.

The information on prices, and/or demand and supply, was used to estimate household real income for each class of households. With the aid of income and price-dependent consumption functions, household consumption was translated into caloric intake. The percentage of households in each class whose caloric intake fell below the minimum norms established by the Food and Agriculture Organization of the United Nations (FAO) for their level of activity and expected per capita calorie deficit were then calculated and summed to derive estimates of the number of people at risk and the degree of seriousness of their food deprivation. The model was stochastic. Agricultural output, internal and international terms of trade, oil prices, and the world price of food are all subject to random fluctuations. These were described in terms of multivariate, correlated, probability distributions of international prices and domestic production that were then transformed by the model into probability distributions of household incomes and household consumptions for each category of households. Finally, for each food security measure, the same distribution of shocks was imposed on the model mean and the variance of people at risk was calculated. The information was then used to evaluate the potential effectiveness of various food security proposals.

The policies simulated fell into three categories: (1) insurance schemes to reduce variance in food prices (world price stabilization and food import insurance); (2) lowering food prices (food aid and grain price subsidies); and (3) development strategies to raise incomes (agricultural development and export-led growth). All policies were calibrated to the same budgetary cost and were assumed to be financed by an equivalent foreign grant. We then calculated the mean-variance frontier for several groups of the population—landless labor, smallholders, unorganized urban workers, organized urban workers, and capitalists. The food security policies were evaluated by their effects on nutritional status as well as by their effects on overall welfare.

In the simulations, none of the policies considered achieved very much in terms of cutting the percentage of malnourished (see Table 11.1). Not surprisingly, we found that the percentage of capitalists suffering from food insecurity was zero under all possible scenarios, and the largest percentage of people suffering from food insecurity was the landless rural workers. The demand for their services fluctuated, reinforcing the results of the shocks. However, overall, the results indicated that (1) policies implemented at the national level were more efficient than those implemented internationally (including grain price stabilization, food import insurance, and food aid) and (2) income-raising strategies (including agricultural development

Table 11.1.   Measures of food security

	Base	World price stabilization	Food import insurance	Food aid	Grain price subsidy	Agricultural development	Export-led growth
				(percent malnourished)			
Small farmers	55.88	55.76	55.91	58.13	56.08	53.23	52.48
Marginal laborers	84.56	85.24	84.51	84.09	58.40	82.12	84.60
Organized labor	13.71	13.56	13.69	12.03	12.60	12.40	15.79
Service labor	20.40	20.42	20.07	18.99	20.40	18.46	20.83
Total population	36.88	36.81	36.81	37.34	36.86	34.79	35.70
			(average daily calorie deficit per malnourished person[a])				
Small farmers	340	338	341	352	340	327	322
Marginal laborers	525	527	524	523	528	495	518
Organized labor	143	143	143	143	138	128	138
Service labor	200	199	199	200	198	188	193

[a] The deficit is measured from 1,930 calories, which is already 10 percent below the norm of 2,200; averages over the 100 replicates.
SOURCE: Adelman and Berck 1991. (Reprinted with permission)

and export-led growth) were the most efficient approaches to reductions in expected malnutrition. Dual price policies, although effective as a food security policy, were more expensive.

The income-raising policies included agricultural development–led industrialization, which centered on investments in productivity to improve investments in agriculture. These generated increases in income in the urban sector through cheap wages and cheap exports. These increases stimulated demand for rural food and raised employment in the labor-intensive industry—the most labor-intensive industry, which is agriculture—and therefore raised the incomes of both producers and consumers of food. Agricultural development led, also, to the lowest percent malnourished. Although it would be desirable to experiment with alternative institutional specifications before drawing definitive conclusions, the group-specific results are likely to be qualitatively generalizable. The major result is that development strategies that raise the rates of growth of incomes of the poor are the most effective approach to reducing malnutrition in the long run.

Notes

1.   Hayami and Ruttan (1985) found that, in the Philippines, when rice prices were high, farm workers were paid in money while, when prices were low, they were paid in kind.
2.   Taiwan was the major exception to this statement in the post–World War II period.
3.   Speech by Robert S. McNamara in 1972, delivered in Santiago, Chile (in Streeten 1985).
4.   Chenery (1974) and more recently Adelman (1979).

References

Adelman, Irma. 1974. Strategies for Equitable Growth. *Challenge* (May–June):37–44.

————. 1979. *Redistribution before Growth*. The Hague: Martinus Nijhoff.

Adelman, Irma, and Peter Berck. 1991. Food Security Policy in a Stochastic World. *Journal of Development Economics* 34:25–55.

Adelman, Irma, and Cynthia Taft Morris. 1973. *Economic Growth and Social Equity in Developing Countries*. Stanford, CA: Stanford University Press.

Adelman, Irma, Cynthia Taft Morris, and Sherman Robinson. 1976. Policies for Equitable Growth. *World Development* 4(7):561–82.

Chenery, H. B. 1974. *Redistribution with Growth: Policies to Improve Income Distribution in Developing Countries in the Context of Economic Growth.* London: Oxford University Press for the World Bank and the Institute of Development Studies, University of Sussex.

Hayami, Yujiro, and Vernon W. Ruttan. 1985. *Agricultural Development.* Baltimore: The Johns Hopkins University Press.

Mellor, John. 1976. *The New Economics of Growth.* Ithaca, NY: Cornell University Press.

Sen, A. 1981. *Poverty and Famine.* Oxford: Oxford University Press.

Streeten, P. P. 1985. *First Things First.* Oxford: Oxford University Press.

Discussant ♦ *Verghese Kurien*

When I received an invitation to participate in the Tenth Anniversary World Food Prize Symposium, I was particularly pleased to have been asked to contribute to the in-depth session on "Food Security and Poverty." Much of my professional life has been devoted to trying to create the former and end the latter.

Although we have come to take it for granted, the fact that there is poverty in this world as we approach the next millennium should be a matter of surprise, if not shame. Very substantial resources have been invested in ending poverty, yet the problem of poverty in much of the world—including the economically advanced countries—appears intractable.

For those who believed the Malthusian prophets of gloom and doom in the 1960s and 1970s, the fact that the world is feeding itself—if barely—might be equally a cause for surprise. Yet for many of the world's nations, food security is far from attained while global food security—if indeed it exists—has been erected on a promising, but fragile foundation.

I count myself at one with Dr. Norman Borlaug when it comes to believing that our world can feed itself. In India, during the early 1970s, there were those who predicted massive famines within a decade. The Green Revolution changed that and, for the moment, we have plentiful stocks of grain available should our agriculture be affected by the shocks of flood, drought, and pestilence. In our own modest way, we have helped to transform India from a nation that was importing almost all of its dairy commodities and where per capita consumption of milk was declining precipitously. Today, our national milk production stands second only to that of the United States and, over the last two decades, we have virtually doubled per capita availability of milk to 200 grams daily. We were on the same path with oilseeds and edible oil when "food security" collided with "comparative advantage." So, on balance, I am confident that this world can feed itself. The question is whether it will.

All of us are aware that the international stocks of grain reached precariously low levels during the past year. As these stocks shrink, prices rise, contributing to two worlds—those who can and those who cannot afford an adequate diet. That this should happen in a world where technology, land, finances, and human capital all exist in sufficient measure to produce bountiful harvests is ironic, to say the least.

If I may be rude to some of those whose profession it is, may I say that we are in our present fix because the theories of economists seem to have replaced common sense, and common decency.

Let me give you an example. I mentioned that India was on the path to self-reliance in oilseeds and edible oil. By adopting a number of measures that are quite commonplace in those countries like the United States, which place a premium on food security, we had transformed that subsector of our agricultural economy from stagnation and decline to vibrant growth. What did we do? We promoted and financed cooperatives to give our producers a voice in a market that, until then, had been dominated by a relative handful of wealthy traders. The cooperatives introduced a competitively priced, good quality vegetable oil in sensitive retail markets. We engaged in some modest but effective price intervention. We moderated imports and ensured that imported oils were not made available at prices with which our farmers could not compete.

Our farmers, assured of a remunerative return, invested in putting more land into oilseeds and used better seed, more fertilizer, plant protection, and so forth. The result was that in a matter of three or four years we went from importing 2 million tons of edible oil to fewer than 200,000 tons.

One would have thought that everyone would have stood and applauded what had been achieved. Instead, the World Bank condemned our support of oilseed cultivation and argued that India lacked a comparative advantage and, therefore, should import edible oils. The fact that our farmgate price had rapidly moved toward parity with the international price for most oilseeds was ignored, as was the fact that the real culprit in the subsector was an archaic and inefficient processing industry. Even more staggering was the use of comparative advantage as a prescriptive rather than a descriptive term. Suffice it to say that no matter what may have prevailed earlier, the world no longer gets its coal from Newcastle.

Now the fact of the matter is that much of India's population cannot afford edible oil and, I'm sure, much the same is true in Africa and in other parts of the developing world. There is more than enough potential demand to keep the soybean farmers of Illinois and Iowa, the palm plantations of Malaysia and Indonesia, and every other oilseed producer in the world busy. The problem is that the demand is constrained by poverty. So is the demand for grain, and for virtually every other agricultural commodity.

What we see today is short-term solutions being prescribed that will ultimately undermine long-term global food security. And we can think in no other terms than of global food security. By global food security, we can mean nothing less than adequately feeding all the world's population—not the situation we have today where tens, if not hundreds of millions remain perpetually hungry, undernourished, and ill from the effects of malnutrition.

Today, the advanced agricultural nations seem to have reached a point where they are convinced that national food security can no longer be at the expense of the taxpayer; rather, it must shift to the expense of the consumer. The invisible hand of the market will bring supply and demand into balance. This will be at the cost of some producers leaving agriculture, some land going out of agriculture, quite possibly at higher long-term prices to consumers, and at the cost of national food security.

Some might applaud this end to intervention in agriculture. Certainly the cost to the taxpayer has been quite high. The subsidies are very substantial. Were I not concerned with the broader impact, I would be relieved that our Indian farmers whose government cannot afford such magnanimity would not have to compete with their brothers and sisters in other parts of the world whose production is so lavishly supported.

But I do have a greater concern. Global food security is, in the end, only the aggregate of national food security. If those nations who, today, account for the major portion of the world's food surplus allow their agriculture to contract, then food will only be available to the highest bidder. And that, too, in a world where many of the world's nations are far from able today to meet the demands of their populations, year in and year out.

Because much of the world is poor, there is enormous pent-up demand for food. To my mind, the goal is to transform that potential into actual demand. That can only be done by addressing the problem of poverty at its roots. Poor people and poor nations cannot buy food. When they can, and then only, will the "market" and terms like comparative advantage truly make sense.

Ironically, we believe that food aid has been a contributor to the looming problem of global food security. Food aid that undercuts domestic prices, that changes food preferences from what is grown locally to foreign products, that creates dependency instead of self-reliance can only contribute to growing poverty—no matter how humane the motives in extending that aid might be.

In our work with India's dairy industry, and then with oilseeds, and now with fruits and vegetables and even with trees, we have tried to use imported and donated commodities in a way to build our own self-reliance. Without the large commodity donations from the European Community, Canada, Sweden, and the United States, much of what we have achieved would not have been possible, or at the very least, would have taken far longer. But, by deploying those commodities at prices that were equivalent to what our farmers would receive and by investing the proceeds in financing the creation of a cooperative infrastructure owned and controlled by those farmers and managed by professionals in their employ, we have been able to energize and ultimately transform our dairy and oilseed industries.

What I would therefore like to offer this symposium is a modest proposal: don't shackle the agriculture of the advanced nations by some curious notions about the market. Instead, use the bountiful harvests that you can produce as an investment in building the agricultural base of the other nations of the world.

In making such an investment, may I urge that we shift from the "development" that is designed, managed, and controlled from international and national capitals to development that is in the hands of producers themselves. If there is a single lesson I have learned from more than 40 years' work it is that only when producers gain control over the resources they create can genuine and sustainable development take place. When this happens, as it has happened in India, those who participate break out of the cycle of poverty. When enough individuals escape from poverty, the nation itself begins to advance. For it is only when a nation's agricultural foundation is strong that economic growth will occur.

What is required is the very small percentage of agricultural production on the margin. The cost is minuscule when compared with the costs if the world cannot feed itself 20 years hence. The financial resources are small, but the human resources that must be invested are great. It is a challenge that requires the best and brightest of my country, of other developing countries, and of the advanced nations.

Some years ago I felt that our world was passing through an era of "enough," but that unless we acted, that era would rapidly end. The early warning signs are already to be seen. Yet, I fear, we lack the collective vision and wisdom to act. Complacency and narrow self-interest are obstacles to actions that are imperative if we are to achieve global food security. I hope that this symposium, which draws together the intellectual and, I hope, moral leadership in the world's agriculture and food industries, can set in motion a movement to attain sustained global food security and an end to world poverty.

Discussant ♦ *Muhammad Yunus*

Let me restate something that has been said many times: poverty and food security are inversely related, and we will assume that if we can remove poverty, we will ensure food security. There may be some variations, but basically that is what we are stating—that if you can remove poverty to ensure greater chances for food security, then food security will increase tremendously. I will not go into how growth and income impact on poverty, because that is a long, drawn-out argument.

My question is: Can poverty be addressed directly? Not in a growth way, but focusing on poverty. We are talking about pushing people above some poverty line in a sustainable way so that they do not slip back again into poverty. My belief is that it is possible, and that to do so would be behind everything I have said today.

Poverty is not created by the poor. If that is acceptable to you, many of the things I will say will become acceptable, too. Poverty is created by the concepts that we have developed along the way, by the political frameworks that we have developed and believed in, and also by the institutions that we have created on the basis of these concepts and political constructs. No matter how we adjust and readjust these concepts, frameworks, and institutions, not much will be achieved in the elimination of poverty.

To me, poverty reduction is a case where you almost have to go back to the drawing board and create concepts all over again and redesign the political constructs and the institutions themselves. So it means a fresh start. Cranking the old stops again and again will not help very much; it will only marginally move back and forth. But the real cleanup of the poverty situation requires an entirely new setup, new approach, and new ambition.

Look at the present situation and compare it with 25 years ago. If, from that point in the past, we had tried to imagine the world 25 years ahead, we would most likely not have been able to visualize clearly what we have today in terms of technology, politics, international relations, and so forth. Now, if we take today and project ourselves 25 years ahead—the world that will be here—I don't think we can predict what the world will be like, no matter what kind of wisdom we can all master together. If we really try to imagine something today, we would have to write about it as science fiction, no less than that. Otherwise, we presumably would not come near to what the reality will be.

Whatever that reality will be, one thing I feel, and you no doubt would agree with me, is that it would be more about bringing people closer than apart—in many different senses. In the sense of ideas, the reality will be people reaching out to other people, easily, quietly, quickly. The reality will be a more or less borderless world. It will be a "distanceless" world because distance would not matter very much. In a future situation like that, to carry on the existing institutional work would be pushing ourselves too hard. New institutions and new ways of doing things will emerge of their own accord, and they will generate their own momentum during the next 25 years. In a future situation of that nature, if we want to eradicate poverty, we will have to do that in the context of a new, emerging world that will be in front of us.

The starting point could be (and I feel this again and again in the work that we and many others do) that we have to adjust the potential in each individual human being. This is apparently the coming message of the future: that individuals will play a much more powerful role in the future than institutions, whether those are multinational businesses or the "mega governments" in many parts of the world. More and more authority, power, and empowerment will come to individual human beings. There is a treasure of potential in individual human beings; whether one is poor or one is not poor, each human being has an enormous amount of creativity and ingenuity. If we look at each individual as a source of tremendous creativity, putting 1.3 billion people out of commission, out of playing an active role in the playground of the world, is a tremendous setback for the rest of humanity because their creativity and ingenuity are not placed in the marketplace, which would have benefited everyone.

So if we look at the situation that way, not only are the poor deprived, but everyone else on the other side of the fence is also deprived in a big way. We would undoubtedly like to get everyone into the playground and be benefited from that. And perhaps that will come about in a way when we recognize the creativity in each individual human being, all the way to the 1.3 billion people that we have talked about.

A second point that has been raised again and again is that not too many people are entrepreneurs, so seemingly we are addressing only the "entrepreneurable" poor people. Our statement right from the beginning— and there is still a big emphasis on it today—is that all human beings are basically entrepreneurs. If someone has absolutely never had a chance to do anything, he or she was only deprived of the opportunity to bring out the entrepreneurial capacity that they had. Given the opportunity and given the environment, he or she can demonstrate that entrepreneurial probability equally. We have also been saying that the fact that any human being is alive is a good demonstration that he or she has skill—that is, the skill for survival. It requires tremendous skill to survive in the old adversities that surround many people. So, if we can support the survival skill itself, a person can move on to build his or her own life.

When we at the Grameen Bank approach an individual woman for the first time in a new area, we suggest to her that Grameen Bank would like to lend her money if she is interested, and if she can come up with an idea. We always hear the answer, "No, I don't need any money. You go and talk to someone else, because I don't need any money." Or the woman will say, "Well, I don't need any money, but you may talk to my husband. He may know what to do with the money, because I never use money myself." We have trained our Grameen staff to know that when they hear such a statement, that woman is the person we are looking for; she is the one who will become an excellent borrower at Grameen Bank. This bank was created for her. And we do not give up; we go on meeting with her and encouraging her. The fact that she said "no" does not mean "no," it simply states the fact of the day. The woman has always been told that she is no good, she is not good for anything, she only bring calamities to the family, so she should stay out of it. So she is trying to avoid troubles for her family, and she is scared to death thinking that she might make some money and invest it in something.

So this is our starting point. To peel off the crust of fears takes time. And ultimately, presumably, sometime the woman will say, "Yes, I would like to take the money." And that day she will be told that she will have to find four other friends to make a group of five, and then she will go around finding a group of five. It will take weeks or months before she can arrange the group. And when her turn comes to take the loan from the bank, she will have a terrifying, sleepless night the night before, because she will not know whether she should go through with such a terrible thing tomorrow morning. And in the morning, she will almost decide not to go through with it; it is too much, she cannot go through with it. She will tell her friends, and they will encourage her not to get so upset, that the bank will do it quickly and nicely. And so she will take the loan and she will tremble at getting this 10 or 15 dollars worth of money. Tears will roll down from her eyes because some institution has trusted her with such an enormous amount of money, and she will promise herself that she will not let them down. And this will be the beginning of her life in a new way. The woman that you see at the beginning of the loan cycle, and the one that you see at the end of that cycle, one year later when she has to pay the loan back, are two different women. This is the experience that one has to go through before we can activate that individual human being.

So, poverty alleviation is not simply giving money or putting somebody on charity. The basic fact of poverty alleviation is to allow each human being to be in charge of the destiny of his or her life. That is the environment we need to create. I believe that environment can be created, and we can create a poverty-free world.

Discussant ♦ *Robert Herdt*

Let me structure my remarks around two ideas: first, that progress has been made; and secondly, to ask what is required to have continued progress.

Progress has been made over the last 25 to 30 years in achieving food security. The proportion of malnourished children in the world has gone down in nearly every developing-country region on average, as Don Winkelmann said, from 35 to 20 percent. World food prices have gone steadily downward, making food more accessible to people. People still need income, obviously, and we still have too many people who are undernourished and malnourished. And as population growth continues, the absolute numbers will rise unless we do better. So we have made some progress, but there is a lot of progress remaining to be made.

The progress that has occurred has been generated through the application of the technologies that we have heard about at this symposium—the Green Revolution. Technologies depend on the distribution of resources, on policies, and on the institutions that govern the societies in which they are used. And the Green Revolution technologies have not been adverse, by and large. This is a point that lay people, the news media, and others do not seem to understand. A couple of examples will make the point. In a 2,000-farm study in Bangladesh in 1975, 80 percent of small, medium, and large farmers all were using high-yielding varieties of rice that came from Bangladesh's relationships with the International Rice Research Institute (IRRI); 80 percent of all groups, whether small, medium, or large. However, in the rainy season only 30 percent of farmers in all size groups were using the new varieties. The big difference between the seasons was what determined the big difference in farmers' adoption of rice varieties. It was not that some were small farmers and some were large. In India, a national study by the National Council of Applied Economic Research showed very similar results in many states. The study, by Peter Hazell and Dr. C. Ramasamy, was documented very carefully in Tamil Nadu in 1973 and then again 10 years later. It found that laborers as well as small farmers benefited from the introduction of technology under the conditions existing in the North Arcot district. So there *have* been gains. But that is not to say that no challenging problems remain.

What things have to be done to make further progress? How can we address the questions of poverty and food insecurity? We have heard some specific ideas. Let me pull back from those a bit and say that we need to have an appropriate balance among the three dimensions: technology, policies, and institutions.

Technology requires the development of varieties, perhaps using biotechnology, of methods to manage soil nutrients and their depletion and other matters. Because we have to have appropriate technology, the emphasis on agriculture research that we have heard about is a necessary emphasis. But the research has to be directed at the appropriate problems, it has to be efficient, and it has to be well-managed. We can waste a lot of money on agriculture research and we can develop technology that does not fit the situation. That is not what we want to do, and that is not where resources should be directed. But technology alone is not going to solve the food security challenge.

Farmers in countries operate under sets of policies that are generated by a mix of forces. We have heard about policies being imposed by the World Bank. Just because the World Bank insists on them does not mean the policies are necessarily correct; but it does not mean they are necessarily incorrect, either. Policies are generated within the national political systems by a complex process. Over time we have seen a lot of policies that involve subsidies. Looking carefully at these policies we see that often the powerful people in a society are the ones who benefit. Subsidies have to be crafted very carefully in order for their benefits to reach the people who really need them.

This is where the third dimension comes in—the institutions. Here the challenge is with the organizations and creative development of new ways of organizing to address the problems. If we take an old system and throw in subsidies and throw in technologies without looking creatively at how those things will interact, the result may be one that we did not anticipate, that we do not like, and do not want. We must have the national will to look at these challenges and to determine that the needs of poor, small farmers in rural areas must be addressed.

The appropriate balance for action among those three broad dimen-sions has to be determined within each situation and each country. Professor Adelman suggested that during the process of going from a low income society to a higher income society, many things change. And the benefit of a given technology or policy depends on whether you are a buyer, whether you are a seller, whether you are a user of technology, whether you are a supplier of labor, and so forth. The kind of analysis contained in Adelman's discussion can help us understand the roles of institutions, policies, and technology in some of the most difficult situations we face, such as in sub-Saharan Africa and in South Asia. That kind of analysis can provide a framework that allows us to see what may happen to the many elements of food security. These are complicated systems, and it is not easy to under-stand what may happen as a result of any initiative, even by looking at what has happened in another country.

The bottom line for addressing the problem of food security, regardless of where it occurs, is the determination by societies to address seriously the needs of the poor and the needs of the food insecure. Only with that serious attention will we solve the problem of food security.

PART III

Special Food Security Issues

CHAPTER TWELVE

Problems of Food Shortage

◆ MODERATOR
◆ RESOURCE PAPER
◆ DISCUSSANTS

Moderator ◆ *Robert L. Thompson*

I would like to address first individual and nutritional security and raise a few points that complement Eileen Kennedy's discussion. She emphasized the fundamental importance of available purchasing power on an individual's ability to access adequate quantities of food, in general, and of specific nutrients, in particular. I feel that in discussions like this, we tend to focus too much on agricultural development and not enough on rural economic development. Certainly, to have food security there must be enough food available. Raising agricultural productivity is essential to ensuring that enough food gets produced. Sufficient food is essential to having global food security. Producing enough food in the world requires finding ways to raise productivity on the fertile, nonerodible soils. The only alternative would be to expand production onto fragile lands or lands that are presently forested.

Poverty is the principal source of individual food insecurity, and the majority of poor people in low-income countries are in rural areas. However, no country in the world has solved the problem of rural poverty in agriculture. Certainly, increasing agricultural productivity raises the incomes and the earning potential of some rural people. However, there is simply not enough land per person employed in farming in most developing countries to permit each individual to grow enough food not only to meet their family's food needs but also to sell to provide a politically acceptable level of family income. The only places that have sustainably reduced rural poverty have created new sources of income from off the farm. These sources may come from cottage industries in the farm home or from off-farm employment.

There are two ways to create off-farm employment: create jobs in nearby rural communities, or move people out of agriculture to faraway cities. An alarming number of cities are growing towards 10 million people. I do not feel that moving a lot of people out of agriculture into cities with populations of multiple millions is a very attractive way to solve the problem of rural poverty. The United States experienced huge social problems in cities of the North in the late 1960s as we moved people out of agriculture in the South at a faster rate than they could be absorbed into employment in distant cities. If low-income rural people see substantially superior earning potential in the cities, many will move. Their motivation is to obtain a better quality of life for their families, including better income, schools, health care, and cultural amenities. I believe that investments in rural development are needed throughout the third world to reduce rural poverty and, in turn, malnutrition and food insecurity—without moving people to the cities. But this will not happen if the pronounced antirural bias in most third world countries' public policies continues. This urban bias is manifested particularly in the allocation of infrastructure investments. Rural roads and communications are deficient, and without them, neither agriculture nor the rural nonfarm sector will develop.

By keeping the price of food artificially low, many developing countries destroy the incentive for their farmers to grow as much food as they could. The agricultural sectors of these countries contribute less to their food security than they could do efficiently. But large investments in rural roads and communications are not sufficient; more investment is needed in rural schools, especially for the education of girls. Fertility rates drop as girls' level of education rises; the quality of care of their families—nutrition, nurturing, and encouragement of education—goes up as more is invested in educating girls. There is a gross underinvestment in public education in rural areas in general; but even where schools have been built, they are often staffed poorly, if at all.

Investments in health care and in drilling wells are also important. A tremendous amount of women's and girls' time in developing countries is wasted fetching water and fuelwood. The girls' time would be much better spent in school than in carrying water or fuelwood. To achieve larger school attendance rates, it may be necessary to reduce the opportunity cost of the time that rural children spend in schools by providing some subsidy to their families. All of these investments are important to increase the attractiveness of rural communities as places to stay or return to. Eliminating the antirural bias in public policy is an essential first step towards reducing poverty and food insecurity.

Most developing countries have abandoned the failed socialist model, but many need a more positive enabling economic environment for the emergence of a thriving private sector. Many developing countries lack an adequate commercial code and fair judicial procedures for resolving contract disputes. Many still have a subtle—or not so subtle—distrust of the private sector. Policies often protect public sector firms, which are often less efficient than the private sector (e.g., supplying seeds and fertilizer and marketing grain). A lot still needs to be done to create a more enabling economic environment for the emerging private sector to thrive.

I agree with several participants' observations that food aid is likely to decline due to the recent changes in domestic agricultural policies in the United States and the European Union (EU). Both the United States and the EU have started decoupling commodity price supports from the volume of production, and have shifted the way they transfer income to their farmers from price distortions to direct payments. Neither is likely to accumulate inventories of agricultural products from price support operations, so there will be less food aid available.

The international market should be able to ensure that enough food is available to the countries that have the purchasing power to pay for it. National, like individual, food security is mainly a problem of adequacy of purchasing power.

If a country cannot efficiently grow enough food to feed itself, national food security becomes dependent on the country's ability to generate foreign exchange earnings. The country must be allowed to export products in which it has a comparative advantage. Unfortunately, many industrialized countries impose protectionist measures that make it very difficult, if not impossible, for developing countries to export the labor-intensive goods in which they have the greatest comparative advantage. These measures put a brake on the growth of both individual income and foreign exchange earnings. Foreign exchange reserves are necessary for countries to import food in times of need or in those countries whose resource endowment is inadequate for them to be efficiently self-sufficient in food.

I believe that the international agricultural system can produce the technologies needed to double food production in the next 30 years without damaging the environment. However, I agree with Dr. Kennedy that the International Food Policy Research Institute's (IFPRI) 2020 baseline paints too rosy a picture by assuming larger investments in agricultural research in the coming decade than at the present time. Investments in agricultural research—national and international—are declining. Without larger investments, it will not be possible to increase agricultural productivity enough to avert environment problems. At the same time, governments need to remove the policy biases that limit farmers from reaching their potential. This is a common problem throughout the third world as well as in many countries of the formerly socialist world.

Finally, there is great potential to increase the efficiency of the world's food system by reducing postharvest losses from rodents, insects, rotting, or other forms of physical destruction of food. In the former Soviet Union, as much as 40 percent of agricultural output is said to be lost between the points of production and consumption; in many developing countries losses may be 20 to 30 percent. This is a tremendous inefficiency in the world's food system. It is immoral to have so much food production disappear after resources have been used to produce it.

Resource Paper ♦ *Eileen Kennedy*

Hope and Reality: Food Security and Nutrition Security for the Twenty-first Century

Introduction

The world has changed dramatically since the 1974 World Food Conference in Rome. Globally, there have been great strides in the past two decades to increase food production and reduce hunger and malnutrition. Nevertheless, hunger and malnutrition continue to be problems of staggering proportions worldwide. Approximately 800 million people, or one in five individuals, have insufficient food (FAO 1995).

The World Food Prize Symposium and the 1996 World Food Summit provided valuable opportunities to reassess the state of food security and nutrition security as we enter the twenty-first century. This chapter reviews what is known about food security and nutrition security as well as identifies potentially effective strategies for reducing hunger and malnutrition as the world enters the next century.

Food Security and Nutrition Security

There are a number of commonly used definitions of food security. Food security is defined here as access by all people at all times to the amount and quality of food they need for an active and healthy life. Nutrition security is a somewhat lesser-used concept that builds on food security but goes beyond food as the only input into nutrition. Nutrition security is defined as food security plus an environment that encourages and motivates society to make food and nonfood choices that promote both short-term and long-term health.

Hunger, malnutrition, and eventually death are the ultimate consequences of failed food security and nutrition security. While hunger and malnutrition manifest themselves at the level of the individual, the causes generally involve a combination of individual, household, community, national, and international factors. Figure 12.1 provides a simplified framework to illustrate the linkages among factors at each level and the impact on the individual.

The availability of a sufficient quantity and quality of food either through domestic production, commercial imports, or food aid is an essential building block for food security. Achieving a sufficient food supply is one part of a strategy to ensure household-level food security (Von Braun 1991). While food availability at the global, national, or community level is one factor that influences household food security, it is not necessarily the most important determinant. For example, it is common to have 20 to 30 percent of a country's population consuming less than 80 percent of caloric requirements even though national-level food availability is at or greater than 100 percent of requirements (The World Bank 1986). Even in developed countries with per capita energy availability far in excess of requirements, segments of the population—usually the poor—can be food insecure. National surveys indicate that in the United States, anywhere from 2 to 4 percent of the population is food insecure based on calorie per capita estimates; if we look at only the low-income population in the United States, this figure jumps to 10 to 12 percent.

Very recently, the U.S. definition of food security has been revised to acknowledge the term as describing problems of food deprivation that result from poverty (Klein 1996). Food security requires access to both an adequate quantity and quality of food obtained in a socially acceptable manner, without resorting to emergency food supplies, scavenging, or stealing. Using this definition, hunger is considered a narrow condition existing within a broader definition of food insecurity (Klein 1996). The indicators of food security, therefore, go beyond energy intake as the sole outcome

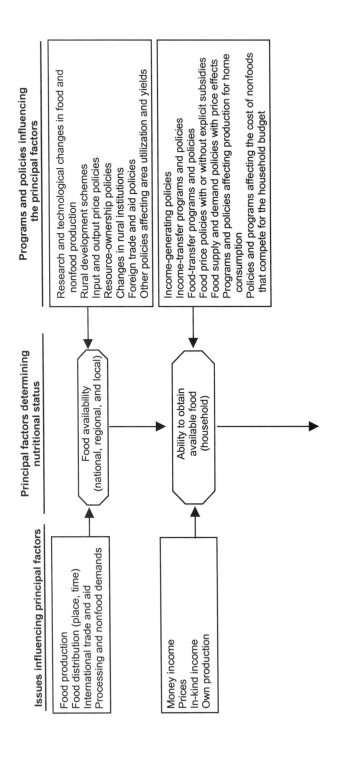

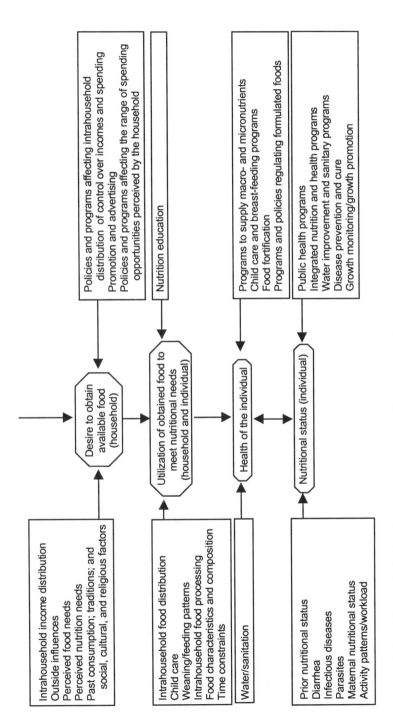

Figure 12.1.   Links to individual food consumption and nutritional status
SOURCE: Kennedy and Pinstrup-Andersen 1983. (Reprinted with permission)

measure and include additional measures such as skipping or cutting meals, not eating for an entire day, and/or going without food because of lack of money. It is access to food—or the household's ability to obtain food—that is critical to ensuring household food security. Policies that increase the access of vulnerable groups to food (either through increased incomes or decreased food prices) will enhance household food security.

An increase in household food intake often is assumed to improve the food intake of each of the individual members. Results from a number of studies now indicate, however, that household food consumption is often a poor proxy for an individual's energy intake (Garcia and Pinstrup-Andersen 1987). This is because the effect of increases in household food access on an individual's food consumption can be modified by a variety of factors including intrahousehold income earning patterns, education of household members, and characteristics of the individual such as gender, age, birth order, and genetic endowment. The strength and direction of each of these factors can vary by sociocultural environment.

For example, in some cultures (mainly in South Asia), boys get preferential treatment in the allocation of food and other resources. But there is little evidence of this gender bias in the allocation of food in Africa (Svedberg 1990). Outside of Africa, however, a number of studies have shown that children and women tend to consume a lower proportion of their caloric requirements relative to other household members (Haaga and Mason 1987; McGuire and Popkin 1989; Piwoz and Viteri 1985).

There are also differences in the allocation of health care among various household members. Again, most of the empirical evidence on intrahousehold gender bias comes from South Asia. A study in the Punjab found that in the first two years of life (years of peak mortality), expenditures for medical care for sons were 2.34 times higher than for daughters (Dasgupta 1977). Dasgupta concludes that gender bias in the Punjab is culturally determined and related to the structure of the rights of asset ownership and decisionmaking, which severely restricts women's authority. Similar findings on maldistribution of medical expenditures were reported for Bangladesh; boys were favored over girls in the allocation of health care (Chen, Huq, and D'Souza 1981).

The education of a child's mother can play an important role in the effect of increased household resources—including food—on the child's consumption and nutritional status. Better-educated women often exhibit behaviors that are more child-centered, which leads to better feeding practices and ultimately to healthier and better-nourished children (Tucker and Sanjur 1988). This maternal education effect on child nutrition has been documented for both formal and informal education of mothers.

New evidence suggests that in certain types of female-headed households where women have more decisionmaking power, mothers allocate a larger share of household food supplies to preschool-aged children (Kennedy and Peters 1992). Other data from Malawi indicate that at low income levels children from some female-headed households are healthier than children from higher income male-headed households. Similarly, recent evidence from Brazil indicates that female-headed households were able to use scarce resources (such as land, labor, capital, and credit) to improve short-term nutritional status for their children (Vosti and Witcover 1990). In view of the demographic shifts occurring in developing countries—in particular, the growing number of female-headed households—it is important to understand how different household structures influence the success or failure of particular policies and programs.

The health and sanitation environment and the nurturing behavior of caretakers may be as important as, or in some cases more important than, food intake in influencing an individual's nutrition status (Von Braun and Kennedy 1994). Caring activities are critically important influences on child health. These include breast-feeding and weaning practices, child care, and other nurturing activities, all of which may be reduced if new agricultural technologies put added demands on women's time. Time allocation studies indicate that, on average, women in developing countries put in more hours per day in nonleisure activities than do men (Juster and Stafford 1991). Not only are women actively engaged in own-farm production and wage earning, but a substantial amount of their day is devoted to the household chores of child care, food preparation, cleaning, and collecting water and fuel.

Many of the health-promoting strategies such as growth monitoring and oral rehydration, which have been advocated as part of the child survival revolution, add to the demands on women's time. The low level of utilization of some of these health strategies may be related to the lack of time available to mothers (Leslie 1989).

In order to design and implement effective strategies to alleviate food insecurity and nutrition insecurity problems, policymakers must have a clear understanding of the underlying determinants of hunger and malnutrition. They must also know how these determinants influence the appropriate policy options and interventions.

Global Food Security

In the late 1960s and 1970s worldwide attention was focused on the problem of food shortages in developing countries. Rapid population growth exacerbated shortages of food that were caused by underlying constraints to food production and food distribution throughout the developing world. These food shortages were related to lack of natural resources, low incentives to produce food, insufficient purchasing power of the poor, and inadequate institutions and infrastructure (Kennedy and Haddad 1992).

Impressive gains in global food supplies were made beginning in the 1970s. The increases in food supplies were related to investments in agricultural research that resulted in technologies that significantly increased food production in developing countries. Dramatic increases in food supplies were most noticeable in Asia and Latin America (Pinstrup-Andersen and Hazell 1985). The increased agricultural production resulted in lower food prices, which directly benefited the urban consumer, the rural landless, and the smallholder producer who was a net purchaser of food (Kennedy and Bouis 1993). In addition, the production increases from the use of improved technology generated higher household income. The technological change in agriculture also generated increased demand for hired labor and thus provided a means of increasing the income of the rural landless.

Today, world food supplies per capita are approximately 18 percent higher than 30 years ago (FAO 1995). However, although globally the food security situation has improved, the progress has been uneven. Significant portions of the population are still food insecure in sub-Saharan Africa (37 percent) and South Asia (24 percent).

This uneven pattern of availability of world food supplies is expected to continue into the next century. Several recent analyses by the Food and Agriculture Organization of the United Nations (FAO) (Alexandratos 1995), the International Food Policy Research Institute (IFPRI) (Rosegrant, Agcaoili-Sombilla, and Perez 1995), and The United States Department of Agriculture (USDA 1996) in preparation for the World Food Summit have looked at projections for global food security over the next 20 to 30 years. The FAO and IFPRI studies are based on similar assumptions: agricultural production will continue to grow but at a slower rate of 1.5 to 1.8 percent. The historically high agricultural growth rates of 2.3 to 3.0 percent during the 1960s and 1970s will not be replicated. Two factors contribute to the slower rate of growth in agricultural production: a slowing of world population growth and, as countries reach an adequate level of calories per capita for the population, a curtailment of further demand for increased food energy. In addition, into the next century, the real price of agricultural commodities will remain constant or decline.

Table 12.1 summarizes the projections for the availability of kilocalories per capita to 2010 or 2020. Worldwide, food energy per person is expected to increase in both developed and developing countries. But here again, progress will be uneven. Despite adequate food supplies globally, serious food deficits will continue to exist in sub-Saharan Africa and South Asia. The data in Table 12.1 indicate it is most likely that in 2010 the regions of sub-Saharan Africa and South Asia will still have per capita food supplies of 2,170 and 2,450 kilocalories—levels that are grossly inadequate for health and well-being. A similar bleak picture emerges for these two areas for 2020 and beyond (Rosegrant, Agcaoili-Sombilla, and Perez 1995). Food insecurity will grow in both of these regions over the next 20 years, resulting in either a higher prevalence of hunger and malnutrition or a higher requirement for food imports and/or food aid.

Table 12.1. Kilocalories per capita per day

	FAO[a]			IFPRI[b]	
	1969/71	1988/90	2010	1990	2020
World	2,430	2,700	2,860	2,773	2,895
Developed countries	3,200	3,400	3,470	3,353	3,532
Developing countries	2,120	2,470	2,730	2,500	2,821
Sub-Saharan Africa	2,140	2,100	2,170	2,053	2,135
South Asia	2,040	2,220	2,450	2,297	2,640
Latin America and Caribbean	2,500	2,690	2,950	2,722	3,026

[a]Alexandratos 1995.
[b]Rosegrant, Agcaoili-Sombilla, and Perez 1995.

The increased demand for food aid comes at a time when resources devoted to food aid are declining. There continues to be a disconnect between food aid resources and actual need. Over the past 15 years, food aid availability has averaged about 70 to 80 percent of need (USDA 1996). If worldwide budgets continue at 1995 levels, the gap between need and resources will grow.

A number of factors have contributed to the insufficiency of food aid budgets. The magnitude and frequency of crises requiring food aid has increased dramatically; the number of people requiring food aid jumped from 44 million in 1985 to 175 million in 1993 (Webb 1995). Humanitarian efforts have been taking a larger share of food aid budgets. This leaves fewer food aid resources to bridge the chronic gap between food production and consumption needs in developing countries.

In addition, the problem of increased demand for food aid resources is compounded by diminishing budgets of donor nations and policies stressing more market-oriented strategies—"trade not aid" has been used to summarize this sentiment.

The worldwide food supply situation projected into the next century looks relatively positive. However, each of the projections for this scenario is based on a very specific set of key assumptions. A fundamental assumption for this positive growth rate in food per capita is that investment in agricultural research and infrastructure will continue at the level of the late 1980s and 1990s (Alexandratos 1995; Rosegrant, Agcaoili-Sombilla, and Perez 1995). As a result of this investment and improved technology, food production will continue to grow at a rate about equal to population growth. Population is expected to grow at the projected UN median rate; per capita incomes will continue to rise; and the trend toward increased urbanization will continue. All of these factors will result in food demand nearly doubling over the next 30 years. If any of these assumptions were to change, the projections of food availability shown in Table 12.1 would also change.

Household Food Security: At-risk Groups

The availability of adequate global and national food supplies is one important link to ensuring household-level food security. However, even where countries have achieved national-level food security, there are typically pockets of the population with inadequate access to food (FAO 1995).

The links among income generation, poverty alleviation, and improved food security are clear. Strategies that increase the incomes of the poor will result in significant reductions in food insecurity and hunger. However, it may take quite large increases in income to bridge the food security gap for low-income households and individuals (Von Braun and Kennedy 1994). In the short to medium term, interventions targeted at low-income households may be needed to buffer vulnerable households against hunger and malnutrition.

In many countries, however, tensions continue between policymakers who favor the use of broad-based economic policies to lift people out of poverty and those who favor more targeted interventions to improve food security and nutrition. However, the reality is that, typically, both types of policies and programs are needed. Even in countries with overall adequate food supplies, pockets of the population—mainly the poor—are at risk of food insecurity and inadequate nutrition. A judicious mix of income-generating policies with programs that provide a nutrition safety net meets the long-term goal of lifting people out of poverty while protecting the health of at-risk groups in the short to medium term.

A good example is the network of food assistance and nutrition programs in the United States. Programs such as the U.S. Food Stamp Program, the Women, Infants, and Children Program (WIC), and school feeding programs were initiated in response to documented problems of underconsumption and undernutrition (Kennedy, Morris, and Lucas 1996).

Results from a number of surveys conducted in the late 1960s and early 1970s indicate that problems of inadequate diets, growth deficits, poor neonatal outcomes, and anemia were significantly more common among low-income segments of the population. Since the implementation and nationwide expansion of the U.S. Food Stamp Program, the WIC supplemental feeding program, and other food assistance programs, the gap between the diets of low-income households and other households has narrowed (Kennedy, Morris, and Lucas 1996). The improvement in energy and nutrient intake in the diets of low-income households has been more marked than changes in the consumption of other income groups. Similarly, nutritional status in the United States as measured by growth, low birth weight, and hematological status has improved. Several recent reviews have credited a part of this improvement to the U.S. food assistance and nutrition programs (Devaney and Schirm 1993; Fraker 1990).

A variety of interventions has been used to improve either household food security or individual nutritional status. The effectiveness of various food security and nutrition interventions has been mixed (Kennedy 1996). A general rule of thumb is that targeted interventions—either to the family or individual household member—are more effective in reaching the vulnerable groups than untargeted approaches (see appendix). Because governments need political support, interventions often cover a group of people larger than those that would be defined as food insecure or nutritionally needy. Although this broader approach might be politically attractive, it can be expensive. The choice of a particular approach should be dictated in large part by the nature of food security and nutrition problems.

Refugees and Displaced Persons

Refugees and displaced populations are the groups most vulnerable to acute food deficits and malnutrition (ACC/SCN 1994). Civil war and political upheaval have replaced natural disasters as the most common causes of the increasing refugee and displaced person populations. As shown in Table 12.2, in 1994, these groups together accounted for more than 40 million people worldwide. Of the 40 million, 18.5 million were refugees or displaced persons in sub-Saharan Africa, adding to the already unfavorable food security environment in the region.

The United Nations' estimates indicate that food availability for refugees—primarily from food aid—ranges on average from 1,500 to 2,000 kilocalories per person per day (ACC/SCN 1994). The range of food availability is wide, with particular concern for individuals surviving on fewer than 1,500 calories per day. A recent report from the Committee on International Nutrition recommended a minimum ration allotment of 2,100 kilocalories in order to provide sufficient food energy for reasonable physical activity (IOM 1995). Clearly this recommendation is well above the caloric ration generally available for refugees and displaced persons and adds to the already looming food aid gap.

Table 12.2. Numbers of refugees and displaced people by region

Region	Refugees	Displaced
	(millions)	
Asia	5.2	N.A.
Africa[a]	6.4	10.1
Latin America	0.1	1.7[b]
Europe	2.6	3.4
North America	1.3	N.A.
Oceania	0.1	N.A.
Former USSR	0.6	1.6
TOTAL	16.3	N.A. but > 25

[a]Refugee Nutrition Informative System No. 3, Jan. 1994, total = 16.5m; 6.4m is United Nations High Commissioner for Refugees (UNHCR) figure for refugees; 10.1m displaced calculated by subtraction; UNHCR figure includes 0.23m in N. Africa, mainly Algeria.
[b]From United States Committee for Refugees 1994.
SOURCE: ACC/SCN 1994.

The Centers for Disease Control (CDC 1992) have documented mortality rates among refugees 80 times higher than normal in one area of Somalia. Outbreaks of scurvy, pellagra, and beriberi have occurred among populations of refugees in camps where diets were severely limited in variety (ACC/SCN 1992). Clearly the mortality, morbidity, and nutrition statistics of refugees and displaced persons speak for themselves.

The total food aid needed to maintain consumption for refugees is projected to be 15 million tons in 1996, increasing to 27 million tons by 2005 (USDA 1996). More than twice this amount would be needed if the consumption target were to meet generally agreed upon minimum energy needs. This is at a time when food aid resources are declining (Shapouri and Missiaen 1995).

All the recent projections of food aid needs into the twenty-first century assume that emergencies will continue. Thus far, history has borne out this assumption. If trends continue, the 20 percent of food aid worldwide used for acute situations would grow to 30 percent of total supplies (USDA 1996). However, at a time when acute food aid needs will be growing, there will be a concurrent rise in chronic food aid needs driven by population growth, slow growth in food production, and sluggish economic growth in many countries.

Nutrition Security

The worldwide nutrition situation parallels the global food security picture. As shown in Table 12.3, for developing countries as a whole, the number of chronically undernourished people based on energy deficits decreased from 36 percent during the period from 1969 to 1971 to 20 percent during 1988 to 1990. However, the improvement in reducing hunger has varied by region; in sub-Saharan Africa, the percentage of the population who are calorie deficient increased from 35 to 37 percent during this same period. The worsening food security situation in sub-Saharan Africa reflects a combination of negative growth in per capita food supplies and declining income.

Table 12.3. Estimates of prevalence of undernutrition

	FAO[a] Chronic Undernutrition[b]			SCN Chronic Undernutrition[b]		SCN Underweight Children[c]		
	1969/71	1988/90	2010	1974/76	1988/90	1985	1990	1995[a]
	(percent)			(percent)		(percent)		
93 Developing countries	36	20	11	33	20	33.8	30.4	30.9
Sub-Saharan Africa	35	37	32			29.2	28.7	31.2
South Asia	34	24	12			55.2	50.1	50.6
Latin America & Caribbean	19	13	6			14.0[d]	12.7	12.2

[a] Alexandratos 1995.
[b] Based on calories per capita.
[c] ACC/SCN 1996.
[d] Middle America and the Caribbean.

Malnutrition rates in children under five years of age declined from 33.8 percent to 30.4 percent during the same period, 1985 to 1990. However, the rate of decrease in malnutrition rates is significantly less than the rate of improvement in energy intakes, reflecting, in large part, the complexity of factors influencing nutrition security. Improved food intake for children is a necessary but not sufficient condition for improving nutritional status (Von Braun and Kennedy 1994).

Unfortunately, preliminary estimates of malnutrition rates from 1990 to 1995 indicate that these rates in children will be stagnant or increase in many developing countries (ACC/SCN 1996). The downward trend in malnutrition noted for 1985 to 1990 is noticeably reversed in sub-Saharan Africa for 1990 to 1995. For seven of eight countries surveyed in the region, a deteriorating nutritional status was apparent during this period (ACC/SCN 1996).

The nutrition security situation in both sub-Saharan Africa and South Asia appears to be worse than was projected in the 1994 World Nutrition Situation (ACC/SCN 1994). Nutrition has improved in Southeast Asia, Latin America, and the Caribbean; however, on average, the estimated number of malnourished children throughout the developing world is rising. At the current rate, the world will not meet the goal set by the International Conference on Nutrition of reducing malnutrition by one-half by 2000. It is important to note that these aggregate statistics on malnutrition rates do not fully incorporate the level of malnutrition among the millions of refugees and displaced persons.

The nutrition security situation appears marginally more promising between 1995 and 2020 (Garcia 1995); each region of the world, with the notable exception of sub-Saharan Africa, will experience a reduction in the absolute number of underweight children by 2020. Malnutrition and food insecurity are likely to remain rampant in most countries in sub-Saharan Africa.

Unfortunately, nutrition security has been a highly visible issue only in times of famine and natural disaster. Once the acute food shortages have subsided, concern about chronic nutrition receives less attention. Food insecurity and nutrition insecurity will continue to be chronic problems that affect large portions of developing countries' populations. Even where national-level food supplies meet requirements, there are typically segments of the population who are food insecure and nutrition insecure. This is true both in developed and developing countries. The subsets of the population who are food insecure are disproportionately represented among the poor.

In addition, the problem of food insecurity has been underestimated. Until recently, research has concentrated on dietary energy as a proxy for overall dietary nutrition quality. A multicountry analysis of consumption patterns of children revealed that an adequate calorie intake in preschool children was not automatically accompanied by an adequate consumption of micronutrients (Kennedy and Payongayong 1991). Deficiencies of vitamin A, iron, and iodine affect a staggering proportion of the world's population. The World Health Organization (WHO) estimates that 40 million children worldwide suffer from vitamin A deficiency and approximately 350,000 infants and children become blind each year owing to vitamin A deficiency (WHO 1989). Iron deficiency is even more prevalent, affecting more than 40 percent of women and children in developing countries;

iodine deficiency disorders are significant public health problems in at least 90 developing countries with approximately 1 billion people at risk (WHO 1989). Only recently has the world focused on this problem of "hidden hunger." As one former World Food Prize laureate noted, "It is a sad fact that shameful and tragic as is the occurrence of famine in today's world, its economic, social and individual significance pales beside the tragedy of hidden hunger that afflicts a majority of the populations of most developing countries" (Scrimshaw 1993). Strategies to deal with food security both from a quantity and a quality perspective need to be identified for the general population as well as for higher risk groups.

Policy Implications for the Twenty-first Century

Great progress has been made since the 1950s in dealing with problems of hunger and malnutrition, in large part because of agricultural research that has resulted in improved agricultural technologies. Technological change in agriculture has markedly reduced poverty in Asia; the effects have been both direct, through increased agricultural employment and higher farm incomes, and indirect through lower food prices and more nonagricultural employment stimulated by increased farmers' incomes. However, the job is not done.

Despite such breakthroughs, the data presented in this chapter indicate that food insecurity and nutrition insecurity are and will continue to be a problem for a large part of the world's people, particularly in sub-Saharan Africa and South Asia. The relatively positive long-term projections for global food supplies emphasize the essential role of increased agricultural productivity in expanding food supplies. These productivity increases will only materialize if public and private sector investments in agricultural research continue. There clearly is a role for both international and national institutions.

Sustaining the gains in agricultural productivity and extending agricultural technologies into untouched areas will depend on continued support for agricultural research. This need occurs at a time when support for both international and national agricultural research centers is plateauing or declining.

The net effect of the technological innovations in the 1970s and 1980s was that, for developing countries as a whole, food supplies increased more rapidly than population growth. At the global level, growth rates in crop yields are declining, yet there is less scope to increase the land area cropped in order to compensate for falling production. Future gains in agricultural productivity will need to come from increases in crop yield per unit of land.

Technological change in agriculture resulting from agricultural research will continue to be the most promising way to increase food supplies. In this second wave of the Green Revolution, however, some agricultural research should be given more priority, including sustainable agriculture, biotechnology, crop diversification, identification of environmentally sound multicropping systems, and technologies that reduce postharvest technology losses.

Projections suggest that the private sector will play a more prominent role in research and development in agriculture. However, an ongoing role for public sector investment in both national and multilateral research will continue, in large part because there are certain areas of research that are unattractive to the private sector. A recent review by the U.S. Department of Agriculture's Economic Research Service indicates that public sector funding will be needed for more basic research where private sector incentives are weak and for research that informs public and private decisionmaking (Fuglie et al. 1996).

Clearly the absence of technical solutions is not the only constraint to improving food and nutrition security. Much more attention needs to be focused on the links between agricultural research at the international level and the means of translating this research into public policy at the national and community levels. The continuing failure of researchers to understand how governments, communities, and households react to innovations in agriculture has been a major impediment to the adoption and success of potentially useful approaches for improving food security and nutrition status. This is not a problem unique to food security and nutrition, but it nonetheless needs to be addressed.

In addition to increased agricultural production, an essential ingredient for improved food security is income growth. Increased income leads to a growth in effective demand for food. In countries where agriculture is a large share of the economy, agricultural growth is highly correlated with overall economic growth. A recent review summarizing the experience of the past 20 years found that developing countries with the most rapid agricultural growth were the countries that also showed the most rapid economic growth (Pandya-Lorch 1994). Clearly, this is another reason for investing in agricultural research that will lead to growth in the agricultural sector.

Countries must continue to pursue policies that increase the incomes of the poor. Since in many developing countries the poor are located disproportionately in the rural areas, growth in the agricultural sector is one of the most promising ways of raising the income of rural households and of simultaneously keeping food prices low by increasing agricultural production. However, even where national macroeconomic policies have been successful in the short to medium term, there often is a need for a social safety net to protect the food security and nutrition of vulnerable groups. Indeed, even in countries where national-level food supplies are adequate, there is still a need for a nutrition safety net to protect households and individuals who do not have access to the food necessary for a healthy and productive life.

Food security, however, is only one of the essential links for nutrition security. Many policymakers assume that the general economic growth resulting from macroeconomic policies, including growth in the agricultural sector, will automatically contribute to improved nutrition. One of the presumed links is that increased national income leads to increased allocation of this income to investments in primary health care and other social services. However, basic services such as access to safe water and investment in health infrastructure do not show a close relationship to national per capita income levels (Von Braun 1991). Thus, we cannot assume that agricultural growth will necessarily translate into improvements in the health and sanitation environment. Deliberate policies aimed at reducing malnutrition are needed in tandem with policies aimed at increasing national and household income. This needs to be done in a way that does not place an undue burden on agricultural policies and programs.

Policies that have been effective in addressing chronic food insecurity problems have few links to the unanticipated acute food shortage situations created by displaced persons and refugees. Most estimates for the twenty-first century suggest that acute food shortages will occur with increasing regularity. Food aid will continue to be an essential requirement for dealing with these disaster situations.

The overall outlook for food supply and food demand in the twenty-first century appears relatively good. But this overall positive picture masks dramatic disparities in food availability in certain regions of the world. For food-insecure countries, short-term assistance, either in the form of assistance for commercial food imports and/or food aid, will likely continue.

The dramatic gains in reducing hunger and malnutrition need to be continued; one of the biggest mistakes at this juncture would be for policymakers and donors to assume that the war to end hunger and malnutrition is over. National and international collaborative efforts need to continue to enhance the gains that have been made.

Appendix

An assessment of food assistance programs

Program	Effectiveness	Constraints
Consumer food price subsidies	There is some evidence that subsidies improve family caloric consumption, but little evidence to suggest that subsidies are able to alleviate preschooler or maternal malnutrition. They are most effective as preventive strategy for improving nutrition.	Subsidies are difficult to implement on a small scale and expensive to implement on a large scale. They are administratively difficult to implement in rural areas. They are most cost-effective when combined with some type of targeting—either to lowest income groups or by use of self-targeting food.
Food stamps	Like subsidies, there is some evidence that food stamps can increase family nutrient intake, but no evidence to date that food stamps are effective in improving maternal or preschooler nutritional status. Focus is preventive rather than therapeutic.	Food stamps are feasible only where households rely on the marketplace for food purchases; in this sense they are prone to urban bias.
	U.S. program has consistent evidence that FSP increases food expenditures and nutrient intake.	In the U.S. urban bias less apparent.
Food-for-work	Information on nutritional effectiveness is limited. Given the focus of most programs (1–3 months participation), it is most effective in alleviating seasonal fluctuations in consumption.	Most programs rely heavily on food aid.
Supplementary feeding	As these programs have been typically operated, they are not very effective in improving preschooler malnutrition. They are most effective when targeted to high-risk individuals. Programs that offer a small ration (200–300 calories) to a large number of people are unlikely to show a measurable impact on growth.	The level of supplementation provided has not taken into account leakage to nontarget group individuals. As a result, net calories consumed by a child are not enough to cover the energy gap and/or improve growth. Also, programs are administratively intensive, requiring moderate amount of infrastructure and logistical support.

(continued)

Program	Effectiveness	Constraints
Supplementary feeding (continued)	WIC program is an exception; strong evidence from a series of studies that WIC participation improves neonatal outcome including decreasing low birth weight and infant mortality. Studies also document improvements in children's growth and/or hematological status.	WIC program capitalizes on strong health infrastructure. Longer participation in WIC equated with more positive impact. Some programs are more effective than others in encouraging early and frequent participation.
Integrated health/nutrition	An appropriate mix of health/ nutrition services is effective in improving maternal and child health. Successful projects have targeted services to high-risk persons, used supplementary feeding selectively, and tailored program components to individual needs.	Program usually requires some health infrastructure and is very labor-intensive.
Formulated foods	Only limited success in improving nutritional status of preschoolers has been observed.	Cost is a primary barrier for commercially available weaning foods. Low consumer acceptability has also limited use of these foods.
Home gardens	Some evidence suggests an impact on increasing micronutrient intake, but the effect on increasing macronutrient consumption appears limited.	Land and labor is insufficient for cultivation of home gardens by the most nutritionally needy families.

SOURCE: Kennedy 1996. (Reprinted with permission)

References

Administrative Committee on Coordination, Subcommittee on Nutrition (ACC/SCN). 1992. *Second Report on the World Nutrition Situation.* Geneva: ACC/SCN.

_____. 1994. *Update on the Nutrition Situation, 1994.* Geneva: United Nations.

_____. 1996. *World Nutrition Report, Update, Preliminary Estimates.* Geneva: ACC/SCN.

Alexandratos, Nikos, ed. 1995. *World Agriculture: Towards 2010, An FAO Study.* Rome: Food and Agriculture Organization.

Centers for Disease Control (CDC). 1992. Nutritional Monitoring in Somalia. Unpublished paper.

Chen, L. C., E. Huq, and S. D'Souza. 1981. Sex Bias in the Family Allocation of Food and Health Care in Rural Bangladesh. *Population and Development Review* 7(1):55–70.

Dasgupta, B. 1977. *The New Agrarian Technology in India.* New Delhi: MacMillan.

Devaney, B., and A. Schirm. 1993. *Infant Mortality Among Medicaid Newborns in 5 States: The Effects of Prenatal WIC Participation.* Princeton: Mathematica Policy Research.

Food and Agriculture Organization (FAO). 1995. *The State of Food and Agriculture: Agricultural Trade: Entering a New Era?* Rome: FAO.

Fraker, T. 1990. *The Effects of Food Stamps on Food Consumption: A Review of the Literature.* Washington, D.C.: Mathematica Policy Research.

Fuglie, K., N. Ballenger, K. Day, C. Klotz, M. Ollinger, J. Reilly, U. Vasavada, and Y. Yee. 1996. *Agricultural Research and Development, Public, and Private Investments under Alternative Markets and Institutions.* Agricultural Economic Report Number 735. Washington, D.C.: United States Department of Agriculture, Economic Research Service.

Garcia, Marito. 1995. *Malnutrition and Food Insecurity Projections, 2020.* IFPRI 2020 Brief 6. Washington, D.C.: International Food Policy Research Institute.

Garcia, M., and P. Pinstrup-Andersen. 1987. *The Pilot Food Price Subsidy Scheme in the Philippines: Its Impact on Income, Food Consumption, and Nutritional Status.* Research Report 61. Washington, D.C.: International Food Policy Research Institute.

Haaga, J., and J. Mason. 1987. Food Distribution Within the Family: Evidence and Implications for Research and Programs. *Food Policy* 12(2):146–60.

Institute of Medicine (IOM). 1995. *Estimated Mean Per Capita Energy Requirements for Planning Emergency Food Aid Rations.* Washington, D.C.: National Academy of Science, Committee on International Nutrition.

Juster, F. T., and F. P. Stafford. 1991. The Allocation of Time: Empirical Findings, Behavioral Models, and Problems of Measurement. *Journal of Economic Literature* 39:471–522.

Kennedy, E. 1996. Intervention Strategies for Undernutrition. Chapter 6 in *Nutrition Policy in Public Health*, edited by Felix Bronner. New York: Springer Publishing Company.

Kennedy, E., and H. Bouis. 1993. *Linkages between Agriculture and Nutrition: Implications for Policy and Research*. Washington, D.C.: International Food Policy Research Institute.

Kennedy, E., and L. Haddad. 1992. Food Security and Nutrition 1971–1991: Lessons Learned and Future Priorities. *Food Policy* 17(1):2–6.

Kennedy, E., P. Morris, and R. Lucas. 1996. Welfare Reform and Nutrition Programs: Contemporary Budget and Policy Realities. *Journal of Nutrition Education* 28(2):67–70.

Kennedy, E., and E. Payongayong. 1991. *Patterns of Macronutrient and Micronutrient Consumption and Implications for Monitoring and Evaluation*. Report submitted to the Agency for International Development, Office of Nutrition. Washington, D.C.: AID.

Kennedy, E., and P. Peters. 1992. Household Food Security and Child Nutrition: The Interaction of Income and Gender of Household Head. *World Development* 20(8):1077–85.

Kennedy, E., and P. Pinstrup-Andersen. 1983. *Nutrition Related Policies and Programs: Past Performances and Research Needs*. Washington, D.C.: International Food Policy Research Institute.

Klein, Bruce. 1996. Food Security and Hunger Measures: Promising Future for State and Local Household Surveys. *Family Economics and Nutrition Review* 9(4):31–7.

Leslie, J. 1989. Women's Time: A Factor in the Use of Child Survival Technologies? *Health Policy and Planning* 4(1):1–16.

McGuire, J., and B. Popkin. 1989. Beating the Zero-sum Game: Women and Nutrition in the Third World, Part 1. *Food and Nutrition Bulletin* 11(4):38–63.

Pandya-Lorch, R. 1994. *Economic Growth and Development*. 2020 Brief No. 1. Washington, D.C.: International Food Policy Research Institute.

Pinstrup-Andersen, P., and P. B. R. Hazell. 1985. The Impact of the Green Revolution and Prospects for the Future. *Food Review International* 1(1):1–25.

Piwoz, E. G., and F. Viteri. 1985. Studying Health and Nutrition Behavior by Examining Intra-household Resource Distribution and the Role of Women in These Processes. *Food and Nutrition Bulletin* 7(4):1–35.

Rosegrant, M., M. Agcaoili-Sombilla, and N. Perez. 1995. *Global Food Projections to 2020: Implications for Investment*. Food, Agriculture, and the Environment Discussion Paper 5. Washington, D.C.: International Food Policy Research Institute.

Scrimshaw, N. 1993. The Challenge of Global Malnutrition to the Food Industry. *Food Technology* (February) 1993:66.

Shapouri, S., and M. Missiaen. 1995. Shortfalls in International Food Aid Expected. *Food Review* 18(3):44.

Svedberg, P. 1990. Undernutrition in Sub-Saharan Africa: Is There a Gender Bias? *Journal of Development Studies* 26(3):469–86.

Tucker, K., and D. Sanjur. 1988. Maternal Employment and Child Nutrition in Panama. *Social Science and Medicine* 26(6):605–12.

United States Department of Agriculture (USDA). 1996. *The U.S. Contribution to World Food Security. The United States' Position Paper for the World Food Summit.* Washington, D.C.: U.S. Department of Agriculture.

Von Braun, J. 1991. The Links Between Agricultural Growth, Environmental Degradation, and Nutrition and Health. In *Agricultural Sustainability, Growth, and Poverty Alleviation: Issues and Policies,* edited by S. Vosti, T. Reardon, and W. von Urff. Proceedings of an International Conference held 23–27 September, Feldafing, Germany. Feldafing, Germany: Deutsche Stiftung Für Internationale Entwicklung (DSE), 73.

Von Braun, J., and E. Kennedy, eds. 1994. *Agricultural Commercialization, Economic Development and Nutrition.* Baltimore: Johns Hopkins University Press.

Vosti, S. A., and J. Witcover. 1990. Income Sources of the Rural Poor: The Case of the Zona da Mata, Minas Gerais, Brazil. In *Income Sources of Malnourished People in Rural Areas: Microlevel Information and Policy Implications,* edited by J. Von Braun and R. Pandya-Lorch. Working Papers on Commercialization of Agriculture and Nutrition No. 5. Washington, D.C.: International Food Policy Research Institute, 47–68.

Webb, P. 1995. *A Time of Plenty, a World of Need: The Role of Food Aid in 2020.* Washington, D.C.: International Food Policy Research Institute.

World Bank. 1986. *Poverty and Hunger: Issues and Options for Food Security in Developing Countries.* A World Bank Policy Study. Washington, D.C.: The World Bank.

World Health Organization (WHO). 1989. *Global Nutrition Status Update, 1989.* Geneva: WHO.

Discussant ♦ *Edward F. Knipling*

Areawide Insect Pest Management by Biological Means

To meet the problem of food security—the thrust of this symposium—agriculture worldwide must become more efficient in the production, distribution, and utilization of food. As Eileen Kennedy told us, there are many factors that contribute to food shortages. I will limit my remarks to a discussion of how crop-damaging insect pests might be controlled in more effective, efficient, and environmentally acceptable ways.

Science achieved a breakthrough in the discovery and development of many highly effective synthetic chemical insecticides that make it possible to control most of the insect pests that attack agricultural production and the comfort and health of people. These developments during the past 50 years have been of great benefit to humankind throughout the world.

But the intensive use of these synthetic chemicals, most of which have broad-spectrum activity, has been a matter of grave concern to many people because of the potential hazards these chemicals create to food and the environment.

Therefore, pest management scientists throughout the world began intensive investigations to develop ways to control insect pests that do not create such hazards. Many scientists concentrated their efforts on safer ways to use the insecticides that were available, while others focused their efforts on alternative but safer ways to control insects. Renewed efforts were made to develop various kinds of insect attractants and to develop resistant host plants. Special attention was given to biological control procedures, such as the use of insect microbial agents, parasites, and predators. Scientists also explored new approaches such as the sterile insect technique.

The Sterile Insect Technique

The spectacular success of the sterile insect technique for the eradication and management of the screwworm, a serious animal parasite, was achieved in the late fifties and sixties, and led to considerable research to develop this new insect control procedure for other important insects. The technique proved to be effective for eradicating or preventing the spread of the medfly and other fruit flies in various parts of the world. Scientists in Japan eliminated the melon fly from Okinawa. The technique is being used in Africa to eliminate tsetse flies after high populations are first reduced by the use of insecticides. The technique is promising for other insect pests. But when viewed from a broad perspective, this environmentally desirable control procedure thus far has not been developed for its full potential against a wide range of insect pest problems.

Integrated Pest Management versus Alternative Control Methods

The same can be said for other alternative control methods. Biologists have obtained a wealth of new information on insect parasites, predators, and insect pathogens. Research on insect attractants has been outstanding. But the use of broad-spectrum insecticides has continued to dominate insect pest control procedures for 50 years. Safer and more judicious use of insecticides has come into being by a supervised system of management referred to as Integrated Pest Management (IPM). But the system that has evolved is largely a defensive-reactive procedure that is used on a farm-by-farm and crop-by-crop basis at the discretion of individual farmers. And, this procedure has fundamental weaknesses. It gives the more dynamic and persistent insect species every opportunity to reproduce successfully until they reach damaging population levels.

Unfortunately, most of the more promising and environmentally acceptable alternative procedures are either impractical to use after the pest populations have reached high levels, or they do not give dependable results because of their slow action or insect movement factors. Thus, for the most part, the advances that have been made on the more economically and ecologically desirable ways to control insects have remained largely stagnated and unapplied.

If a list were made of the insects that were of major concern to agriculture before the new insecticides came into being and a similar list were drawn up for today, most of the same insects would be on both lists. In effect, relatively little progress has been made in reducing the threat that insects pose for agriculture in the United States. The current list would not include the screwworm and the boll weevil in the southeastern United States. Significantly, these are the only species that have been subjected to fully organized and coordinated areawide management programs.

Areawide Preventive Control Procedures

I have long had the view that preventive measures would offer more rational ways to deal with many of our major insect pest problems if practical and safe methods of regulating total insect populations could be developed. The sterility technique offered one possibility, but its density-dependent suppression action and the necessity to use the technique on an areawide basis were regarded by many scientists as serious limitations. The pest management community, in general, and many of the agricultural and environmental leaders in both the public and private sectors were unable to see the necessity of undertaking large-scale areawide preventive programs as long as farmers have the means of dealing with most insect pest problems when they occur by the use of insecticides. Thus, the concept of areawide preventive control procedures for the most part has not been accepted. Nevertheless, I have continued to analyze the dynamics of many important insect pests and have made appraisals of the suppression characteristics of various methods of insect control when used alone and when integrated. The results of these investigations have convinced me that we already have or can readily develop all the basic technology needed to manage populations of many of the major insect pests rigidly, on an areawide basis, by environmentally benign and economically sound procedures. I have references to such persistent and damaging pests as the corn earworm, European corn borer, sugar corn borer, boll weevil , Colorado potato beetle, codling moth, tropical fruit flies, and many other species that have cost agriculture billions of dollars and have required the application of hundreds of millions of pounds of ecologically disruptive insecticides in our agro-ecosystems each year for 50 years.

Advantages of Sterile Insect and Parasite Augmentation Techniques

The sterility and the parasite augmentation techniques, in particular, offer outstanding opportunities for managing insect pest populations on an areawide basis at low costs. The two techniques have a number of similar suppression characteristics. They both involve the use of mobile biological organisms, which virtually rules out their practical use on a small scale in unisolated areas. But this is one of the most desirable characteristics when used on an areawide basis. Both techniques have density-dependent suppressive actions. This characteristic limits their usefulness when pest populations have already reached their damage levels, but makes them highly effective and practical for maintaining pest populations below significant damage levels. This ability is the very objective of preventive pest management systems. Most importantly, both techniques are virtually target-pest specific. To achieve pest management by pest-specific means has been the primary goal of pest management scientists since the new broad-spectrum insecticides came into being; it has also been the hope of those concerned over the environmental consequences of nonselective pest control methods. Also of major importance from an economic standpoint is the fact that when fully perfected, the proactive areawide management system based on the use of parasites and sterile insects should achieve nearly complete control of many major insect pests, at annual costs that are unlikely to exceed 10 percent of the losses the pests cause under current management procedures.

Outlook for the Future

The time will come—and very soon—when agriculture worldwide will be faced with the monumental task of producing enough food for the expanding world population. And this will have to be accomplished on diminishing agricultural lands. Nevertheless, people will—and should—demand that this be accomplished without risks to the safety of the food we eat and the safety of the environment in which we live.

More effective insect pest control will be one of the ways to meet the increased demand for food. To achieve more effective insect pest control, we must change from localized, reactive pest management procedures based largely on the use of insecticides to low-cost, areawide proactive preventive systems designed to keep total pest populations below significant damage levels. The two biological control procedures proposed will require development of efficient ways to mass-produce and distribute the needed biological organisms throughout their pest ecosystems. Fortunately, nature has already given us the basic tools. It has created hundreds of parasite species that can be used in augmentation systems and the pests themselves that can be modified genetically to bring about their own destruction. The development of efficient ways to eventually mass-produce perhaps up to 100 different kinds of insects by the hundreds of millions or even billions per year will be a major challenge for insect-rearing scientists. However, based on their past achievements in rearing such insects as the screwworm, medfly, melon fly, and even the tsetse fly, I am confident they will experience few obstacles, given appropriate resources. The investment in such research will be minimal compared with the large economic and environmental benefits that can be realized.

In conclusion, developing effective, efficient, and safe ways to control many of the world's insect pests is one of the challenges that agriculture must meet in the years ahead. Based on years of research and investigations into the principles and concepts of insect pest control procedures, it is my conviction that science already has, or can readily develop, the information and basic technologies that can accomplish this in an effective, environmentally safe, and economically sound manner.

Discussant ♦ *Jeanne Downen*

Eileen Kennedy rightly pointed out in her discussion that both food availability and access to food are important to food security, and that food security is often not equivalent to nutritional security due to intrahousehold distribution problems. She also said that food aid needs are greater than what donors will support. This is a twenty-first century reality that nongovernmental aid organizations are very concerned about. There has been a dramatic increase in wheat prices in the past year alone, which has had a negative effect on the amount of food aid available.

In the United States in particular, increased prices for food, the removal of agricultural policies that encouraged large food surpluses, and a reduction in the foreign assistance budget mean that barely one-third of the food aid needed by developing countries may be available by the year 2005. It is an irony that as early warning systems to identify potential disasters have improved, our ability to respond to them has declined.

Dr. Kennedy emphasized that "hidden hunger" or micronutrient deficiencies are serious problems. We agree that one cannot focus only on the need for sufficient calories and proteins.

Although it emphasizes the importance of agricultural research and the need for supporting funds, Kennedy's discussion does not adequately stress that agricultural research needs to focus on issues of access as well as availability. Small farmers need a menu of options appropriate to their agro-ecological zones. A few high-yielding varieties are not enough; we also need high-yielding varieties that do well with the limited inputs that poor farmers with few resources can access. Varieties that can provide more stable yields despite annual fluctuations in weather are also needed. Research agendas need to ensure that both national and international research centers include these options for small farmers to manage risk.

Although Kennedy supports investment in social services and safety nets, a more holistic approach is needed for sustainable food security, one that will look systematically at the range of constraints faced by poor households in meeting food needs and that will enable us to better target the causes of food insecurity.

Absolute poverty has three dimensions: (1) the availability of essential resources to meet basic needs; (2) the financial and other means of households and individuals to access these essential resources; and (3) the physical, social, and cultural status and position of households and individuals that influence their access. An effective poverty alleviation strategy that allows for sustainable food security must address all three dimensions.

When essential resources to meet basic needs are not readily available, and a household's means are limited, households may be forced to make difficult trade-offs in the satisfaction of different needs. For instance, if a household does not have access to water, to schools, or to cash income, its members may cut down on the amount of food they eat. But if resources for education were saved by reducing food expenditure, hunger may hamper the success of the education program.

Thus, when calculating the minimum income required for meeting basic needs, it is important to take into account these differences in availability of essential resources and means. By looking at their needs holistically using a household livelihood security approach, we can understand the linkages between problems and identify the key areas to address that affect food security. It also helps us to understand what we cannot do and to look for other ways to address needs. Household livelihood security has been defined in general terms as the adequate and sustainable access to income and other resources to enable households to meet basic needs (including adequate access to food, potable water, health facilities, educational opportunities, housing, time for community participation, and social integration).

The risk of livelihood failure determines the level of vulnerability of a household to income, food, health, and nutrition insecurity. The greater the proportion of resources devoted to food and health service acquisition, the greater the vulnerability of the household to food and nutrition insecurity. Thus, food and nutritional security are subsets of livelihood security. Food needs are important; but there are other needs or aspects of subsistence and survival that are also important to households.

Livelihood systems in many areas of the world are likely to become more structurally vulnerable due to one or a combination of factors: (1) increasing population growth outstripping local resources; (2) recurrent droughts; (3) loss of economic opportunities during transitional periods of market liberalization; and (4) complex emergencies in politically unstable areas. A number of communities are experiencing a progressive erosion of their basis of subsistence, leading to the further degradation of their natural resource base to compensate for these shortfalls. Community-level buffers against periodic income and food shortages are beginning to disappear. At the same time, the allocation of government resources to social services, food transfers, and agricultural development has been significantly affected by structural adjustment measures and by resource allocation to emergency or drought relief operations.

Recognizing that poor households are not static in their ability to make a living, a range of intervention options is needed to deal with the various circumstances that face poor populations. We must view relief, rehabilitation/mitigation, and development interventions as a continuum of related activities, not separate, discreet initiatives (such as livelihood provisioning, protection/rehabilitation, and promotion). Projects undertaken by CARE target people who are unable to meet basic needs. Of the estimated 1.3 billion people today who are unable to meet basic needs, 800 million are food insecure. CARE aims to improve household livelihood security so people can become food secure.

Nutritional status is a good indicator of household livelihood security, but it does not identify the key causal factors that prevent households from achieving adequate nutritional status. We can determine key leverage points for reducing inadequate nutrition and poverty by looking across the broad spectrum of water, health, sanitation, education, and other issues that affect the household.

Kennedy did not make explicit the existing and potential interlinking partnerships among institutions to address the problem of food insecurity. Governments, bilateral and multilateral institutions, nongovernmental organizations, community-based organizations, and the private sector can play various roles and assume different responsibilities to promote greater food security by addressing the issues of availability, access, and utilization.

Governments also face problems in the allocation of scarce resources. When funds are limited and governments are organized vertically in strong sector lines, interventions may focus on the development of selected resources only. For example, the development of primary education may compete with the restoration of basic health services. And who will be responsible for long-term safety nets if governments do not have the resources or the will to maintain them?

Finally, Kennedy's discussion hit relevant points on the relief to development continuum but did not specifically mention disaster mitigation and rehabilitation activities. A growing number of food-insecure people can be targeted through food-for-work and cash-for-work activities that will help prevent them from selling off their productive assets in times of hardship. The cost of emergency response and long-term safety nets is far higher than long-term development assistance that addresses the causal factors of poverty and food insecurity.

CHAPTER THIRTEEN

Food Processing and Distribution

♦ MODERATOR
♦ RESOURCE PAPER
♦ DISCUSSANTS

Moderator ♦ *Richard L. Hall*

This session is about what happens—what must happen—to food between harvest and consumption. Only a small portion of the world's food is produced at the time and place it is to be consumed. Food must, therefore, be stored and transported to consumers and stabilized and protected for storage and transport. Except for the cereal grains, most raw foods are not naturally stable and deteriorate rapidly after harvest or slaughter. These losses in developing countries may reach 50 percent of the food originally produced—a situation we should not tolerate and cannot afford. The solution is certainly not to compensate by producing more, with the waste of increasingly scarce resources this would entail. Proper storage and packaging protect the processed food against contamination and attack. Processing also, and very importantly, renders many foods safe, nutritious, easy to consume and digest, and attractive, when the same food unprocessed would lack these necessary qualities. Packaging not only protects, it informs. A complex organization of food control laws and agencies, public and private, has been created to ensure that processing, storage, and packaging are effectively and honestly performed. These and other postharvest issues are the subject of this session.

There are five basic categories of food processing techniques:
*   heat processing, including cooking, blanching, and retorting;
*   cold processing, including chilling and freezing;
*   dehydration;
*   chemical processing, including fermentation, salting, smoking, and additives; and
*   food irradiation.

Modern food processing consists of variations and combinations of these five basic process categories. Cooking is doubtless the oldest form of food processing. Fermentation, dehydration, salting, and smoking also predate history. Refined over centuries, the associated products have become popular ethnic identifiers—sauerkraut, Kim-chi, fermented fish sauce, soy sauce, and air-dried beef.

The first use of refrigeration may well have been by our distant ancestors who eked out an uncomfortable existence near the edges of the glacial ice sheets. Refrigeration and freezing only came into common use when electric power became widely distributed, refrigeration units and insulation became compact and efficient, and Clarence Birdseye discovered the *vastly* higher quality provided by quick-freezing.

The first great advance in food processing to date only from historical times, we owe to Napoleon. To free his armies from the need to live off the countries they invaded, he offered a prize of 12,000 gold livres to anyone who could devise a broadly useful method for preserving food. That prize was won by Nicolas Appert, the inventor of canning, and our first modern food technologist.

Only one other category of food processing is more recent. Food irradiation, preservation of food by exposing it to ionizing radiation, has been developed only since World War II. Its application has been slow because of the need to gain regulatory and consumer acceptance, and because it typically displaces existing, acceptable, and often more economical processes.

Cereal grains, and also some dehydrated food products, lend themselves to storage and distribution if they can be protected against rodents, insects, and moisture. The costs and uncertainties of moving them, however, were such that only in the last two centuries has distant distribution become feasible for any but the most expensive commodities, such as spices. For most foods, distribution beyond the community of origin had to await the methods of preservation already mentioned. The combination of these methods with protective packaging and vastly improved transportation has led to the variety and year-round availability more developed countries now enjoy. Accompanying benefits have been the reduction in the percentage of family income spent on food and greatly improved nutritional status.

Our task now is not only to continue to improve the processes we have but also to extend them to those who do not have them, and usually can currently neither afford nor implement them. This task requires far more than just food technology. To lead us in discussions of these complex and urgent problems, we have a distinguished resource paper "Food Processing and Distribution" by Colin Dennis and a panel with Robert Smith, Verghese Kurien, and Joseph Hulse.

Resource Paper ♦ *Colin Dennis*

Food Processing and Distribution

Within the next three decades the world's population will grow by nearly half, from 5.7 billion to about 8.5 billion people. Although the precise projections vary according to source, there is no doubt that future generations will face the task of feeding a much larger population. These predictions indicate that each year we have to find a way to feed an entirely new country that is roughly 1.5 times the population of the United Kingdom. This population increase provides a major challenge as well as an opportunity for the food processing industry. In addition, economic growth is changing the nature of food demand. When people have more income, one of the first things they do is improve their diets. As much as 40 percent of additional disposable income in the developing world goes to improving the family diet. This use of income has enormous implications for the food processing and distribution industry.

Much of the population growth we anticipate will occur in the developing world, including some areas facing the prospect of continued strong economic growth. In three decades, almost two of every five people in the world will reside in either India or China, and one in five will live in Africa. The developed world will see its population remain static at best and decline as a percentage of the total global population. The rate of growth in gross domestic product (GDP) will be higher in the developing world than in the developed world, largely due to strong performance among major Asia-Pacific nations.

Data provided by the Organization for Economic Cooperation and Development (OECD) show that the five largest economies in the world today are the United States, Japan, China, Germany, and France. By 2025, four of the five largest economies in the world will be Asian—China, Japan, India, and Indonesia; the fifth economy being the United States. This transition is very important to the food industry. Increasing population and higher incomes translate into demand for more meat, dairy, fish, vegetable oils, and other protein sources. Given the means and the opportunity, people prefer chicken to rice.

The OECD estimates that if China, India, and Indonesia maintain an economic growth rate of 6 percent per year, 700 million people in these three countries alone will have annual incomes equal to those of Spain today. At present only about 100 million have such a level of income. These developments indicate a stronger global market pull for food that can satisfy a rising demand for more higher-value food products, including animal proteins and the feed grains and oilseeds needed to produce them.

The Global Food Business

The sheer size of the food industry, £2,500 billion (pounds sterling) in 1995, far exceeded a number of leading consumer products and durables businesses. The size of the organized food industry, however, still represents a small fraction of the total amount of food produced and consumed in the world. In spite of its impressive size, the complexities of the business are often understated, together with the vulnerabilities and opportunities these represent. From a business perspective there is every reason for optimism about the future of the food industry. World demand for food continues to grow. It has been said that the food business is the only business that wakes every morning with a quarter of a million new potential customers, whether we do anything or not!

The leading food companies of the world operate in Europe and in the United States, and the majority of these are transnational corporations. Philip Morris in the United States and Nestle in Europe are by far the largest food companies, followed by PepsiCo, ConAgra, and Unilever. In spite of the impressive size of the major companies in the United States and Europe and the restructuring that has taken place in the past 10 years, the industry is fragmented and characterized by a large number of regional and local operations. This trend is more marked in Europe than in the United States. The strong national and ethnic preference for food compared with other consumer products represents the sum of a number of complex issues.

Leading brands of snack foods, soft drinks, and certain other beverages have found greater universal acceptance than basic meals and ingredients, which are still much more influenced by local preferences, although this relationship is beginning to change. In addition, the rapid growth of economies in the Pacific Rim countries and the changes in Eastern Europe offer enormous potential to manufacturers. However, compared with other consumer products, the nature of these opportunities and the ways and means to fulfill them are probably less well understood. Much of the processed food market in the developing economies is dominated by a large number of small and local operators, but the potential for certain popular brands and other products adapted to local tastes is significant. The major difficulties include poor data on consumer tastes and habits, the ability to upgrade ethnic products by technological and multinational innovation, and the availability of trained managers and technicians capable of interpreting and developing the opportunities.

The Main Consumer Trends

In the West there have been significant changes in the social, economic, and demographic environment. Population growth is virtually static and the percentage of older people is increasing. Households are becoming smaller with a much higher proportion of one-person households. The proportion of women in the workforce is increasing. These demographic trends are accompanied by changes in so-called "lifestyle." Formal family meals are becoming rarer as snack eating increases. There is much greater concern with health and appearance accompanied by a demand to know more about the food that we eat, a willingness to pay for quality, and an increasing concern about the environment. To this we have to add the complication of consumer, media, and government reactions to food scares, real or imagined.

The implications of these changes in lifestyle for the food industry and for food processing are varied and far reaching. There is no doubt that for food manufacturers to survive in what will be an essentially static total food market, they will have to provide consumers with choice, convenience, quality, and information. In contrast, the growth of the GDP in large parts of the developing world, the growing middle class and urban population, and the globalization of communications offer significant scope for the growth of processed food.

In terms of GDP per capita, Japanese consumers are well known for their very high purchasing power, but entry of modern processed food from the West into Japan's enormous market remains relatively small. On the other hand, the Japanese food industry has achieved impressive and rapid growth by modernizing traditional Japanese products, many of which are derived from fermentation and advanced mild preservation technologies. Japan is also at the forefront of certain leading-edge developments in creating novel foods and health-related products. There is much to be learned by the rest of the food industry from Japan.

Although in GDP per capita terms Eastern Europe and the rest of the world may appear to be low, this situation grossly underestimates the opportunities these markets represent in terms of absolute population and GDP. In addition, in parts of the developing world and in the transition economies of Eastern Europe, the processed food industry is vestigial and dominated by local operators who, in many instances, use obsolete technologies that inevitably result in the high cost of processed food products as

well as high levels of wastage due to spoilage and underprocessing. This
state of affairs, when seen along with the rising proportion of the middle
class in these countries, indicates the potential opportunity. Although the
choice of technologies may be somewhat restricted, an understanding of
market dynamics, ethnic tastes and habits, and the role of local entrepreneurs in these markets will be of prime importance in future development.
These developing markets will likely grow much faster in the first two decades of the twenty-first century in terms of absolute volumes and value
than the mature markets of the West.

The Technology Challenge

Food processing can be viewed as the conversion of agricultural products
to products that have particular textural and organoleptic characteristics,
by using economically feasible methods. Traditionally, processing of the
more highly perishable agricultural products also has served to preserve
them and thus extend their availability well beyond the normal harvesting
season and indeed, in many cases, to offer an all-year-round supply. More
recently, food processing has provided greater variety, choice, convenience,
and enhanced nutritive value through added-value products.

The general development of food and drink consumption in the developed world is towards more elaborate and complex foods with a high
technological and service element. There have been five dominant forces
driving demand in the food and drink markets:

- higher quality or perceived higher quality with respect to taste
  (flavor, texture) and appearance;

- healthier foods that are "fresher," "more natural," and nutritionally improved (high fiber, low fat, fewer additives);

- products having greater convenience; meals, microwaveable
  products, chilled sandwiches, snack products, and individual
  packs;

- greater variety of products; ethnic foods and "new" flavors; and

- safer foods, primarily as related to microbial pathogens, but also
  fewer chemical contaminants such as pesticides and mycotoxins,
  as well as chemicals migrating into food from packaging
  materials.

Postharvest, processing, and distribution technology has had to be developed to meet these consumer and marketplace demands.

In general, foods can be "preserved" by any process that inhibits, destroys, or removes microorganisms and/or prevents their subsequent entry into food. In addition, it is necessary to prevent deleterious biochemical, chemical, and physicochemical changes to preserve nutritive value and organoleptic properties. Most of these deleterious changes, in common with microbial growth, require water; it is fortunate that conditions that inhibit microbial growth also retard most of the other changes. No preservation process can completely stop all changes in a food and, therefore, none can effect an infinite storage life.

Of the many techniques used to preserve food only a few rely on killing of the relevant infecting microorganisms. Thermal inactivation is still the most widely used process and is unlikely to be replaced in the foreseeable future, although irradiation is becoming more widely accepted, particularly for specific uses such as decontamination of spices. Other processes such as superclean or aseptic systems rely on physical exclusion of microorganisms after filtration or sterilization. Most procedures, however, merely rely on slowing the process of food spoilage and the growth of poisoning organisms. Of course, it is possible that, with extended storage of these foods, the death rate of microorganisms will exceed that of growth and numbers will ultimately decline. Physical procedures such as chilling, freezing, drying, concentrating, or modifying the headspace gases in the packaging (vacuum, modified atmosphere packaging, oxygen scavenging) as well as chemical techniques such as acidification, fortification (addition of alcohol), or chemical preservatives are all examples of sublethal preservation. Alternatively, acid and alcohol can be produced in the product by fermentation preservation processes.

In practice, to ensure successful food preservation irrespective of the technology used, an appropriate combination of raw materials, preparation and processing, packaging, storage, distribution, retailing, and consumer handling is necessary. Each of these aspects is interdependent in providing product shelf life and other characteristics important to the consumer (e.g., perceived quality and safety, composition, value for money, and convenience).

Process technologies for the preservation of food by heating, chilling, freezing, drying, concentrating, and a range of combination treatments will continue to exist for the foreseeable future and well into the twenty-first century. Each will have a unique role in meeting the demands of an ever increasingly critical consumer who demands safety, quality, variety, and convenience. Their relative importance will change in accordance with new technological developments and marketplace needs.

Within these generic preservation and processing technologies, there have been significant developments in the last two decades; undoubtedly, further developments and refinements will occur in the early part of the twenty-first century. These developments will be designed to meet the marketplace needs for safer, healthier, high quality, convenient added-value products. Thus, technologies that result in less severe processing are required (e.g., less severe heating, minimal overheating, minimal freeze damage), along with less use of artificial additives, more use of natural preservation systems, and lower levels of fat, salt, and sugar. These objectives can be achieved by the application of rapid and novel forms of heating or by alternative nonthermal processes involving electrical heating (ohmic, microwave, and radio frequency heating), ultrahigh hydrostatic pressure, pulsed energy processing, ultrasonication, and irradiation.

Rapid and Novel Heating Techniques

Thermally processed foods (pasteurized or commercially sterilized) constitute a substantial category of foods available to consumers throughout the world, especially for the developed economies. Thermal processing extends the shelf life of foods anywhere from a few weeks to a few years, depending on the type of food, packaging, and processing used. However, traditional in-container heat processing (canning, bottling) produces distinctive products due to the long processing times required to ensure that the slowest heating point in the container receives sufficient heat treatment to achieve commercial sterility.

The advent of continuous thermal processing via plate, tubular, and scraped-surface heat exchangers, or by direct steam injection or infusion in combination with aseptic packaging systems, has overcome the overprocessing effects of in-container sterilization. Developments in packaging technology were undoubtedly the key to the increased commercialization of ultrahigh temperature (UHT) processing and aseptic packaging. Rapid progress in the development of multilayer packaging by lamination, co-extrusion, and various molding methods led to a wide range of pouches, bags, bottles, cartons, and pots. Business frameworks based on aseptic technology continue to be of interest to the food processor for the production of ambient stable food. The UHT milk and pasteurized fruit juices, which followed soon after, were among the first commercially successful products to be sold in modern flexible, semirigid packaging with barrier materials made of thermoplastics and aluminum foil. In the last 10 years the range of aseptically packaged products has increased dramatically and now includes dairy products such as creams and custards, and wine, tomato products, soups, sauces, jellies, rice, and chocolate desserts. The major breakthrough was to achieve sufficient process control for low-acid and particulate products. However, large-particulate food materials still present a major challenge both with respect to ensuring commercial sterility in all particles and to aseptic filling. Although different experimental particulate systems have been developed, commercialization has not yet occurred, except for the ohmic heating system.

The diversity of packaging types and forms available for use in aseptic systems, together with the heat process being independent of container size, have been major advantages in the commercialization of aseptic technology. This technology is especially suitable for intermediate commodity products transported in bulk (e.g., tomato paste, fruit pieces, pulps, and purees) as well as recipe products for institutional, restaurant, and hotel catering. Thus, aseptic technology has the potential to provide a convenient and cost-effective way of distributing partially processed products for use by the global food industry.

The commercial use of extrusion cooking has increased considerably in the last 10 to 15 years and has provided new options for the processing of starch-rich or protein-rich, low-moisture food mixes. Extrusion cooking is a continuous thermal-plus-mechanical process, and the extruder can be considered as a continuous chemical reactor for the modification of biopolymers. Twin-screw extruders provide considerable flexibility in screw configurations for processing different regimes. Together with options for external heating and die designs, these processes enable applications such as the production of snacks and biscuits co-extruded with a filling, and of crisp bread, breakfast cereals, cereal bars, bread crumbs, and intermediate products (precooked flavors, starches, chemically modified starches). Twin-screw extrusion also has been used as an alternative to ethylene oxide treatment to sterilize spices. The Masterspice process subjects the dry powders to high-temperature shear and pressure. These modified extrusion processes have the potential to be extended to other products.

The most recent and major advances in thermal processing have been in the development and commercialization of electrical heating systems both directly (ohmic heating) and indirectly in the form of electromagnetic energy (microwave and radio frequency systems). These heating systems offer the advantages of more rapid and more even heating of food products that in turn, offer opportunities for improved product quality and reduced processing times.

Ohmic Heating

The ohmic heating system exploits the principle that if an electric current flows through a substance of suitable conductivity, heat is generated within the product. The extent of heating depends upon the electrical properties of the food, power input, and residence time in the heater. Many food products conduct electricity sufficiently well to be suitable for ohmic heating—fats and oils being notable exceptions. The key is that the heating is uniform throughout the volume of the product, with no heat transfer surfaces. This principle is not new; patents for simple ohmic heating devices were filed early in this century, but the inventions failed in practice because of the limitations of available electrode materials and control technology.

This principle was translated into a commercial, truly continuous heating method for pumpable food products by APV (Aluminium Pan and Vessel) in the late 1980s. Commercial ohmic heating columns are available for use in aseptic sterilization systems for low-acid foods, pasteurization of fruit and other high-acid products, preheating, and the hygienic production of pumpable particulate ready meals. Particles in suspension in these systems heat at the same rate as the carrying fluid (e.g., sauce or gravy) if the conductivity of each of the two components is the same. Thus, the ohmic system provides rapid, even heat treatment of liquid and particles with minimum heat damage. This system has the potential to offer higher levels of product safety due to minimum residence time differences between liquid and particles. There is also the reduced risk of plant fouling due to the absence of hot heat transfer surfaces, thus enabling longer run times. Low maintenance costs owing to the reduction in moving parts and the easy-to-control fast start-up and shut-down times offer further potential benefits. Ohmic heating combined with aseptic filling offers flexibility in package type and size ranging from retail to bulk packaging of intermediate or finished products for global distribution at ambient temperatures. Commercial ohmic systems have been established in Japan and Europe for low-acid ready meals and high-acid fruit products.

Microwave and Radio Frequency Heating

Microwaves and radio frequency (RF) are forms of electromagnetic energy classified within the electromagnetic spectrum. Specific frequency bands are allocated by the government for the operation of industrial microwave and RF equipment to avoid interference with communication channels. If a material of suitable composition is placed in an electrical field alternating at radio or microwave frequencies, heat is generated throughout the mass of the material by the rapid reversal of polarization of individual molecules. The selection of microwave or RF heating will depend on the physical properties of the product and the required process conditions for a particular application.

The application of microwave technology to cooking, heating, and processing of foods is rapidly becoming an integral part of food preparation and represents a technological innovation that will readily serve current needs as well as future ones. Commercial use of microwave processing has considerable potential but the current major uses are limited to tempering and drying. Microwave processes such as blanching, baking, pasteurization, and sterilization are likely to increase in the future.

In recent years there has been considerable interest in microwave heating for pasteurizing food products. Industrial applications of microwave pasteurization have occurred primarily in Japan and continental Europe, for products such as chilled pasta and chilled ready meals. The use of microwave technology for sterilizing food recently has been commercialized in Belgium for pasta-based ready meals. The major advantage of the microwave sterilization process is the high temperature, short time process needed to preserve the taste, texture, and other organoleptic properties of food, including the *al dente* quality of pasta.

The application of microwave energy to food processing has been recognized as a promising technology for several years. The major challenge has been achieving uniform heating. This type of heating is influenced by food composition and geometry, packaging geometry and composition, and by the microwave energy feed system. With an improved understanding of microwave food technology gained from the design of food products for reheating in microwave ovens, a better use of the beneficial effects of microwave heating can be obtained. Food products can now be designed to heat more uniformly in a microwave field, achieving both the required degree of microbiological inactivation and a high-quality product. Microwave processes often occur in combination with conventional heating treatments, and the most successful applications will be those that use the beneficial effects of each.

The largest application in the food industry for RF heating is in the finish drying or post-baking of biscuits and other cereal products. Other applications being developed are the drying of products such as expanded cereals and potato strips. Most recently, RF cooking and pasteurization equipment for pumpable meat products has been developed. This process involves pumping a meat product into a plastic tube placed between two electrodes shaped to give uniform heating. Excellent uniformity of heating has been demonstrated in these applications for the continuous cooking of products such as luncheon meat.

The system will revolutionize the processing of section-formed meat products by replacing time-consuming traditional batch processes with a rapid and hygienic continuous process. This system is capable of delivery reductions in production costs, combined with improved product quality, higher product yields, and reduced risk of bacterial contamination.

The various frequencies of the electromagnetic spectrum each have advantages and limitations with respect to depth of penetration, uniformity of heating, capital, and running costs. In general, electrical heating methods offer several major advantages:

- direct heating of foods with limited energy losses to the surroundings and the heat processing equipment;

- energy efficiency often in the 80 to 90 percent range;

- clean without creating excessive heat in the working environment; and

- easily controllable.

The major disadvantages are the following:

- relatively high capital cost;

- need for training personnel and modification of other parts of the manufacturing line; and

- relatively high cost of electrical energy in many countries.

Nonthermal Technologies

In addition to the novel approaches to developing more rapid and uniform (hence less severe) heating technologies, significant developments also have occurred in alternative approaches to the processing and preservation of food by using novel technologies. These developments have included the use of ultrahigh hydrostatic pressure (UHP), pulsed energy processing, and ultrasound. These approaches provide the potential for greater commercial applications in the food and drink industry in the twenty-first century.

Ultrahigh Hydrostatic Pressure Processing

Nearly a century ago the principle was established that high isostatic pressures in the range of 500 bar to 10 kbar had a lethal effect on microorganisms at room temperature. Changes in the three-dimensional structure of proteins and polysaccharides also were observed. However, it was only in the late 1980s as a result of work in Japan that interest in the industrial use of UHP in the food industry was created. This type of processing was further emphasized in 1990 when the first UHP product, fruit jam, was launched in the Japanese retail market. A year later the product range was extended to include fruit yogurt, jelly, salad dressing, fruit sauce, and citrus juice. Ultrahigh hydrostatic pressure processing is generally a batch operation, although semicontinuous systems are available for citrus juice. The food or drink is packaged in a flexible container, sealed, and the container placed in a chamber filled with a fluid. The fluid acts as a transmission medium; the mechanical pressure applied to the fluid is transmitted to the container and then to the food or drink. Due to the nature of the process, very high precision is needed in the manufacturing equipment. The food or drink being processed must be packaged so as to minimize headspace in the container. If possible, the foods should be deaerated before sealing the containers so that the pressure may be transmitted uniformly through the product. Flexible and semirigid packaging materials (plastic and aluminum foil laminates) are ideally suited for in-container UHP processing. Any package with enough flexibility to compensate compression of the food (about 12 percent at 400 MPa) can be UHP treated without suffering damage. Barrier properties (oxygen permeability, water vapor, transmission rate) and mechanical properties (ultimate tensile strength and elongation) as well as seal strength and integrity of multilayer plastic and aluminum films are not affected by UHP treatment.

Ultrahigh hydrostatic pressure processing alone is only suitable for the preservation of high-acid products because only vegetative bacterial cells are inactivated by UHP. Processes of 20 to 40 minutes at 4,000 atmospheres or a few seconds at 8,000 atmospheres at room temperature are required. Inactivation of bacterial spores requires treatment at 6,000 atmospheres for 20 to 40 minutes at 60°C.

Enzymes also are known to be very pressure resistant and cannot be completely inactivated by pressures that are industrially available. Thus, a combination of pressure and moderate heat may be suitable. If this treatment is not acceptable for food quality, enzymic degradation can be retarded by chill distribution. The UHP-pasteurized, high-acid foods such as jam, fruit juice, and fruit desserts are already commercially available in Japan and are distributed at chill temperatures to retard enzymic degradation, giving a shelf life of two months. The UHP pasteurization and sterilization of low-acid foods such as meat and fish products, dairy products, and ready meals have been suggested. In such products UHP has the potential to yield preserved foods of unique nutritional and sensory quality.

The UHP treatment also has the potential to be applied in food preparation based on its ability to change the structure of proteins and polysaccharides. Possibilities include gelation of proteins, meat tenderizing, improved starch and protein digestibility, and chocolate tempering. For example, pressure-induced protein gels are known to maintain their natural color and flavor and have greater elasticity than heat-induced gels. Technology involving UHP offers the food industry a unique opportunity to develop foods of high nutritional and sensory quality, novel texture, greater convenience, and increased shelf life. One of the challenges of the commercial applications of UHP technology in the food industry is the identification of applications for which the higher UHP processing cost (as compared with thermal processing) is justified by unique and superior food product properties.

Pulsed Energy Processing

Pulsed energy processing for food applications can be divided into three distinct approaches: pulsed electric field (PEF), pulsed light (PL), and pulsed magnetic field (PMF). Most work has explored and developed PEF and PL in relation to food processing. Although it is known that oscillating magnetic fields can be used to inactivate microorganisms in a manner similar to PEF, there is still limited data on the use of PMF to inactivate foodborne microorganisms.

Pulsed electric field technology was developed in Europe, Japan, and North America in the late 1980s and early 1990s with specific reference to milk, fruit juice, and other beverages. The lethal effect of electric pulses on microorganisms is a result of the induction of an electric potential across cell membranes causing a charge separation in the cell. When the electric potential exceeds a critical value, irreversible changes in the cell membrane result in cell death. The critical potential gradient depends on the type of cell and its size, shape, and growing and environmental conditions. Yeasts are killed by much lower electrical field strengths and energy inputs than bacteria. Vegetative bacteria are killed only above an electric field strength of 20 kV/cm. A recent observation indicates that microbial inactivation is significantly higher if pulses of alternating polarity are used as opposed to repetitive, same-polarity pulses of the same intensity, frequency, and duration.

The limited amount of data available suggests that PEF does not inactivate bacterial spores or enzymes. Thus, similar to UHP, a combination of heat and PEF would be necessary to sterilize foods and ensure enzyme inactivation. Commercial-scale equipment has been developed by Pure Pulse Technologies, Inc., in the United States in which fruit juice, milk, liquid egg, and beer have been processed (CoolPure™ Technology).

The PL system uses a pulsed power system to drive an inert gas lamp that produces intense, broadband white light pulses. The intense flashes of light have a duration of 10^{-6} to 10^{-1} of a second and an energy intensity approximately 20,000 times that of sunlight at the earth's surface. This technology is very effective in inactivating microorganisms at the surface of food, in transparent foods, and on packaging and other food contact materials. The inactivation is a result of both photothermal and photochemical mechanisms and is effective on vegetative bacteria, spores, and molds.

The PL technology is being developed commercially by Pure Pulse Technologies, Inc., in the United States (Pure Bright™ Technology) in collaboration with Tetra Laval for sterilizing packaging materials and reducing pathogens on food contact surfaces. Pure Bright Technology provides an attractive potential alternative to hydrogen peroxide for decontaminating packaging in aseptic packaging systems.

Ultrasound

High-intensity ultrasound can produce a number of changes within food, mainly through the effects of cavitation. The main effect of ultrasound on microorganisms is physical, owing to the rapid alternating pressures and cavitation. However, the energy released also can bring about chemical reactions by creating free radicals and by causing cell membranes to become more permeable and sensitized to harmful substances. Ultrasound has found several application niches in the food sector where, when combined with existing technologies, it has led to improved processing efficiency and product quality. Examples of such areas are crystallization, hydrogenation of oils, accelerated aging of alcoholic beverages, accelerated filtration, dehydration, and heat transfer.

Ultrasound also has been used in cleaning operations. It has a number of advantages over conventional cleaning systems in that ultrasound is claimed to kill microorganisms as well as remove dirt and fouling films from inaccessible places without the use of chemical detergents or expensive downtime. Ultrasound in combination with heat can significantly reduce the survival of bacteria. Further work is in progress to explore the feasibility of using such a combination in food processing.

Irradiation

Irradiation is not a new technology; its use in food processing has been approved by the Food and Agriculture Organization/World Health Organization (FAO/WHO) for more than 50 types of food since 1981. However, this technology is not widely used commercially owing largely to consumer resistance and also to economic and logistical concerns. The technology involves the use of ionizing radiation, from a variety of sources, to inactivate microorganisms and pests and retard physiological and biochemical changes (sprouting and ripening) in fruits and vegetables. The commonly used sources of radiation are gamma rays (usually cobalt 60), x-rays, and high-energy electrons from an accelerator. Each type of radiation source has benefits and drawbacks. The gamma irradiation from cobalt 60 has the greatest penetration and is most effective in treating products with a wide range of densities and packaging dimensions. However, unlike x-rays and electron beams, gamma rays cannot be turned off, they require a highly secure facility (a concrete-, lead-, or steel-lined chamber), and the cobalt source has to be lowered into a water storage pool when not in use.

Electron accelerators can deliver highly energetic particles to the food quickly, but the product thickness that can be treated is limited by the food's density. Thus, electron beams are usually used to treat low-density products in thin layers at high throughput. The use of pulsed accelerator technology to produce a high-power, high-volume x-ray irradiation food processing facility is currently being developed in the United States. X-rays have greater penetration depth (60 to 70 cm) than electrons but are capable of being produced from similar equipment and are therefore controllable.

Irradiation is considered to have minimal impact on the nutritional value of food and has the potential to be used in combination with other technologies, especially refrigeration and drying, to improve the security of our food supply. This potential has increased recently in view of the public's demand for zero tolerance of pathogens in foods and no nutrient degradation as a result of processing.

New and Emerging Drying Technologies

For many centuries food products have been dried to prevent spoilage and aid storage and distribution. The twentieth century has seen the introduction of a diverse range of mechanical drying equipment such as tray, drum, and spray dryers. The emphasis in recent years has been on methods designed to dry products while allowing maximum retention of their original flavor, color, and nutrients. There also has been a constant refining of mechanical methods, more recently in combination with other technologies such as microwaves or dehumidified process air, to minimize process times.

Current laboratory research shows the potential for applying methods such as osmotic and electrohydrodynamic discharge drying to produce high-quality dried food. Studies of emerging technologies in other industries show the potential of technology transfer for processes such as acoustic/ultrasonic and infrared drying. Heat pump dryers and steam dryers, originally applied to nonfood products such as wood and textiles, are being used increasingly for energy-efficient food drying. In addition, recent advances in mathematical modeling and the availability of computational fluid dynamic (CFD) programs are starting to allow designers of drying equipment to experiment with dryer design at the pilot level before scale-up. This approach further enhances product quality and allows more compact dryers. For example, packages are available for the design and scale-up of spray dryers. The following are some examples of the developments currently in progress.

Heat Pump Dryers

Heat pump dryers offer the ability to dry products at lower temperatures than common mechanical systems and potentially promise an economical alternative to energy-intensive methods such as vacuum freeze drying for temperature-sensitive food products, particularly granular foods. Rehydration ability, taste, and color compare favorably with those of freeze-dried products. Heat pump drying has the added advantages of maintaining higher mechanical strength within food products compared with vacuum freeze-drying, and of being considerably lower in energy consumption.

Osmotic Dehydration Processes

Osmotic dehydration processes involve water-rich foods, such as fruit, being immersed in aqueous solutions (usually highly concentrated salt or sugar solutions). This treatment causes an outward flow of water from the product, a transfer of the solute to the product, and a slight loss of the product's own solutes to the aqueous solution.

The use of osmotic drying has been restricted to a limited range of dried fruit products. In practice, osmotic dehydration is usually followed by drying in heated air to produce the final product. Osmotic drying offers the potential to improve food quality while saving energy as the process can be carried out at ambient temperatures. However, the cost of concentrated solutions into which the food is immersed and the difficulties of scale-up, including the treatment of large amounts of waste-concentrated solutions, need to be examined. Microbiological validation of osmotic processes also needs attention, particularly as there is now interest in using this technique for low-acid foods such as meat and fish.

Linking osmotic dehydration with conventional air drying, freezing, and compression with vacuum packaging has resulted in several patents on the novel technology of dehydro-freezing. This technology is used for the manufacture of non-freeze (NF) and compressed non-freeze (CNF) products. Patents for this process were developed by Byron Agriculture Co. Pty. Ltd., Australia; several licenses have been taken out for different products in Europe, the United States, Australia, New Zealand, and Sweden.

The patents describe methods used to partially dehydrate, compress, and quick-freeze a wide range of vegetables, fruits, herbs, and spices. This technology has been developed to reduce the high energy costs associated with the freezing, packaging, transporting, warehousing, distribution, retail storing, and displaying of frozen foods.

The NF vegetables have been partially dehydrated to such a level that the products, when cooled to $-20°C$, do not freeze and have a water activity below 0.9. Time for dehydration is short, and consequently costs are extremely low; high-quality NF vegetable products with rapid cooking times and very good flavor, color, and texture are obtained.

Partially dehydrated NF vegetables can be readily compressed either by exerting an external pressure or by vacuum packaging. The vegetables remain dry and pliable to touch and can be packed into a range of pack sizes. Energy costs are greatly reduced as are packaging, transportation, and storage costs. One kilogram of CNF sliced beans when rehydrated occupies the space of 12 kg of normal quick-frozen beans. Recent work also has extended this technology to new product areas such as meat and fish.

Microwave Technology

Microwave technology enables rapid drying of food products because of its ability to remove moisture by generating heat deep within the product with selective absorption of microwave energy by the bipolar water molecules in the food. This technology is in contrast to conventional drying processes in which moisture evaporates at the outer surface of the product with the dry layer slowly progressing towards the center. In practice, microwave energy is usually used in combination with other drying methods such as forced convection employing heated air to achieve optimum product drying.

Microwave energy also has been used to improve other drying techniques, such as vacuum drying, where applying microwaves uniformly supplies the power to vaporize water rapidly from the product, while the vacuum environment permits water to rapidly vaporize at significantly lower temperatures than at atmospheric pressures. This technique maintains the color, flavor, and shape of the product. A system patented under the name *MIVAC,* and developed in the United States, uses combined microwave and vacuum technology to produce high-quality dried fruit.

Microwave energy also is particularly effective in freeze-drying because the ice, rather than the dry solids in the product, selectively absorbs the microwave energy. Thus, the energy for subliming the frozen water is distributed throughout the product; it is not limited by heat transfer through the dried porous layers that insulate against conductive and convective heat transfer.

Dehumidifiers

Dehumidifiers based on the desiccant principle have been used for many years, although their use in the food industry has been limited. Recently, this process has become more economical and easier to apply, often in combination with an existing mechanical system, due to the commercial availability of compact units with a desiccant (typically lithium chloride)-impregnated wheel that is continuously loaded with moisture followed by the removal of that moisture by using heat. The process air is passed over the wet product where it picks up moisture. This air is then passed through a desiccant-impregnated wheel that is rotating slowly. Through both sorption and adsorption, moisture is collected in the honeycomb structure of the rotor leaving the outlet air considerably drier. This dry air can be recirculated over the product to pick up additional moisture before returning to the wheel. The moisture that has been deposited on the wheel is subsequently driven off using hot air, leaving the wheel dry and ready for more moisture loading.

Examples of recent applications in the food industry include the modification of drum dryers to combine them with desiccant dryers for drying tomatoes, apples, and potatoes. This combination enables reduced drying temperatures and thus reduced thermal damage to the products, as well as increased production capacity. Similar modification of blast and spiral freezers for use in combination with desiccant dryers to dry the process air has reduced ice buildup on conveyers and floors, and, in turn reduced expensive downtime. The freeze/defrost cycles on conventional cooling coils also cause fluctuations that lead to variable product cooling rates and affect the quality of a product. Bypassing this operation by using very dry air can resolve this problem.

Steam Drying

The use of superheated stream for drying food products avoids the high energy cost incurred for drying large quantities of air. The moisture that is removed also may be used as a source of heat for other processes. Steam drying also is characterized by relatively short residence times, often making this process suitable for sensitive products. A recent development in steam drying known as *airless drying* has been patented. Superheated steam is formed from the moisture driven off a product as it travels through the process on a belt feed. This steam is retained within the drying chamber by a gastight steam/air stratification layer. Although this process has been targeted for textile and paper drying, the process warrants investigation for its application to foods.

Acoustic/Ultrasonic Drying

The use of acoustic/ultrasonic energy has been shown to accelerate the transfer of heat and materials. It is claimed that by using these techniques food products can be dried with rises in temperature of 1°C or less when using ultrasonic energy sources. Research has shown that foods with high thermal sensitivity could be dried using low frequencies in combination with air at 60° to 93°C for a much reduced processing time compared with equivalent hot air drying. The use of acoustic low frequencies to dry tomato paste, orange juice, and coffee resulted in negligible loss of flavor, color, and nutrients.

Electrohydrodynamic Drying (EHD)

Electrohydrodynamic drying, involving the creation of an *electric wind*, enables accelerated drying of products and offers advantages in drying heat-sensitive foods and in reducing processing times. However, additional work is necessary to extend the knowledge of EHD drying characteristics to food products before the method can be used commercially.

Refrigerated Food

The dramatic development in the volume of chilled or refrigerated food in response to consumer demand for added value, choice, convenience, freshness, and minimal processing has required:

- the development of ultrahygienic preparation and processing areas;

- sophisticated and well-managed refrigeration, retail, and distribution systems;

- novel packaging materials and systems; and

- innovation in product formulation to give the necessary quality and shelf life of such products.

Improved quality and extended shelf life have been achieved in many product categories (meat; fish; bakery, including products with cream; fruits and vegetables; cheese; pasta; and ready meals) by the adoption of modified atmosphere packaging (MAP). This technology modifies the atmosphere within the pack either by gas flushing before sealing, by the natural respiration of the products (fruits and vegetables), or by the use of gas absorbers or emitters. The use of MAP usually involves changes in the levels of oxygen, carbon dioxide, or nitrogen although for some applications the use of other gases is being considered. For example, argon and nitrous oxide may be used for the MAP of fresh produce.

Apart from red meat and non-oily fish, where oxygen concentrations of 60 to 85 percent and 20 to 30 percent are used, respectively, oxygen is usually absent from MAP packs except for fruits and vegetables where low oxygen concentrations (3 to 5 percent) are used to prevent anaerobic respiration. The very high oxygen levels for red meats are used to ensure maintenance of the red color by the continued presence of oxymyoglobin. Recent research work has, perhaps surprisingly, shown beneficial effects from packing prepared green vegetables in high oxygen (70 to 100 percent) atmospheres. Preliminary results indicate reduced microbial development and inhibition of enzymic browning caused by polyphenol oxidases. It is postulated that high oxygen levels may cause substrate inhibition of the polyphenol oxidase enzymes or, alternatively, the high levels of colorless quinones that are formed may cause feedback product inhibition of the polyphenol oxidases. High concentrations of carbon dioxide (15 to 100 percent) are used to retard microbial development in all product sectors except for fresh fruits and vegetables, where levels of 3 to 5 percent are used.

Nitrogen is used as the filler gas in all cases. Fresh fruits and vegetables continue to respire in MAP packs and, unlike other food products, require the packaging material to allow some exchange of oxygen and carbon dioxide. This exchange needs to prevent the creation of anaerobic conditions while ensuring sufficient modification of the atmospheres to bring about shelf life extension. The gas exchange required depends on the type of produce (different respiration rates) and the storage temperatures. Further developments are still required in relation to changes in gas permeability with changes in temperature. The ratio of gas permeability in plastic materials is usually about three to one whereas the ratio of respiration is one to one.

At least two technologies have been proposed to permit the package material to recognize the temperature and respond to it by increasing gas permeability. In one, a co-extruded material contains deliberate microcuts that open with increasing temperature because of differential thermal expansion of the two film components. The other involves altering the molecular structure of the base polymer to add a branch that changes physical properties with increasing temperature.

Technology involving MAP has been a natural extension of controlled atmosphere fruit storage that was developed in the 1920s and 1930s and that was used for apples and pears throughout the world by the 1960s. More recently, controlled atmosphere technology has been applied to shipping containers to provide optimum temperature and gas conditions for preserving fresh produce such as lettuce, strawberries, and raspberries. In fact, the rapid rise in international trade in fresh produce is, in part, attributable to the use of containers in which the temperature as well as the oxygen, carbon dioxide, water vapor, and ethylene concentrations are controlled throughout the transporting process.

Generic Technologies

Apart from the potential developments in postharvest, processing, and distribution technology referred to here, it also is apparent that developments in other technologies will undoubtedly influence our ability to improve processed food security in the twenty-first century. For example, developments in information and sensor technology will enable greater and more effective use of modeling and simulation techniques in the design and control of unit processes and whole-food processing and packaging lines. Developments in materials technology will allow the use of more hygienic materials in equipment and processing areas as well as further applications of intelligent packaging. The incorporation of materials in packaging that monitor storage and distribution conditions and indicate abuse are feasible developments.

The impact of the products of modern biotechnology on food processing are also varied and far reaching. These include providing raw materials and ingredients from genetically engineered crops, making available modified or novel ingredients, and creating processing aids from fermentation or other biologically mediated reactions.

Genetic Engineering

Genetically engineered crops, ingredients, enzymes, and microorganisms are now cleared for use in the food industry. The genetically engineered crops being produced commercially include tomato, squash, potato, maize, rapeseed, and soybean; tomato paste, protein, oil from soybean, and oil from rapeseed have been cleared for use as ingredients in the food industry. Similarly, commercial sources of bovine chymosin are now available from *Aspergillus niger, Kluyveromyces lactis,* and *Escherichia coli* K-12. Genetically engineered *Saccharomyces cerevisiae* is used for faster liberation of carbon dioxide (baker's yeast) and greater ability to break down dextrins and starch (brewer's yeast). Commercialization of genetically engineered food crops has been generally slow. It has been accompanied by substantial information being made available to the public in the hope that they will accept materials from genetically modified crops and organisms. Providing this is the case, then genetic modification offers considerable potential for producing raw materials with improved functionality. This, however, does depend on food scientists and technologists being better able to specify the required properties and subsequently understanding the biochemical and physiological consequences of gene modification. In fact, the real power of genetic modification initially will be to provide the tools to enable improved specification of functionality of raw materials and ingredients.

Modern biotechnology has already played a major part in the development of improved diagnostic and analytical tools used by the food processing industry in their quality assurance and quality control programs. This technology is perhaps most marked in development of microbiological methods but also has included methods for analyzing chemical composition of food and detecting contaminants. Developments have included the application of adenosine triphosphate (ATP) bioluminescence, immunological approaches, gene probes, polymerase chain reactions (PCR), and genetic fingerprinting methods.

ATP Bioluminescence

Perhaps the most evident use of ATP bioluminescence in the food processing industry is for hygiene monitoring. An analysis is conducted for both bacterial and food ATP, because food material left on the production line after cleaning indicates poor hygiene and could be a focus for microbial growth and further contamination. For hygiene monitoring, swabs can be taken from production areas and analyzed in less than two minutes, allowing for very rapid action if poor hygiene is indicated. Current developments for the rapid detection of specific microorganisms are based on either immunological detection procedures or on the use of nucleic acid–based detection procedures (gene probes).

Immunological Approaches

Immunological approaches are based on the use of antibodies that are raised to react to specific microorganisms. Antibodies cannot by themselves be used to detect an organism. They must be attached to suitable labels that allow the binding of the antibody with the organism to be visualized. Most of the early immunological methods that were aimed at detecting foodborne pathogens used enzyme labels and were known as enzyme linked immunosorbant assays (ELISA).

Currently available ELISA systems are aimed at pathogens such as *Salmonella, Listeria,* and staphylococcal enterotoxins and usually reduce detection times from four to five days down to 48 hours. Fully automated ELISA systems are now available. The newest form of immunoassay for microorganisms is based on the use of dipstick technology first developed for pregnancy testing. In this case, food samples go through an enrichment, followed by analysis with the dipstick unit. This form of immunoassay is easy to use in the laboratory and requires no automated equipment.

Gene Probes

The cell nucleic acids deoxyribonucleic acid (DNA) and ribonucleic acid (RNA) define the specific characteristics of any organism and can, therefore, be used as targets to very specifically detect microorganisms. *Nucleic acid hybridization probes,* or gene probes, have been produced that can detect a number of food pathogens such as *Salmonella, Listeria, and Listeria monocytogenes.* As with immunological procedures, they can only operate if a suitable label is attached to the probe. This label is usually either an enzyme or luminescent marker. Probe tests can reduce test times for pathogens from four to five days down to 48 hours, or allow the confirmation of isolated *Listeria monocytogenes* colonies in 30 minutes rather than in two days.

PCR System

One of the newest developments in rapid microbiology has been a chemical amplification system called the polymerase chain reaction (PCR). This method can amplify the microbial cell DNA in about three hours to levels that are very easily detectable. The PCR technique has now been developed into pathogen detection kits for the food industry. The BAX (Qualicon, Inc., United States) and Probelia (Sanofi, France) both reduce test times for *Salmonella* down to 24 hours, thus providing one of the most rapid procedures available for pathogen detection in food.

Genetic Fingerprinting

Very recently, food microbiologists have become aware of a range of genetic fingerprinting techniques that can be used to identify isolated organisms to a very fine degree. These methods are so discriminatory that it is possible to trace the epidemiology of a contamination of a food product, i.e., to show where a particular contamination occurred during production. These methods can be used to trace and prevent future contamination of a product. Until recently, these procedures were confined to specialist laboratories; however, through the production of fully automated systems it is now possible to operate them routinely in the food industry. The RiboPrinter™ (Qualicon, Inc., United States) is a fully automated instrument used to ribotype bacteria automatically in eight hours, producing a genetic fingerprint that can be used in epidemiological investigations. This type of rapid, automated system will considerably aid the food microbiologist if serious problems of food contamination occur, pinpointing where the contamination occurred, and allowing prompt, directed corrective actions.

Conclusions

Food processing and distribution will undoubtedly continue to play a key role in food security in the twenty-first century because the ability to meet rising food demands will depend increasingly on providing people in need with access to available supplies. For the processing and distribution part of the food chain to achieve its full potential in feeding an increased world population and an increasingly discerning consumer in many parts of the world, it has to be viewed within the context of the whole food chain.

The challenges of the twenty-first century will continue to be the following:

- ensuring the availability of cost-effective raw materials;

- ensuring the safety and quality of raw materials;

- improving management and control of current methods of manufacturing, distribution, and supply;

- improving product quality and efficiency by innovative methods of manufacturing, distribution, and supply;

- reassuring the safety of food;

- matching products to consumer requirements;

- improving understanding of the food chain by the general public, through improved education and provision of balanced information;

- understanding the links between diet and health; and

- providing food with specific health benefits.

Many of these needs for the early twenty-first century will be met by the continued use of existing preservation, processing, and distribution technologies although further refinements and modifications will occur in meeting the needs of the market and ensuring a commercially viable industry. The new technologies described here will have a part to play both directly in the processing of food and indirectly in support of this part of the food chain, either by providing raw materials or by improving the security and integrity of processing and distribution operations.

However, the speed of modification of existing technologies and the adoption of new technologies will be influenced by the complex interactions of several factors, including:

- advances in science and technology;

- social and fashion trends;

- relationships between diet and human health;

- demography and economic prosperity;
- energy and environmental factors;
- consumer awareness and perceptions; and
- financial investment and return.

In addition to these interactions, Whitney MacMillan, Chairman Emeritus, Cargill, Inc., in a recent paper emphasized one of the most important factors that will shape the future of the global food industry—liberalized food trade. He stated:

> Our ability to meet rising food demand depends in large part on providing people in need with access to available supplies. It sounds so simple, but sometimes we make it so difficult.

> Expanded, liberalised world trade helps create better markets for producers, and it provides customers and consumers with a more stable, reliable supply of products. It allows supplies to move quickly and inexpensively from surplus to deficit areas. In other words, it expands economic opportunities for farmers, and it promotes exactly the kind of food security consumers value so highly.

> I might also add that it promotes the overall economic growth that will fuel rising demand for the goods and services provided by our industry. It promotes agricultural productivity in developing countries. It frees people from subsistence economies to participate in industrialisation. It contains food costs, allowing people to upgrade their diets while generating savings to spend on other goods.

> The president of the North American Export Grain Association recently projected that by the year 2025, annual trade in wheat, coarse grains, oilseeds, and their products could be worth $100 billion. That reflects exactly the kind of robust global demand that spells opportunity for our industry (MacMillan 1996, 6).

Reference

MacMillan, W. 1996. What's Ahead for a Global Food Industry. 18th Annual Campden Lecture at Campden & Chorleywood Food Research Association, Chipping Campden, Glouscester, England.

Background References

Betts, R., M. F. Stringer, J. Banks, and C. Dennis. 1995. Molecular Methods in Food Microbiology. *Food Australia* 47(7):319–22.

Brody, A. L. 1996. Integrating Aseptic and Modified Atmosphere Packaging to Fulfill a Vision for Tomorrow. *Food Technology* 50(4):56–66.

Broomfield, P. L. E. 1996. Microbial DNA Fingerprinting. *International Food Hygiene* (August):5–7.

Campden & Chorleywood Food Research Association. 1995. New and Emerging Drying Technologies. *New Technologies Bulletin* 11.

_____. 1996. Emerging Technology Implications for Frozen Foods. *New Technologies Bulletin* 12.

Campden Food & Drink Research Association. 1994. Genetic Modification in Food Production. *New Technologies Bulletin* 9.

Day, B. P. F. 1996. Novel MAP for Fresh Prepared Produce. *European Food and Drink Review* (Spring):73–80.

Dunn, J., T. Ott, and W. Clark. 1995. Pulsed Light Treatment of Food and Packaging. *Food Technology* 49(9):95–8.

Ledward, D. A., D. E. Johnston, R. G. Earnshaw, and A. P. M. Hasting. 1995. *High Pressure Processing of Foods*. Nottingham: Nottingham University Press.

Tempest, P. 1996. *Electroheat Technologies in Food Processing*. Crawley, West Sussex, England: APV Processed Food Sector.

Discussant ♦ *Verghese Kurien*

Food processing, packaging, and distribution in India are still in the early stages of development if we compare ourselves with the developed world. From an economy in which almost all food was sold unprocessed and unpackaged and, I might add, therefore untransported, India has progressed a long way. The result has been that loss through wastage has been reduced, making more food available for a burgeoning population.

My experience has been mainly in the dairy industry. We have moved from the local milkman who poured out unpasteurized, adulterated milk, a liter at a time, into domestic containers preferred by metropolitan housewives, to pasteurized milk sealed in bottles and plastic sachets. This milk came from those areas in India that produced it, and it was sold in the cities where there was a constant demand for it. In the process, we transported the milk into the cities instead of transporting the buffaloes and cows from which it came. This transport became possible because of the pasteurization and chilling plants that were set up near the source of supply.

Keeping the animals where they breed best and milk best, owned and looked after by millions of small farmers who also own the processing, transport, and distribution facilities in producer cooperatives, ensured the prosperity of the individual farmer and the democratization of capitalism. It also ensured that the fresh quality of milk was maintained. And, true to the capitalist ethic, the farmer breeds more buffaloes and cows to sell more milk thus garnering for himself increasing profits every day. This system has resulted in India, a nation never in its history known for its dairying prowess, becoming today the second largest producer of milk in the world. We expect to be able to produce more milk than the United States in another couple of years, some two years ahead of the millennium, a target we set ourselves.

The distribution network for milk has now evolved into an India-wide National Milk Grid that involves the transport, in refrigerated tankers, and distribution of milk to cities all over India. For example, Calcutta in eastern India gets milk that originates in Anand in Gujarat near the west coast—a distance of some 2,000 kilometers. The milk goes in refrigerated railway tankers often amid summer heat during which temperatures can reach 45°C.

The increasing availability of milk has resulted in being able to set up facilities for processing milk into products such as milk powder, baby food, ice cream, cheese, and Indian milk sweets, all owned by the same small farmers as part of the great cooperative endeavor. Today, the Indian market is dominated by products made by cooperative milk marketing federations. These federations have now developed an infrastructure of cold chain distribution networks for the distribution of perishable foods that is of immense importance in a subtropical country such as India. The cooperative federations also have set up normal merchandising channels in towns and cities all around the country with a retail universe of more than half a million outlets.

Multinational corporations such as Nestle, Lever, and Cadbury were established in India even before the policies of the Indian government changed to give multinationals a more open market. The cooperatives have succeeded in spite of their presence. The present policy of the government of opening up agricultural markets in India to multinationals is not something we approve. But we have proved we could do it better in the past and we can prove we can do it even better in the future. Our disapproval of multinationals entering the food processing and distribution industry is based on actual experience, not just the desire to keep the market to ourselves.

For example, many years ago the government in Delhi was concerned about the supply of condensed milk. At that time it was considered an important product and there were no manufacturers in India. A multinational was approached and requested to help with making the product in India. The response was so ill-mannered that I will not mention the multinational's name. They said in effect: This is not a technology you natives can manage. Only we can do it. This was not the type of response that pleased the government, which then asked us (the cooperatives) if it was possible for us in India to make condensed milk. It was a challenge, but I knew that I had some of the finest scientists and technologists in the dairy industry working with me. So I said we would try. We succeeded. Once we succeeded it was discovered by the multinational that it could, in fact, manufacture condensed milk in India.

A little later the government became concerned with the need to manufacture baby food in India. Once again they contacted a multinational. This multinational gave a different reason for protecting its import into India. It said that there were too many bacteria in India's air to make such a sensitive product. The government contacted our cooperative and, once again, we took up the challenge. Today, most of the baby food produced in India is produced in the cooperative sector owned by Indian farmers. Needless to say, once we had succeeded, the bacteriological contamination of Indian air miraculously disappeared and the multinational got to work making baby food in India.

Not long ago edible oil was only sold unprocessed, unbranded, and unpackaged. This led to adulteration and the import of edible oil at prices that hit Indian farmers the hardest. The effort made by the cooperative farming industry and the market intervention operation started by the Dairy Board of India changed the face of the oil scene. The cooperatives have a collaboration with a firm called Tetra-Pak. Together they own a plant where they manufacture the laminated paper that is used to make Tetra-Pak containers. When the government asked them to become involved in edible oil they knew that it was important to sell oil in small quantities in tamperproof packaging. The Tetra-brik seemed a good option. So the cooperatives asked their collaborators to help. "No," they replied, "the packaging cannot be used for a product like edible oil." Well, these cooperatives were not about to take "no" for an answer. Their scientists and technologists began looking at the problem. And they solved it. Today there is more edible oil sold in India in Tetra-briks than in any other type of packaging.

Our reasons, therefore, for being wary of multinationals stem from the hard experience of being at the receiving end for the last four decades. I am not willing to believe that they will change their behavior overnight just to make us happy. No country willingly permits its food security and its capacity to process or distribute food to go into the hands of people whose only loyalty is to their shareholders who live in Switzerland and not to the country in which they operate.

We also have tried our hand at other food processing and distribution targets. We set up a system to procure fruits and vegetables from cooperative growers all over India for distribution in metropolitan Delhi. The system works in tandem with the Delhi Mother Dairy's sale of milk through strategically located booths in the city. This system has resulted in farmers who grow vegetables getting better prices, while consumers in Delhi get better vegetables at lower prices. As the supply of vegetables has improved, we have set up processing units for blast freezing to make them available elsewhere and on a timely basis.

Today, India has the capability to process, preserve, transport, and distribute food products as basic as milk and vegetables over periods of time and distances that would have been unimaginable just two short decades ago. When the Bovine Spongiform Encephalopathy (BSE) crisis hit Britain, newspapers in India suggested that India should donate packaged milk to Britain as they were to slaughter all their dairy cattle. It did not come to that. But it does show the kind of confidence that we have developed in India in dealing with our food industry.

Discussant ◆ *Joseph H. Hulse*

Postproduction Systems and Urban Food Security

Rene Prudhomme, an economist at l'Institut des Affaires Urbaines in Paris, observed that economists seem generally more interested in what is produced and how it is produced than in the means and systems by which what is produced is best delivered to those who want or have need of it. Whether or not Prudhomme has reason, I cannot judge, but the main burden of this paper relates to the preservation and processing of agricultural and fisheries products and to their delivery to markets and consumers, with particular concern for rapidly expanding urban communities in many countries of Asia, Africa, and Latin America.

Professor Dennis has elegantly reviewed recent developments and progress in food science and technological processes as they relate to food preservation and transformation, many of which are most evident among more affluent societies; various of which, appropriately adapted, will in time find application among developing nations. This paper will attempt no comprehensive review of modern food science and technology but will offer a brief account of the essential elements of postproduction systems (PPS): all that exists or must exist between farms, fish landings, and urban markets and consumers.

Agro-industries: Engines of Economic Growth and Diversification

Of all industries, food production, processing, and distribution are the most vital to human survival, health, and welfare. Agro-industries, using technologies that profitably converted materials from harvested crops and livestock into useful food and fiber, were the primary engines of economic development in Europe, North America, and Oceania. Agro-industries are among the fastest-growing industries in many nations of Asia, Africa, and Latin America. With shipments valued at over $400 billion per annum, food processing is the largest industrial sector in the United States. Food industries among the European Community (EC) employ more than 2.5 million people and process two-thirds of all farm produce with annual sales value in excess of $300 billion. Although barely 3 percent of Canada's workforce is employed on farms, close to 25 percent earns its livelihood in the postproduction food sector: in processing, marketing, distributing, catering, and related service industries.

More than 2 million people work in Indian food factories, proportionately the largest component of the industrial work force. Since 1971, Malaysian food processing industries have grown annually at approximately 11 percent and comprise 21 percent of all industrial activity. Of the Thai food processors, who constitute 60 percent of the industrial economy, almost 90 percent are classed as "small-scale industries" that may broadly be interpreted as local and labor-intensive.

Despite their critical economic and social importance, PPS have received relatively little systematic support from donor or development agencies. A committee convened at the Food and Agriculture Organization of the United Nations (FAO) in 1990 by the Consultative Group on International Agricultural Research/Technical Advisory Committee (CGIAR/TAC) reported that barely 0.7 percent of CGIAR center budgets was invested in postproduction systems research (PPSR). The committee also reported that support for postproduction systems research and development (PPS R&D) among development and donor agencies was fragmented and uncoordinated. This support was frequently dedicated more to the export of donors' technologies, goods, and services than to a systematically determined assessment of the needs and opportunities of developing nations, or to a rational stimulation of their food processing and distributing industries. It is not unusual to be informed by representatives of development agencies that food processing and distribution are matters for "the private sector." But is not the same true of farming? Nowhere have nationalized farms been notably successful.

Private sector enterprise, whether it be crop and livestock production, or food processing, distributing, and marketing, requires a congenial politico-economic climate in which to function effectively. Agricultural production and postproduction entities both rely upon essential rural and urban infrastructures and services, and on stimulative and congenial food, agriculture, science, and technology policies that only governments can enact and enforce.

It is, therefore, difficult to understand development strategies that provide technical and economic support and assistance to agricultural production systems while denying comparable amenities to enterprises and services upon which the safe, economic, and efficient preservation and distribution of food depend. Agro-industries in developing economies stand as much in need of technical assistance and a favorable political climate as do farmers.

Postproduction Systems: Nature and Purposes

The 1990 committee report referred to above broadly defined the purposes of food PPS as follows:

*   to protect and preserve the products of harvested food crops, livestock, and fisheries so they may be safely stored and conveyed from regions and seasons of surplus to those of scarcity; and

*   to process and transform the raw materials of agriculture and fisheries into foods that are acceptable, affordable, convenient, nutritious, and safe, together with by-products of commercial, social, and economic utility.

All products of agriculture and fisheries are in varying degrees perishable. Various cereal, legume, and root crops, and livestock and fisheries products are indigestible and/or unpalatable in their raw state. Crop species being seasonal and the carcass of a large animal being too much for one family meal, the need to preserve food has persisted since humans evolved from herbivorous primates to omnivorous hunters, gatherers, and scavengers. Professor Dennis's paper describes the many sources of spoilage after harvest and slaughter. One might add to the biological causes, physical and solar heat damage and contamination during handling, transportation, and processing operations, much of which could be obviated by good housekeeping and by systematic control of handling, storage, factory processing, and distribution practices. All of these potential causes of spoilage argue for the provision of technical advisory services to PPS, comparable to the "extension services" provided to farmers.

Cereals: Properties and Processes

Cereals now and well into the future will provide most of humankind's food energy. The earliest cereals eaten around the Mediterranean were the wild seeds of emmer and einkorn. The earliest "naked" wheat, free of a tough outer hull appeared about 8000 B.C., possibly the progeny of a promiscuous outcross between emmer and some other wild grass. In most primitive societies where the harvested cereal grains come in many shapes and sizes, it is convenient to pound them into a coarse meal to be cooked in water to a porridge. As economies expand and diversify, diets of cereal porridge give way to cereals processed into a variety of structured foods, distinctive in form and texture, and more convenient to transport and eat. Leavened and unleavened bread, pastas, biscuits, and cakes are typical examples.

About 6000 B.C., in the Middle East, pounding in primitive pestles and mortars gave way to saddle-stone mills that over time progressed from simple grinding of whole grains to techniques by which the outer seed coats were separated from the endosperm. Herringbone grooves incised into the surfaces of the millstones broke open the grain to release endosperm fractions, the bran being separated by hand-winnowing over woven fiber screens.

Modern break-and-reduction wheat flour mills use essentially the same principles. The grains are broken by counter-rotating incised steel rolls; vibrating gravity tables combined with woven textile screens, from which the bran is removed by suction fans, are the mechanical equivalents of hand-winnowing. In common with many other "modern" food processing technologies, flour milling, baking, and brewing depend on the same basic principles as were employed in Ancient Egypt and Republican Rome.

Until relatively recently, progress in food processing was most evident in the replacement of human hands by machines, animal power being sequentially superseded by energy from water, wind, steam engines, and electric motors. In India and many other "developing" countries, grain milling technologies ranging from the primitive pestle and mortar and rotary quern to modern electrically powered mills are all still in use. Because grain milling and many other industrial food technologies began as domestic and artisanal crafts, most are independent of scale and location. They may be highly mechanized or labor-intensive, located wherever the essential resources and access to markets are conducive. Near the center of Rome, there is an integrated complex of a modern flour mill, a bread bakery, and a pasta and biscuit factory. The milling operations and transfer of milled products to the processing units are controlled from a central console in the flour mill. In each of the processing factories, the weighing, moving, and mixing of ingredients and the packaging and conveyance of final products are entirely controlled electromechanically. Close by this hands-off mechanized facility one finds several small enterprises where bread, pizza, pasta, and flour confectionery are produced commercially almost entirely by skilled artisans' hands.

In recent years, small rural grain mills have appeared in several African countries, their primary purpose being to relieve rural women of tiresome hours of pounding grain. Evident is a growing interest in milling African grains for secondary processing into bread, pasta, biscuits, snack, and infant foods, many of which are now processed from imported grains. According to the International Plant Genetic Resources Institute (IPGRI), the CGIAR centers have accessioned more than 630,000 food crop genotypes, of which the agronomic and phenotypic properties are recorded comprehensively. There is less evidence across this immense range of genetic diversity of any systematic examination that relates to functional utility: to the properties of interest and concern to millers and other food processors. It is not proposed that CGIAR centers establish food science and technology laboratories, but there is evident need for systematic cooperation between international plant breeders, food scientists, and food processing industries to study and evaluate across this vast germ plasm collection the biochemical and biophysical properties that are potentially useful to food processors; how such properties vary among genotypes; and by what genetic or technological modifications properties congenial to efficient processing could be enhanced.

The properties of every processed food depend upon the properties of the raw material ingredients and how these are modified during transformation. Grain millers, bakers, and pasta processors require grain types that are of consistently desirable quality. If the harvest yield of a grain is increased by 5 percent and the miller's subsequent grind out yield drops by 10 percent, the net result is an economic and postharvest loss. Food crops consistent in desirable functional properties are essential to food processors. Food processing industries are of increasing, indeed critical, importance to future food security, particularly to the expanding urban populations of Asia, Africa, and Latin America. The rapid growth in numbers and changing patterns of food demand may be regarded as an awesome challenge or as an exciting opportunity for rural agriculture, agro-industries, and employment. Failure to meet the challenge by a systematic, creative response will result in excessive wastage and food insecurity for many poor urban people.

Urban Growth and Food Security

Predicting the size of the global population and how many persons the planet can sustain 50 or 100 years from now, though intellectually diverting, is pragmatically futile. Of more pressing urgency are the consequences of rapidly changing demographics and rising disposable incomes among nations that, until recently, were regarded as poor and underdeveloped. The World Resources Institute (WRI) states on page 3 of its publication *The Urban Environment*: "Between 1990 and 2025, the number of people who live in urban areas is expected to double to more than 5 billion...almost all of this growth...[about] 90 percent will occur in...the developing world." More than 70 percent of Latin Americans already live in urban areas, and African and Asian urban communities are increasing at a rate of roughly 4 percent per year. Fifteen of the 19 cities with a population above 9.5 million are in developing countries.

Although some food can be produced in cities, kitchen gardens, and small allotments, most urban communities rely on food brought in from rural farms and coastal and inland fisheries. Safe and efficient distribution of crop and livestock products from rural farms and coastal fisheries to urban markets and consumers involves economically effective integration of preservation, processing, packaging, storage, transportation, and marketing. The greater the distance in time and space between the rural producer and the urban consumer the more complex is the postproduction system, logistically and technologically. For various reasons, the demand for industrially preserved and processed foods will increase as urban populations expand and their patterns of food consumption change.

The literature on global food security in general and for the World Food Summit in particular deals predominantly with the need progressively to increase agricultural and fisheries production, a need that is not in dispute. Development agencies' programs for food security, however, give relatively scant attention to PPS: all that is entailed in delivering foods safely and economically from rural farms to urban markets. Inefficient and uneconomic PPS result in urban food prices that the urban poor cannot afford.

Urban food security requires consistent access to a safe, adequate supply of acceptable, affordable foods, and demands these factors:

- technologically efficient systems of preservation and processing, preferably located close to the place of harvest or slaughter;

- logistically efficient and economic means of storage and transportation;

- reliable market outlets that retail foods acceptable in quality, unit quantity, and price that the diverse segments of urban communities can afford;

- market and technical advisory services to ensure that what is produced is what is required by consumers, processors, and the markets that serve them; and

- national food and agricultural policies that provide the infrastructures, services, and socioeconomic and regulatory conditions essential to the above demands.

Preservation and Processing

Professor Dennis's paper consummately reviews the principles relevant to food preservation and related contemporary technological progress. Suffice it here to state that most developing countries have tropical or near tropical climates that are conducive to food spoilage. Food and waterborne microbial contamination are difficult enough to control in temperate regions, but in these tropical ecologies, food spoilage and foodborne infections can be of egregious proportions with meat, fish, milk, and fruits and vegetables high in water content being notably susceptible.

Among affluent nations, low-temperature processing and storage is extensively used as an effective means of preservation, a means dependent upon electrical power. Food processing, the largest industrial sector in the United States, consumes barely 2 percent of the total electrical power generated. Nonetheless, the supply is, for the most part, consistent and reliable. Although demand for electrical power in many developing countries is rising rapidly, disruption because of overload and poor maintenance is not unusual. Refrigeration reliant on public power is, therefore, a restricted option particularly in rural areas where food processing is most effectively carried out.

Processes of preservation and transformation proposed and practiced must be applicable and sustainable under the conditions that prevail. There would seem to be an urgent need for systems of cryogenic preservation that are independent of public power supplies. A brewery in Zimbabwe converts carbon dioxide from its fermentations into "dry ice" that it sells to ice cream makers. An oil refinery in Thailand supplies food processors with liquid carbon dioxide that permits food in insulated containers to be transported over long distances.

Professor Dennis describes opportunities for packaging and storage under modified atmosphere and the progress made in genetic modification of yeasts that produce carbon dioxide by fermentation of carbohydrates. A particular advantage to fermentation industries in tropical climates would be the use of yeasts genetically modified to ferment efficiently at 45°C rather than at approximately 25°C, which is, I believe, the near optimum for *S. cereviseae.*

One could recite a melancholy litany of misadventure in the transfer of food technologies by donor agencies and their advisers seeking to impose methods established within their home territories without first assessing comprehensively the resources, conditions, and constraints existing in the developing communities concerned. Farming systems methodologies developed at the International Rice Research Institute (IRRI), and the diagnostics and design techniques devised by the International Centre for Research in Agroforestry (ICRAF), demonstrate the wisdom of understanding first what is in place before seeking to impose novel technologies.

Demand for Livestock Products

The FAO, the Winrock Foundation, and various other organizations predict future rising demand for meat, milk, eggs, and other livestock products in Asia, Africa, and Latin America. Consumption of livestock products in India is growing by more than 10 percent per year; demand in China and other Asian countries is predicted to rise even faster. In 1991, 90 billion hamburgers were bought from fast-food outlets, roughly 18 for every living person. Over 40 percent of McDonalds' $30 billion hamburger sales was outside the United States. McDonalds expects to open 2,000 new outlets outside the United States every year. The fast-rising demand for livestock products and for meat in "fast foods," many sold by street vendors, could present serious consequences for agricultural production, land use efficiency, ecological conservation, and public health.

Generation of fibrillar and textured proteins by extrusion technologies has made possible a variety of meat extenders from soybeans. Selective extraction of soluble and water-dispersible proteins has produced substitutes for milk and milk derivatives from soybeans and groundnuts. There is an apparent need for research on tropical legumes to identify proteins that, by genetic or technological transformation, could be structurally modified and converted to meat and milk-like analogues or extenders. Production of tropical food legumes has not kept pace with the increased harvests realized from the high yielding varieties of cereal grains. Innovative research to expand and diversify the industrial utility of tropical legumes, particularly as meat and milk extenders, would stimulate production and ameliorate arable land use efficiency.

Rural and Urban Food Industries

Rural food industries are essential for rural and urban food security, while offering such economic and social benefits as employment for skilled, semiskilled, and relatively unskilled people, thereby discouraging rural to urban migration. Perishable foods are best processed soon after slaughter or harvest. Most small industries benefit more from operations research and advice—to improve the efficiency and utility of existing processes—than from research to devise new food products. Apart from financial constraints, rural agro-industries in developing countries labor under diverse, discouraging difficulties, including:

* inadequate, often nonexistent technical advisory services;

* poorly maintained rural infrastructures and transport services; and

* unreliable access to market information concerning prevalent and changing patterns of urban consumers' food demands and needs.

Government and academic food scientists and research facilities are usually found in large cities, remote from rural industries, and generally are more disposed to the invention of novelty than to providing useful, practical advice to small industries. Among the developing nations, one finds countless instances of rural grain mills, abattoirs, and fish, fruit, and vegetable processors that suffer excessive wastage for want of hygienic housekeeping, systematic equipment repair, and maintenance. Some years ago, with support from Canadian food industries, the FAO provided mobile technical service units to a dozen countries in Asia, Africa, and Latin America. After two weeks of demonstration to small rural canning industries in Chile, operational efficiency and profitability in all of the factories showed remarkable improvement.

Rural and Urban Markets

As urban populations expand, the proportion of food expenditures given to marketing services, transportation, and distribution rises significantly. As disposable incomes rise, so do marketing functions. In a favorable political climate, food markets provide valuable social and economic services. Yet support for food marketing, transportation, and rural infrastructures seems only a minor consideration among many governments and development agencies.

In urban centers of developing nations, food retail outlets are of three broad categories: public markets and street vendors, neighborhood stores, and self-service supermarkets. A correlation between disposable income, the category of market patronized, and the types of food purchased is evident in most urban communities. Affluent consumers tend to visit a supermarket once or twice each week, whereas the poorer citizens buy from public markets or local small stores daily. Consequently, package sizes for the supermarket trade tend to be larger than those for public markets, street vendors, and local stores. Inconsistent access to specific quantitative and qualitative urban market information severely constrains profitable planning and production by small rural industries.

Present and predicted expanding demand for livestock products and other perishable foods by urban populations in many lands is unlikely to be reversed. Safe and reliable satisfaction of these demands offers exceptional opportunities for rural food industries, provided the constraints to technical operational efficiency, access to urban market information, reliable transportation, and consistent raw material supply can be overcome.

Food Transportation

A 1993 World Bank report by C. D. Creightney titled "Transport and Economic Performance: A Survey of Developing Countries" (Technical Paper 232) states: "The Bank has largely ignored the crucial role of transport in improving economic performance;...infrastructures...are more important than prices in the balance of supply and demand." Transport systems and their relative efficiencies influence agricultural production, as they affect access to essential inputs and markets, and consumer prices, particularly as the latter relate to seasonal fluctuations.

Food distribution is inhibited by poorly maintained roads, particularly evident among feeder roads that purport to connect farming communities with rural industries and their urban markets. It seems sadly evident that politicians are more favorably disposed to construct sports stadia and conference centers than to invest in road maintenance. Deterioration of newly paved roads begins slowly but accelerates rapidly after three or four years under heavy truck traffic. The World Bank study indicates that more than $40 billion is now needed for urgent road repair because of long neglect in 85 developing countries.

In tropical climates, refrigerated trucks carrying frozen foods require heavier insulation than in temperate regions, not only on the roof and sides but also underneath, because of heat radiated from black road surfaces.

Across the planet, motor vehicles are increasing proportionately more rapidly than human populations. Throughout much of the developing world, road vehicles are appearing faster than the roads needed to carry them. By 2010, over 60 percent of the planet's oil production will be consumed by motor vehicles. The resultant cost of oil and transportation and the consequent effects on urban food prices should be matters of serious concern to international agencies and to government planning and policy departments.

Planning and Policy Formulation

There are no global panaceas for urban food security. But if disorder is not to deteriorate into catastrophic chaos, provident planning and policies, based upon scientific analyses of total food production and delivery systems, are an urgent necessity. Simulation models of transport and distribution logistics that take into account essential and contingent factors within cities, and among rural producers, processors, and urban markets, could contribute usefully to diagnostics and design. Total food systems analyses would indicate the best sites for urban markets and storage depots as well as optimum locations for rural food processing in relation to agricultural and fisheries production sources—all in precise relation to the markets and consumers to be served.

Recognizing the limited capabilities among many developing nations to formulate cogent, systematic strategic plans and policies to ensure future food security, there is evident need for an international advisory service, staffed by systems analysts and other essential disciplines, to assist developing nations' governments in defining policies and systematic plans for urban food security. Such a service could draw upon an immense reservoir of knowledge and experience gleaned over many years by long-established food processing, distributing, and processing industries, and by men and women recently retired from such industries who are willing and able to offer sound practical advice. It is suggested that provision of such an advisory service should be seriously considered by the FAO and the World Food Summit, in cooperation with the International Food Policy Research Institute (IFPRI), as a critical component of the Organization's Global Food Security and Food for All initiatives.

The eighteenth century poet William Cowper wrote in Book One of *The Task* that God created the countryside but man made the towns. On present evidence, God was demonstrably more presciently aware than man of what he was about.

Summary and Recommendations

1. Predictions of an expanding global population and the consequent need to increase crop and livestock production are extensively documented and unquestionably necessary.
2. Investments in plant breeding and farming systems research have facilitated increased food crop production in many countries.
3. Increasing agricultural and fisheries productivity alone cannot guarantee food security for rapidly expanding urban populations.
4. Access to an adequate, affordable food supply for diverse urban communities requires that agricultural and fisheries production systems be complemented by economic and logistically efficient PPS.
5. Despite their economic and social importance, PPS have attracted less than adequate study and support by development agencies.
6. An international working group convened at FAO in 1990 reported that barely 0.7 percent of CGIAR centers' resources were then invested in PPS R&D; and that bilateral support for PPS was fragmented, uncoordinated, and, in many instances, dedicated more to the export of donors' technologies, goods, and services than to systematic assessments of the needs and opportunities for, and constraints to, the design and development of desirable and effective PPS.
7. The working group defined two primary purposes of PPS: to protect and preserve products of food crops, livestock, and fisheries to enable safe storage and conveyance from regions and seasons of surplus to those of scarcity; and to process and transform agricultural and fisheries products into foods that are accessible, affordable, convenient, and nutritious.
8. The greater the distance in time and space between the place of harvest and urban markets and consumers, the more complex PPS become.
9. In varying degrees, all foods are perishable and are most effectively preserved and processed close to the place of harvest or slaughter.
10. Expanding urban populations and their diverse food demands present both a challenge and an opportunity for rural food processing and marketing industries and for rural employment.
11. To be effective and profitable, rural food industries require a favorable political and economic environment with essential supporting services. Many appear severely constrained by poor

access to financial resources and services; inadequate technical advisory services; poorly maintained rural infrastructures and transport services; and unreliable access to market information concerning changing demands of urban consumers.

12. Although no global panaceas for urban food security can be offered, certain courses of action seem necessary. The PPS must be assessed and developed systematically by methods analogous to those of farming systems research and the diagnostics and design of agro-forestry. Also, to assist developing nations in urban food security planning, the FAO with the International Food Policy Research Institute (IFPRI) and the World Bank should consider the establishment of a food security policy advisory service to be staffed by systems analysts and other essential experts. Experienced men and women from established food processing and marketing industries could provide helpful, complementary advice.

Discussant ♦ *Robert E. Smith*

Colin Dennis has certainly provided an in-depth summary of the available processes that have potential to extend the availability of processed foods and to control microbiological activity in processed foods. In doing so, he has underscored the number one concern of all food processors; that is, in the manufacturing and handling of foodstuffs, we need to control the microbiological and toxicological aspects of our food.

While conventional processes keep our food safe, we need to conduct more research on emerging pathogens of all types, in the areas of identification, process control, and irradiation. The recent incident of mad cow disease, I think, is just the tip of the iceberg. The fact that we do not know anything about the etiology of that pathogen, and how to measure and stop it, is rather terrifying.

Whenever we talk about food security, we must talk about food safety. Further to this point, no food security symposium would be complete without mentioning the quality control process Hazard Analysis and Critical Control Point (HACCP) for controlling food manufacturing. Although some food industries were using a HACCP program in the early 1970s, it has become more popular lately, as regulatory agencies seek to find simpler oversight methodologies. Suffice to say, this procedure examines all aspects of the food processing chain and identifies those areas where a health hazard exists and then clearly outlines the appropriate control or remedies. One word of caution, however, is that the use of HACCP, certainly from an industrial point of view, should be relegated only to controlling the safety aspects of food and not necessarily the quality aspects of food, because it could be taken over by government oversight. We usually do not do too well when the government gets involved in regulating the quality of food. They should be interested only in the safety of food.

Another issue of major importance is the safety and quality of the water used in food processing. I am astounded that this issue has been mentioned only a few times in the symposium, mostly in the context of the lack of water available to grow crops. You cannot imagine the problems we have in food processing when we do not have potable water. This issue is going to become serious in the future, especially in the developing nations.

Dealing further with food quality, it is essential that we do more to educate populations with regard to quality control. This education is particularly important as we move into developing countries and into the developed countries where foods are becoming less processed as retail outlets take on more ready-to-eat and fresh foodstuffs. Also, because almost 50 percent of meals in developed countries are eaten outside the home, we must ensure that restaurant and food service handlers are well versed in safety and quality matters.

Moving somewhat beyond food safety and quality in the classical sense, a matter of overall food integrity is appearing on the horizon in the developed nations, where we now have the capability of removing macro-ingredients from foods without knowing the full impact on the population of diets consisting of only these modified foods. For example, we have the capability of producing fat-free diets. The selection of only fat-free foodstuffs could have a negative impact on the population. Similarly, we are moving into the neutraceutical arena without a good appreciation of the overall long-term impact of many of these components on the population at large. This is not to say that such emerging foodstuffs are inherently bad; it is merely brought up to emphasize that it is essential for nutritional education to go hand in hand with these process breakthroughs. It would certainly be a disaster if, in the midst of plenty of food, we were to manufacture food that led to nutritional imbalances.

Food standards cannot be ignored when we talk about food processing and distribution. Food will undoubtedly become the economic linchpin in future country-to-country relationships. And the only way we can avoid food sanctions in such an environment is to have in place meaningful food standards based on sound scientific principles. It is essential that a body such as the Codex Alimentarius Commission be supported as the recognized worldwide definer of food standards. And as stated by Dr. Nevin Scrimshaw, the 1991 World Food Prize laureate, we need to have nutritional integrity in our foods, and probably the only way to achieve this end is with classical fortification programs. Again, we need fortification programs that are accepted worldwide.

While my background has been in the process food industry, I also am concerned about the decrease in public and private funding for agriculture and postharvest technology, with particular emphasis on preservation and distribution improvements. One example of a joint initiative between the U.S. government, academia, and the private industry that involves food and agriculture projects shows that, if these joint projects are to be exploited properly and extended worldwide, we need to have more cooperation instead of competition among these three areas.

Dr. Dennis's brief remarks about biotechnology and those made by Dr. Khush need to be underscored. If we are to increase food production, make it safer, reduce chemical exposure to the environment, increase the nutritional content, and reduce the cost of foods, I believe that it can only be done, on the scale required, through biotechnology. Furthermore, we must all find ways to make the marketing of such products as easy as possible without compromising safety.

While on the subject of the difficulty of marketing products, we need to find better ways to communicate with the consumer and particularly with consumer advocates, who see death lurking behind every new technology. There is no doubt that radiation preservation is essential for some crops and foodstuffs; yet, we have been unsuccessful in selling this concept as safe to the consumer. We need to find the answer to favorably selling new technologies such as biotechnology to the consumer before foods produced through biotechnology reach the market.

We also need to be cognizant of the impact of our marketing on supplier nations. We need only remember the impact of the tropical oils removed from foods in the United States, when tropical oils came under attack. Such marketing decisions are a major challenge for the food industry, academia, and government, and all of us must stand up and be counted when some of these issues arise.

Certainly, as Dr. Hall has indicated, this session needs to pay attention to food processing and distribution in developing countries. After all, these countries are precisely the ones where food tends to be perishable, and processing might make it more transportable or stable. We need to find out how to avoid food losses, which have been estimated at 25 to 50 percent postharvest. Is it possible to do more to develop technologies that are effective, inexpensive, and locally available for preserving food and making it more transportable? I think the panelists in this session have indicated that if we can work more with the people in the countries involved, such technologies can be achieved.

Finally, I would like to say, with full knowledge that it may invoke the wrath of Dr. Kurien, that the commercial food industry will not participate in world security without a profit motive. This is a fact of life. Dr. Yunus hinted at this symposium that even the Grameen Bank has to show a financial return. We need to spend time exploring ways to provide incentives to the developing and developed food industries to get them to participate in all levels of the food production and development chain.

Although I have mostly mentioned issues that we are facing in the developed world, it is with the understanding that developing nations, knowing these pitfalls, can avoid many of these issues, and hopefully, will be able to concentrate on transferring the more valuable technologies and lessons already learned. Beyond that, they can move towards implementing those technologies that will be valuable in the future.

CHAPTER FOURTEEN

Geographic Food Security

- ◆ MODERATOR
- ◆ RESOURCE PAPERS
- ◆ DISCUSSANTS

Moderator ◆ *Charles E. Hess*

In this section, we address food security issues in two parts of the world that will provide great challenges for the future. As Dr. Mandivamba Rukuni describes in his paper, the number of undernourished people on the African continent is projected to increase from 100 million in 1996 to 300 million in 2020. China also poses a major challenge for geographic food security. Tremendous progress was made in China in the early 1980s following the cultural revolution, when agricultural productivity increased dramatically and raised China from a nation of food shortages and periodic famines to a nation that was not only self-sufficient in basic food for the first time in modern history but also an exporter of food and fiber. Former Minister of Agriculture He Kang, the 1993 World Food Prize laureate, provided key leadership in the Chinese version of the Green Revolution. But, as Vice President of the Chinese Academy of Agricultural Sciences Yang Yansheng tells us, feeding 21 percent of the world's population on 7 percent of the world's cultivated land will present a challenge to all of us, particularly as the population grows and the availability of cultivated land decreases.

One of the challenges for Africa, China, most developing nations, and even the developed nations is how to make it attractive for people to stay in rural areas. Agriculture is one essential component but not alone sufficient to meet the challenge. Agricultural and food policy also must ensure that rural farmers receive fair returns on their investments. There is also the need to encourage the development of other industries in rural areas. Minister He Kang was a real advocate for the development of rural industry, which now accounts for a substantial component of China's gross national product. The policy in this area was "it is all right to leave the farm but do not leave the village."

Finally, all of us must do a better job of helping policymakers in areas other than agriculture, such as finance, supply, and trade, to understand the key, fundamental role of agriculture in development. We have that challenge here in the United States, where the funding for international agricultural research has been on the decline even though there is convincing evidence that it is in the national interest to make such investments. Recent studies published by the International Food Policy Research Institute (IFPRI) have shown that the gains in wheat and rice production in California alone more than pay for the International Maize and Wheat Improvement Center (CIMMYT) and the International Rice Research Institute (IRRI) on an annual basis, as well as building the potential of future trading partners and achieving the humanitarian goal of improving the quality of life for all people. Hopefully, this symposium will help bring attention to the challenges and opportunities associated with geographic food security.

Resource Paper ◆ *Mandivamba Rukuni*

Food Crisis: The Need for an African Solution to an African Problem

Introduction

The failure to give priority to public sector investment in agriculture by African governments has been arguably their most serious error in political judgment. In the twenty-first century, Africa will be the region where the number of undernourished people will have increased from 100 million in 1996 to 300 million in 2020. In a search for African solutions, it is now necessary to view this food crisis within the broader context of good governance, politics, and democracy. This broader context is appropriate because selected emerging positive signs of recovery suggest that it is unnecessary, particularly as the twenty-first century approaches, to have growing numbers of hungry people on the African continent.

Hunger and malnutrition in Africa have three major causes: pervasive poverty, famine from natural disasters, and man-made famine due to civil unrest. The Great African Famine of 1985 was the eruption of a crisis that had been in the making for more than two decades. Since then the international community has had a better understanding of the effect and interaction between poverty and famine. The experience in the 1980s and 1990s has demonstrated the inseparable relationship between peace, stability, good governance, politics, and democracy on one hand and food security and human rights on the other. The persistence of chronic hunger and famine, therefore, must be seen as morally outrageous and politically unacceptable (Sen 1981).

The Food Crisis

The fact that even the most optimistic projections into the twenty-first century show the need for massive food aid to Africa is evidence of the depth of the food crisis (Table 14.1.1). Several authoritative projections (FAO 1996; USDA 1995; The World Bank 1995) indicate that food aid needs in sub-Saharan Africa will double by 2005. Sub-Saharan Africa will account for 55 percent of all food aid, and food aid needs of the region will exceed the projected global supply for food aid (USDA 1995). For Africans to maintain 1996 levels of food consumption per capita, food aid needs will increase from 5 million to 12 million metric tons in 2005 (Table 14.1.2). A total of 26 African countries are projected to need food aid for the entire period to 2005, and the Greater Horn of Africa has the most severe deficit.

The African food crisis needs to be addressed on both sides of the food security equation: availability and access to food. Per capita food availability in sub-Saharan Africa is 2,300 calories compared to 3,500 in Western Europe and 3,600 in North America (FAO 1996). On the access side, Africans at the family level continue to lose capacity to purchase food on the market, and at the national level continue to lose capacity to import enough food to cover the national food deficits. The net cereal deficit for sub-Saharan Africa is expected to almost double to 50 million metric tons by 2010.

Africa's population has grown at just below 3 percent per year for the last two decades, while food production has grown at about 2 percent. Between 1980 and 1996, for instance, population increased 53 percent, while food production grew by 45 percent. Over the period, food imports were 50 percent higher, while food exports were 25 percent higher (FAO 1996). The bottom line is that an estimated 217 million Africans, 35 percent of the population, are currently chronically hungry and malnourished.

Table 14.1.1.   The World Bank food gap scenarios in Africa: 1990 to 2020

	1990	2000	2010	2020
Case I				
Population (millions of persons)				
(with constant fertility)	500	700	1,010	1,500
Food production (mtme)				
(at current trend growth rate of				
2 percent a year)	90	110	135	165
Food requirement (mtme for				
universal food security by 2020)	100	160	250	410
Food gap (mtme)	10	50	115	245
Case II				
Population (millions of persons)				
(with constant fertility)	500	700	1,010	1,500
Food production	90	135	200	300
(at 4 percent annual growth)				
Food requirement (mtme for universal				
food security by 2020)	100	160	240	410
Food gap (mtme)	10	25	50	110
Case III				
Population (millions of persons)				
(with total fertility rate declining by				
50 percent to 3.3 percent by 2020)	500	680	890	1,110
Food production				
(mtme at 4 percent annual growth)	90	135	200	300
Food requirement (mtme)	100	150	220	305
Food gap (mtme)	10	15	20	110

NOTE: mtme = millions of tons of maize equivalent.
SOURCE: The World Bank 1989. (Reprinted with permission)

An important characteristic of Africa's food crisis is its chronic nature. Poverty, therefore, is a more important and serious cause of hunger than civil strife. War leads to acute food shortages and famine, and this tends to divert attention from the underlying problem, a lack of economic development. Poverty is concentrated in rural areas where the bulk of the population lives. Compared to their urban counterparts, the rural poor are considerably poorer. For instance, an estimated 92 percent of the poor in Uganda are rural, as are 86 percent of the poor in Kenya (Donovan 1996). Agriculture employs about 67 percent of the labor force, earns 32 percent of the gross domestic product (GDP), and makes up 40 percent of foreign exchange earnings of the sub-Saharan African economy.

Amartya Sen's moral outrage is based partly on the realization that poverty can be successfully attacked from a number of fronts if political will prevails (Sen 1981). The enhancement of "entitlement" can be brought

Table 14.1.2. Projected chronic grain food aid needs, status quo consumption

Region	High commercial imports			Low commercial imports		
	Commercial imports	Food requirements	Food aid needs	Commercial imports	Food requirements	Food aid needs
			(million metric tons)			
Sub-Saharan Africa						
1996	6.8	61.0	4.8	6.0	60.8	5.4
2005	7.1	79.6	11.8	3.6	79.5	15.2
East Africa						
1996	1.4	21.6	1.9	1.4	21.6	1.9
2005	0.7	27.8	4.7	0.2	28.0	5.4
West Africa						
1996	3.7	26.0	1.1	3.1	25.9	1.6
2005	4.8	34.6	3.6	2.5	34.6	5.9
Southern Africa						
1996	1.2	10.5	1.6	1.1	1.5	1.7
2005	1.0	13.1	2.7	0.6	13.0	3.0
Central Africa						
1996	0.5	3.0	0.2	0.4	2.9	0.2
2005	0.5	4.0	0.8	0.4	3.9	0.9
North Africa						
1996	15.7	38.9	1.6	15.7	38.9	1.6
2005	20.1	46.5	2.9	19.6	46.5	33.4
Africa Total						
1996	22.5	99.9	6.4	21.7	99.7	7.0
2005	27.2	126.1	14.7	23.2	125.9	18.6
Latin America Total						
1996	3.7	12.2	1.6	3.6	12.2	1.7
2005	4.9	14.6	1.8	4.4	14.4	2.1
Asia Total						
1996	11.2	289.9	2.3	11.1	289.9	2.4
2005	15.6	360.3	4.9	14.8	360.3	5.7
Total (60 countries)						
1996	37.4	402.0	10.3	36.4	401.8	11.1
2005	47.7	501.0	21.4	42.4	500.7	26.4

SOURCE: USDA 1995.

about by expanding income earning opportunities for rural people. Although solutions to poverty and enhancing entitlements appear largely economic in content, political and social factors often determine the associated possibilities. The estimated negligible annual growth in real per capita income, 0.3 percent for Africa to 2000 (Table 14.1.3), is a reflection of the early stages of development in the region's political and economic institutions. Comparable growth estimates are 5.7 percent for East Asia, 3.1 percent for South Asia, and 2.2 percent for Latin America (The World Bank 1992).

Table 14.1.3. Annual growth in real per capita income by region, 1980 to 2000

Region	Period 1980 to 1990	1990 to 2000
	(percent)	
Sub-Saharan Africa	-0.9	0.3
East Asia	6.3	5.7
South Asia	3.1	3.1
Latin America	-0.5	2.2
Middle East and North America	-2.5	1.6
Developing countries	1.2	2.9

SOURCE: The World Bank 1992. (Reprinted with permission)

The Urgent Need to Get Agriculture Moving

W. Arthur Lewis's adage of 1955 still applies: Most Africans are still farmers, and it follows that raising the productivity of smallholder agriculture is a necessary condition for raising the African standard of living and reducing poverty (Lewis 1955). The "food production/population imbalance" forces African agriculture to grow at ambitious levels of 4 to 5 percent annually for at least 20 years, if the food gap is to be closed. These are rates that these nations are not likely to achieve given the less than adequate prior investment in their rural economies.

African scholars and leaders, to their credit, have known for some time that an African solution is required to halt the decline of the agricultural economy. The Lagos Plan of Action in 1980 is the prime example of several unfulfilled declarations which, in fact, did identify the problem squarely. The lack of political will and/or sacrifice to see a shift from urban focus to rural priorities is, in my judgment, a function of a distorted political system rather than a lack of intellectual capacity to internalize the problem.

The Structural Adjustment Programmes (SAP) in the 1980s and 1990s confirmed that poor macroeconomic policies and management contributed to the stagnation of agriculture. By the 1980s most economies were experiencing high inflation, declining real interest rates, budget deficits ranging from 10 to 20 percent of GDP, and deteriorating terms of trade. It would appear, however, that the SAP fell short in their ability to realign the investment patterns with rural priorities. In summary, there is a need to go beyond current reforms to get agriculture moving. Markets have been liberalized and exchange rates devalued; now we need strategies that will achieve a better supply response by small farmers. Public sector investment rates of about 10 percent of agricultural GDP will be woefully inadequate, if this supply response is to be forthcoming. Nearly 30 years after his seminal work, W. Arthur Lewis reflected on the frustration of the stagnation of agriculture by concluding that in the developing world, the failure of agriculture has mainly been at the political level, because the small producer carries little political weight (Lewis 1984).

The prime movers' development framework lays the ground for a strategy that puts the solution in the hands of Africans. Six integrated prime movers must work in tandem to achieve sustainable development:

- public and private investment in land development and agrarian and land tenure reform;

- new technology to meet the changing needs of farmers through sustained investments in research, both public and private;

- human capital at all levels, through long-term investments in education and training facilities for farmers, professionals, managers, and technicians;

- physical and biological infrastructure such as roads, dams, irrigation systems, grain storage, and improved livestock herds and plantations;

- effective institutions, particularly research, extension, credit, and marketing that work as a system to support farmers; and

- an enabling political and economic environment, with top priority for budgetary allocation for agriculture and rural areas, and economic policies that promote private investment in rural areas.

No single prime mover on its own can get agriculture moving on a sustained basis. For instance, improved producer prices through market reforms, while positive, have been unable to achieve a significant supply response. Public sector investments have to cover the six prime movers on a sustained basis. Compared to similar developing regions of Asia, Africa needs greater investment in rural transportation and other infrastructure (Table 14.1.4).

Table 14.1.4.   Selected indicators of surface transport for Africa and Asia, 1990

Region/Country	Rail and road mileage per 1,000 persons	Rail and road mileage per 1,000 hectares of cultivated land	Number of motorized vehicles per mile of paved road
Africa			
Benin	0.17	0.36	9.0
Kenya	0.30	1.09	19.2
Malawi	0.28	1.55	17.0
Senegal	0.44	0.60	12.2
Tanzania	0.15	0.09	14.2
Togo	0.37	1.26	13.5
Zimbabwe	1.09	4.11	35.7
Asia			
Bangladesh	0.07	0.75	47.5
India	0.68	3.58	49.0
Pakistan	0.73	3.71	42.5
Philippines	1.65	7.38	51.8
Korea, Republic of	0.58	11.41	67.2

SOURCE: Plateau 1990, with data from Ahmed and Donovan 1992.

The Political Economy of Food and Agriculture in Africa

Powelson (1994) examines various paths of economic growth in Japan, Europe, and North America and compares these with the developing regions of the world. He concludes that the *power diffusion process* is essential for sustained economic development. The "new" political economy approach in Africa is demonstrating that smallholder farmers have an insignificant voice in the political process. Bratton (1994) uses the case of a smallholder farmers' union in Zimbabwe to demonstrate how neopatrimonial political systems prevalent in Africa make it difficult to achieve a bottom-up development based on the farmers' voice. The minority urbanites monopolize political power, while the rural majority have no viable and/or voluntary farmers' associations or special interest groups that have the power to get their governments to listen to the rural citizens.

Rural Economic Institutions: Putting the Farmer at the Center

The performance of agricultural research and development (R&D) organizations such as research, extension, and credit has been disappointing in Africa. Africa's early stages of institutional development may explain the slow pace of technical change and agricultural transformation. To facilitate transformation, rural economic institutions must function as a system with a common vision and agenda (Bonnen 1990). Such a system of institutions, adequately decentralized and crafted to suit local needs, has the farmer at the center, and not at the bottom as is generally the case. A demand-driven system of rural institutions has the capability to empower and develop farmer and traditional organizations.

Agriculture R&D organizations in Africa need to develop the capacity to act in the economic interest of farmers. Change, therefore, is a constant requirement for these organizations. Institutional reforms are needed to address the decay of African R&D organizations. There is a need to improve the associated value and belief systems, and there is a need for dynamic leadership to steer key institutions through the badly needed reforms. These reforms include the capacity to work within political markets, and require knowledge of cultural factors through enhanced organizational learning. The pursuit of social objectives will be the primary source of institutional change (De Capitani and North 1994)

Political economy issues arise on both the supply and access sides of the food equation. Food access issues bring to the forefront the inseparability of politics and food security. The prevalence and continued occurrence of famine in Africa has instructive lessons for needed political-bureaucratic reform. Famine, due to poor weather or war (or both), has similar devastating effects. The fact that famine is often a result of a cumulative number of bad seasons or of protracted civil conflict reflects the lack of preparedness or political will to address the problem before the disaster. The result is usually a massive loss of life, social disruption, economic chaos, displacement of persons and livestock, distress sales, crime, and a general decay in the moral and customary codes of behavior (Mellor and Gavian 1987).

Botswana, a tiny desert nation of 1.5 million in Southern Africa, is arguably the best example in Africa of government taking the lead in building capacity for famine preparedness and a permanent safety net for the poorest of the poor, and averting famine in this dry country that experiences four droughts every five years. The role of the state in providing good governance has been captured by Timmer (1993, 37): "By and large, markets are good at giving us growth, but they are not very good at giving us stability and equity, and we cannot sustain the growth process unless we have a balance among the three—growth, stability and equity."

Agricultural Markets and Trade

During the period 1979 to 1994, 35 sub-Saharan African countries undertook SAP supported by the World Bank and the International Monetary Fund (IMF). In total, the Bank invested about $13 billion in

SAP, and the IMF an additional $16.4 billion (Donovan 1996). With an additional $3 billion from bilateral sources, SAP represented a $33 billion investment, impacting primarily on agriculture through liberalization of product and factor markets as well as balance of payments impacts due to exchange rate devaluations. The impact on African agriculture has been somewhat mixed, with the jury still out on the long-term effects. It is important, however, to examine the impact in the areas of product and factor markets, food markets, trade, and food aid.

Food Markets

The liberalization of food markets in Africa was arguably the single most important achievement of SAP. Government intervention in food markets has been largely eliminated. In Eastern and Southern Africa, for example, grain markets (particularly maize) are relatively more efficient, with benefits to both producers and consumers (Jayne et al. 1994). The governments in Eastern and Southern Africa had traditionally controlled grain markets while markets for tubers and root crops in Western and Central Africa had been largely uncontrolled. Subsidies to grain markets were, however, largely for consumers aimed at lowering prices of urban food. In the late 1980s, for instance, maize subsidies accounted for 17 percent of the total Zambian government budget (Howard, Chitalu, and Kalonge 1992).

One important research finding on food markets involves the large proportion of rural Africans who rely on food markets for income and food security (Weber et al. 1988). The myth of the self-sufficient, or the Robinson Crusoe-type, African farmer is now far from reality because in many countries a high proportion of rural households are net buyers of basic staples or rely on the market during certain periods of the year. Reducing transaction costs of rural food markets is, therefore, a major contribution to household food security. Deregulation of maize markets in Zimbabwe, for instance, led to a significant increase in low-cost maize meal processing by rural and semiurban hammer mills, capturing the market from large, urban industrial mills that produce refined maize flour.

The opening of food markets has also promoted a more positive linkage between food and cash crops. As opposed to traditional tradeoffs between export/cash crops and food crops, Dioné (1989) found strong links in Mali between cotton production and food crop production stimulated by farm-level and community-level capital formation. Reliable rural food markets strengthen this positive linkage and diminish the historical paranoia over cash crops in Africa.

Factor Markets

Improved seed, fertilizer, and credit are important inputs but have had limited impact on increasing the productivity of African farms. Both the seed and fertilizer industries are in their infancy in sub-Saharan Africa. In the 1990s, the seed industry appeared to have received a positive boost from privatization efforts. The fertilizer industry, however, appears to have suffered from escalating costs due to the devaluation of local currencies. The collapse of the public sector credit system in many of these countries has exacerbated the decline in fertilizer use.

On average, sub-Saharan African farmers applied 15 kilograms per hectare of fertilizer in 1992 to 1993 (Donovan 1996), an increase from 11 kilograms per hectare during 1988 to 1990 (Alexandratos 1995). The annual growth rate in fertilizer use is estimated at 3.3 percent (Table 14.1.5). However, compared to other developing countries, which applied between 60 and 90 kilograms of fertilizer per hectare during 1988 to 1990, sub-Saharan Africa remains well below the average in terms of fertilizer use. The region requires a more positive fertilizer development policy as part of its transformation strategy.

Table 14.1.5. Fertilizer use per hectare in developing countries, excluding China

Region	Fertilizer in kg per hectare of harvested land[a]		Annual growth rate during the period
	1988 to 1990	2010	1988/90 to 2010
			(percent)
Developing countries (excluding China)	62	110	2.8
Africa (sub-Saharan)	11	21	3.3
Near East/North Africa	89	175	3.3
East Asia (excluding China)	79	128	2.3
South Asia	69	138	3.4
Latin America/Caribbean	71	117	2.4

[a] Manufactured fertilizer in kg of nutrient content (nitrogen, phosphate, potash).
SOURCE: Alexandratos 1995. (Reprinted with permission)

In addition to fertilizer, the seed industry and rural financial services are important areas for market development in sub-Saharan Africa. Not only are these key to increasing production, but rural growth linkages are fostered largely through more efficient financial and labor exchange. The seed industry is making most significant strides in the maize-growing regions of Eastern and Southern Africa. Maize promises to be a Green Revolution crop in this region. Unfortunately, the market is severely underdeveloped for the other grains and oilseeds, particularly millet. The struggle of developing the fertilizer, seed, and rural financial markets signifies the undeveloped relationship between the public and private sectors. The industrial capacity of nations also determines the viability of these markets. The general collapse of public sector credit systems can also be related to the narrow supply-side approach of these rural financial services (Mabeza-Chimedza 1994). Greater emphasis will be required on savings mobilization and developing low-cost microfinancing systems that are not only responsive to the production needs for seed and fertilizer, but offer households real options and opportunities to bridge their financial needs, including those for food.

Trade and Food Aid

The signing of the General Agreement on Tariffs and Trade (GATT) in 1993 was expected to usher in a new era of international trade relations for Africa. The partial liberalization of grain trade is expected to hurt Africans because of increased world grain prices. This has an impact on both commercial imports and available food aid (FAO 1994). In addition, Africans are losing on the preferential trade agreements such as the Lome Convention. The reduction in most favored nation tariffs by 30 percent by the European Union (EU) alone is expected to cause the loss of export revenues of $70 million a year for the countries of sub-Saharan Africa.

Over the last three decades, sub-Saharan Africa has suffered the steepest decline in the agricultural terms of trade. Over this period, Africa has lost world market share for all major agricultural products (Table 14.1.6). Between 1970 and 1990, a span of 20 years, sub-Saharan Africa's market share for cocoa beans dropped from 60 to 41 percent, coffee from 30 to 23 percent, palm oil from 19 to 2 percent, cotton from 16 to 14 percent, and bananas from 6.5 to 2.5 percent (Donovan 1996). It is instructive to note that sub-Saharan Africa lost this market share almost entirely to the other developing regions of the world: coffee and bananas to Latin America, cocoa and palm oil to Asia (Malaysia, Indonesia), and cotton to Pakistan.

The United States and Canada traditionally have been the largest food donors. Both GATT and domestic policies, however, mean that food aid from these nations will likely decline significantly. The change in North America toward the reduction in government-induced grain surpluses and public stockholding is expected to increase the cost of food aid, and reduce the volume of program food aid. The short-term implications for sub-Saharan Africa are expected to be negative, on balance, given Africa's poor capacity for supporting commercial food import. In the long term, however, it is possible to institute food aid policies that address its negative impact on local food markets. Opportunities for the monetization of food aid, for instance, may be enhanced. It has also been argued (DeRosa 1995) that the Uruguay Round reforms to agriculture will affect the agricultural exporting countries favorably through increased market share and better prices.

Table 14.1.6.   Market shares for agricultural exports, selected years and projections to 1995

Commodity	1969–71	1979 to 1981	1990	1995 (proj)	1970 to 1990 (growth rate)
		(million dollars)			(percent)
Cocoa					
Sub-Saharan Africa	979	907	1,298	1,386	1.4
World	1,638	2,052	3,190	3,401	3.4
Coffee					
Sub-Saharan Africa	987	900	1,126	1,500	-0.3
World	3,261	3,649	4,869	4,595	1.6
Palm Oil					
Sub-Saharan Africa	186	96	193	190	-2.2
World	1,002	3,230	7,884	11,417	10.4
Cotton					
Sub-Saharan Africa	627	372	699	815	1.3
World	3,929	4,558	5,035	6,820	1.5
Bananas					
Sub-Saharan Africa	389	246	234	236	-4.5
World	5,929	6,900	9,330	10,068	1.3

SOURCE: The World Bank 1992. (Reprinted with permission)

Food aid, therefore, may continue as a controversial form of development assistance. Some of the damage may be difficult to reverse, particularly in the case of the entrenched tastes for wheat and rice products. The reversal of the effects of the massive flows of food aid, representing an annual $1 billion to Africa, or over 30 percent of the region's cereal imports and about 8.5 percent of all official development assistance (The World Bank and World Food Programme [WFP] 1990) will require a reallocation of significant proportions of this development assistance.

The challenge, therefore, is integrating food aid into the national economic strategies for development. This, of course, requires greater capacity by African nations to rationalize and to accept this kind of development assistance. The monetization of food aid requires a functioning food market and an efficient bureaucracy.

Regional Integration and Cooperation

Regional cooperation has yielded varying results in the three major sub-Saharan regions of West and Central Africa, East Africa and the Greater Horn, and Southern Africa. It is more evident today that regional cooperation cannot be sustained simply through establishing or trying to establish common economic markets. The notion of political markets or politically viable communities is allowing Africans to build the capacity to assist regional neighbors on matters of civil conflict or war. The Southern Africa Development Conference (SADC) is an example of a loose regional grouping of originally diverse political and economic nations. Sixteen years after the establishment of SADC, these nations appear to have converged their perceptions and aspirations in economic and political ideology. Nations are assisting one another economically and politically. The chances of regional integration are greater with the entrance of South Africa into the community.

In 1992, Southern Africa faced the worst drought in recorded history. The region required a record over 10 million tons of imported cereal in one year. On the African continent a food deficit of this size would normally mean massive starvation and civil upheaval. A catastrophe was averted mainly because the 10 regional nations, over a period of a decade, had crafted a crude framework for regional food security to provide early warning and open up regional transportation routes. When disaster struck, it took a few months for a regional commission to smooth the bureaucratic process, mobilize local and foreign assistance, and move grain into the interior of land-locked countries. Zimbabwe, importing in excess of 2 million tons in that season, allocated an African record of more than 1 million tons to commercial imports in one season. This Southern Africa case, with lessons that require greater analysis and dissemination, is an example of slow but steady progress toward regional integration, taking into account their major differences in historical, political, economic, and colonial histories.

Science and Technology

Can science and technology provide solutions for African farmers? The intensification of smallholder agriculture, and its transformation to more science-based production systems, is now an urgent requirement if Africa is to meet its food needs as well as raise incomes and create jobs. It can be argued, however, that Africa has yet to develop a capacity for research and development consistent with the region's human and natural resources. Whether or not technology exists on the continent to transform African agriculture has been a difficult issue to agree upon, largely because of the gap between research and extension, and also because of a poor interface between public sector institutions and farming communities.

Agriculture R&D organizations, therefore, will continue to have a limited impact, unless indigenous scientific and intellectual leadership make it part and parcel of their political agenda that they become more demand driven. There is a need to balance the current largely "supply-side research" driven by national scientists and international centers with "demand-driven research" through empowerment of and support to smallholder economic interest groups. The financial cuts during the 1980s and 1990s have further incapacitated public sector R&D organizations. The rapid buildup of research staff during this period, for instance, was not matched with growth in financial resources (Pardey, Roseboom, and Beintema 1995). Spending per scientist declined drastically during the 1980s, although the decline has continued for the past 30 years.

A number of studies during the 1980s and 1990s have shown a generally high rate of return to public investment in agricultural research for selected commodities in Africa. But there have not been many adoption studies to focus on institutional blockages to technology development and transfer. Because of the lack of adequate focus on institutional innovation, R&D organizations have not developed the capacity for change and reform that is aimed at building capacity for institutional innovation; that is, proactive, anticipatory, or creative actions that reduce the cost of technology development and transfer. African R&D organizations should regard institutional innovation as a product of organized research and analysis, just as biophysical innovations are a result of good science. While both biophysical and institutional innovations may have a serendipitous quality, scientists have to proceed with a conscious effort to achieve both.

The poor client-research-extension linkages, therefore, represent a dubious and artificial separation between research and extension. The linear sequential view of knowledge and information has contributed to the poor participation of farmers in the technology generation and transfer process. Scientists have yet to develop the capacity to talk with and listen to farmers who generally have practical skills and ideas that could assist scientists with problem-solving research.

Agriculture R&D institutions need to be crafted and reformed in total, through the pragmatic approaches to learning by doing and through institutional experimentation. This is suggested because institutional development is location-specific and depends on the nation's capabilities and developmental state. Leadership of these organizations is extremely important because growth and development of institutions depends on the learning and adaptive capacity of its leader. Institutional reforms to date have focused more on issues of management than on leadership. Support to National Agricultural Research Systems (NARS) by International Service for National Agricultural Research (ISNAR) and similar international organizations and donors has of late emphasized issues of planning and priority setting, as well as management of national research systems. The numerous efforts by most African nations to institute research master plans, either voluntarily or in anticipation of donor funds, do not appear to have yielded positive results (Maredia et al. 1996).

Leadership is important in providing vision, values, ideology, and doctrine necessary for effective institutions. Strategic planning and priority setting are the sum result of values and a vision to do the right things. Leadership, as distinct from management, is important when organizations are undergoing change and/or crisis. Formal priority setting exercises and research master plans prepared so far by African NARS have demonstrated that these are not good substitutes for visionary leadership and scientific maturity. Even without formal priority setting, effective scientific organizations have in the past been able to determine problems of scientific importance and identify those with feasible solutions.

Finally, African R&D organizations, particularly research, extension, and credit, need to develop the capacity for technical change and the ability to act in the economic interest of small farmers. Their clientele has to be consciously and explicitly expanded to include farmers, scientists, manufacturers of agricultural products, lenders, agribusiness firms, nongovernment organizations (NGO), users of research results, and administrators and politicians who make budgetary decisions. More courageous approaches to institutional reform are now required. It is not enough to tinker with the "soft" part of the research system, addressing issues of priority setting and research master plans. The time has come to address the people-related issues of leadership, values, client orientation, scientific maturity, and motivation to tackle significant problems that will make a difference to poor farmers' lives.

Research and development organizations have to tune into both economic and political markets as the sources of institutional innovation. These organizations have to support and empower farmer organizations and special interest groups as part of their mandate. Leadership is required that is able to identify and engage in an appropriate and revolutionary form of science in poor societies where technology that works is more important than simply trying to do good science. As research funding becomes more scarce, NARS must actively seek the patronage of their own farmers as opposed to donors. The artificial separation of research and extension, and the linear top-down approach that this encourages, must be dispensed with and replaced by a "farmer at the center" approach where R&D organizations develop common agendas with the farmer.

Conclusion

An African solution to the food crisis is needed and is feasible. The painful process of economic reforms that most sub-Saharan countries have gone through brings about the realization that political reforms are even more painful yet necessary. Food security is more than an economic issue; the fight against poverty, hunger, and malnutrition now requires solutions that bear on politics and governance. If an acceptable decline in hunger is to be achieved, Africa at best requires major investments for the next 20 years to get agriculture moving. Poverty and hunger are long-term problems requiring long-term solutions.

The development of rural economic institutions and the empowerment of farmers and rural communities is a prerequisite for developing effective public sector support services and organizations. Food markets and factor markets are beginning to function more efficiently, but the lack of technology and a poor social infrastructure do not permit greater production and distribution. The development of local markets is necessary before Africa can recapture lost world trade markets.

Developing indigenous capacity for science-based agriculture that is consistent with African resources and social structure is now an urgent need. In addition, institutional reforms that give public sector organizations the entrepreneurial skills to exploit rural economic opportunities is required. Good governance and effective leadership should be consciously factored into development assistance. Governments must demonstrate care and take responsibility for the nutritional well-being of their citizens through effective food policies, safety nets, and economic policies that incorporate food aid into mainstream development activities, income generation, and job creation.

References

Ahmed, R., and C. Donovan. 1992. *Issues in Infrastructural Development: A Synthesis of the Literature.* Washington, D.C.: International Food Policy Research Institute.

Alexandratos, N., ed. 1995. *World Agriculture Towards 2010. An FAO Study.* New York: FAO and John Wiley and Sons.

Bonnen, James T. 1990. Agricultural Development: Transforming Human Capital, Technology, and Institutions. In *Agricultural Development in the Third World,* edited by C. K. Eicher and J. M. Staaty. 2d ed. Baltimore: Johns Hopkins University Press, 262–79.

Bratton, M. 1994. Micro Democracy? The Merger of Farmer Unions in Zimbabwe. *African Studies Review* 37(1):9–37.

De Capitani, A., and D. C. North. 1994. Institutional Development in Third World Countries: The Role of the World Bank. Draft paper discussed at The World Bank Seminar, March 11.

DeRosa, D. A. 1995. *The Uruguay Round Agreement on Agriculture and Sub-Saharan Africa in Re-Establishing Agriculture as a Priority for Development Policy in Sub-Saharan Africa.* Edited by Abdulai and C. L. Delgado. Washington, D.C.: International Food Policy Research Institute; Zurich: Swiss Federal Institute of Technology, and Swiss Development Cooperation.

Dioné, J. 1989. Policy Dialogue, Market Reforms and Food Security in Mali and the Sahel. In *Food Security Policies in the SADCC Region; Proceedings of the Fifth Annual Conference on Food Security Research in Southern Africa,* edited by M. Rukuni, G. Mudimu, and T. S. Jayne. Harare, Zimbabwe: University of Zimbabwe/Michigan State University Food Security Research Project, 143–70.

Donovan, G. D. 1996. Agriculture and Economic Reform in Sub-Saharan Africa. AFTES Working Paper No. 18. Washington, D.C.: The World Bank.

Food and Agriculture Organization (FAO). 1994. *The State of Food and Agriculture 1994.* Rome: FAO.

_____. 1996. *World Food Summit: Food Security Situation and Issues in the Africa Region.* Nineteenth FAO Regional Conference for Africa. Ouagadongouu, Burkina Faso, April 16–20.

Howard, Vulie, George Chitalu, and Sylvester Kalonge. 1992. The Impact of Investments in Maize Research and Dissemination in Zambia: Preliminary Results. Paper presented at the Symposium on the Impact of Technology on Agricultural Transformation in Africa, Washington, D.C.: USAID.

Jayne, T. S., D. L. Tschirley, J. M. Staatz, J. D. Shaffer, M. T. Weber, M. Chisvo, and M. Mukumbu. 1994. *Market-Oriented Strategies to Improve Household Access to Food: Experience from Sub-Saharan Africa.* International Development Paper No. 15. East Lansing: Department of Agricultural Economics, Michigan State University.

Lewis, W. Arthur. 1955. *The Theory of Economic Growth.* London: George Allen and Irwin.

_____. 1984. Development Economics in the 1950s. In *Pioneers in Development,* edited by G. Meier and D. Seers. New York: Oxford University Press.

Mabeza-Chimedza, R. M. 1994. *Rural Financial Markets in Zimbabwe's Agricultural Revolution,* edited by M. Rukuni and C. K. Eicher. Harare, Zimbabwe: University of Zimbabwe Publications.

Maredia, M. K., D. Boughton, J. A. Howard, D. Karanja, M. C. Collion, O. Niangando, and T. Bedinjas. 1996. *No Shortcuts to Progress: Case Studies of the Status and Impact of Strategic Agricultural Research Planning.* Draft paper. East Lansing, MI: Michigan State University, Department of Agricultural Economics.

Mellor, J. W., and S. Gavian. 1987. Famine: Causes, Prevention, and Relief. *Science* 23(5):538–45.

Pardey, P. G., J. Roseboom, and N. M. Beintema. 1995. *Investments in African Agricultural Research.* EPTD Discussion Paper No. 14.

Plateau, J-P. 1990. The Food Crisis in Africa: A Comparative Structural Analysis. In *The Political Economy of Hunger,* edited by J. Dreze and A. Sen. Vol. 2. Oxford: Clarendon Press.

Powelson, John P. 1994. *Centuries of Economic Endeavor: Parallel Patterns in Japan and Europe and Their Contrast with the Third World.* Ann Arbor, MI: University of Michigan Press.

Sen, Amartya K. 1981. *Poverty and Famine.* Oxford: Clarendon Press.

Timmer, C. Peter. 1993. Keynote Address: Setting the Stage. In *Agricultural Transformation in Africa,* edited by D. Seckler. London: Heinemann.

United States Department of Agriculture (USDA). 1995. *Food Aid Needs and Availabilities; Projections for 2005.* Economic Research Service Report. Washington, D.C.: USDA, Economic Research Service.

Weber, M., J. Staatz, J. Holtzman, E. Crawford, and R. Bernsten. 1988. Informing Food Security Decisions in Africa: Empirical Analysis and Policy Dialogue. *American Journal of Agricultural Economics* 70(5):1044–53.

World Bank. 1989. *Sub-Saharan Africa: From Crisis to Sustainable Growth.* Washington, D.C.: The World Bank.

_____. 1992. *Market Outlook for Major Primary Commodities.* Vol. II. Washington, D.C.: The World Bank, International Economics Department.

_____. 1995. *A Strategic Vision for Rural Agricultural and Natural Resources Activities of the World.* Washington, D.C.: The World Bank.

World Bank and World Food Programme (WFP). 1990. *Food Aid in Africa: An Agenda for the Nineties.* Washington, D.C.: The World Bank; Rome: World Food Programme.

Background References

Busch, L., and W. B. Lacy. 1983. *Science, Agriculture, and the Politics of Research.* Boulder, CO: Westview Press.

Covey, Stephen R. 1989. *The 7 Habits of Highly Effective People.* New York: Simon and Schuster.

Eicher, C. K. 1985. Famine Prevention in Africa: The Long View. In *Food for the Future.* Proceedings of the Bicentennial Forum held at Philadelphia on November 6. Philadelphia: Society for Promoting Agriculture, 82–101.

_____. 1990. Building African Scientific Capacity for Agricultural Development. *Agricultural Economics* 4(2):117–43.

Governments of Canada and the United States. 1996. *World Food Summit: Food Security Situation and Issues: A US/Canada Perspective.* Washington and Ottawa: Joint Canada and U.S. release.

Jones, S. 1989. The Impact of Food Aid and Food Markets in Sub-Saharan Africa. A Review of the Literature. Working Paper No. 1. Food Studies Group. Oxford: University of Oxford.

Rosen, S. 1989. *Consumption Stability and the Potential Role of Food Aid in Africa.* Washington, D.C.: USDA, Economic Research Service, Agriculture and Trade Analysis Division.

Rukuni, M. 1994. Report of the Commission of Inquiry into Appropriate Land Tenure Systems. Vol. I, Main Report; Vol. II, Technical Reports; Vol. III, Methods Procedures, Itinerary, and Appendices. Harare, Zimbabwe: Commission of Inquiry.

Rukuni, M. Ackello-Ogutu, H. Amani, P. Anandajayasekeram, W. Mwangi, H. Sigwele, and T. Takavarasha. 1994. *Getting Agriculture Moving in Eastern and Southern Africa and a Framework for Action.* Discussion paper for the Eastern and Southern Africa Conference of Agriculture Ministries held in Harare, Zimbabwe, April 12–15.

Staatz, J., T. Jayne, D. Tschirley, J. Schaffer, J. Dioné, J. Oehmke, and M. Weber. 1993. *Restructuring Food Systems to Support a Transformation of Agriculture in Sub-Saharan Africa: Experience and Issues.* Staff paper No. 93–96. East Lansing, MI: Department of Agricultural Economics, Michigan State University.

Resource Paper ◆ *Yang Yansheng and Jikun Huang*

Agricultural Development and Food Policy in China: Past Performance and Future Prospects

Introduction

China, with more than 1.2 billion people at the end of 1995, is the world's most populous country. Since 1978 when China opened up and gradually began to broaden the role of the market mechanism in its economy, it has enjoyed spectacular economic growth. The growth rate of the gross domestic product (GDP) during the reform period since 1978 is roughly double that of the immediate pre-reform period, 1970 to 1978 (Table 14.2.1). The GDP has expanded at an average annual rate of more than 9 percent over the last 15 years. This strong economic growth has brought about a significant improvement in the living standards and a substantial reduction in absolute poverty in China (The World Bank 1992; MOA 1995).

Growth of the agricultural economy has provided the foundation for the successive transformations of China's reform economy. China's agricultural performance over the past two decades has been more impressive than that of other countries in South and Southeast Asia. Although the growth rates of agricultural production have declined in China (Table 14.2.1), they have also declined in all other Asian countries due to the declining comparative advantage of the sector with general economic growth. China, in fact, was one of the countries with the highest growth rate in agricultural production and trade in East, South, and Southeast Asia during the 1980s.

Table 14.2.1.   The annual growth rates of China's economy, 1970 to 1995

	Pre-reform	Reform period	
	1970 to 1978	1978 to 1984	1984 to 1995
	(percent)		
Gross domestic products[a]	4.9	8.5	9.7
Agriculture	2.7	7.1	4.0
Industry	6.8	8.2	12.8
Service	NA	11.6	9.7
Foreign trade	20.5	14.3	15.2
Import	21.7	12.7	13.4
Export	19.4	15.9	17.2
Population	1.8	1.4	1.4
GDP per capita	3.1	7.1	8.3
Consumption per capita	1.6	8.0	6.8
Rural	0.6	9.2	4.9
Urban	3.7	5.2	7.6

NOTES: Growth rates were computed using regression method. GDP and per capita consumption growth rates refer to values in real terms.
[a] The figure for GDP for 1970 to 1978 is the growth rate of national income in real terms.
SOURCES: SSB, *Statistical Yearbook*, various issues; *Statistical Survey* 1996.

This economic growth is a notable accomplishment by historic standards. Moreover, the foundations for continued expansion of the Chinese economy appear to be in place. The government intends to maintain growth rates near current levels throughout the decade. The sheer size of China's economy, its rapid growth, its gradual progress toward a market-oriented economy, its increasing integration into the global economy, its urbanization, the shifting of comparative advantage of economic sectors, dietary diversification, diminishing agricultural land, and population growth will require a more dynamic agriculture in the coming years.

The Role of Agriculture and Agricultural Sector Performance

The Role of Agriculture in the Economy

Over the past few decades, Chinese agriculture has made important contributions to the development of the national economy in terms of gross value added, employment, capital accumulation, welfare of urban consumers, and foreign exchange earnings. Even though its share in the national economy has declined, agriculture directly contributed more than 20 percent to both GDP and export earnings in the early 1990s. The agricultural sector has also played an important role in industry development in terms of capital accumulation and market expansion as well as in providing industrial raw materials. Currently, two-thirds of the light industry outputs depend on agricultural products as raw materials. Including processed agricultural exports, the share of agriculture was about 40 percent of total exports in the early 1990s. Moreover, the sector provided more than one-half of the country's employment, and about 75 percent in rural China in 1995.

It is noteworthy that the decline in agricultural comparative advantage has not been reflected in agricultural self-sufficiency for China (Table 14.2.2), indicating the critical role of the government's food self-sufficiency policy. Rice and coarse grains self-sufficiency levels have declined by only about 3 percent in the last four decades. The wheat self-sufficiency level has increased over this period.

Table 14.2.2.   Self-sufficiency ratios for grain products and cotton in China

Year	Rice	Wheat	Coarse grain	Cotton
1961 to 1969	103	83	99	94
1970 to 1979	104	94	103	94
1980 to 1984	101	86	100	89
1985 to 1989	100	87	96	102
1990 to 1995	100	91	96	95

SOURCES: Computed from SSB, *Statistical Yearbook,* various issues; MOFERT, various issues.

Agricultural Production and Trade Growth

China is a highly land-scarce country. Total cultivated land was about 95 million hectares in the early 1990s. However, measured on a per capita basis, cultivated land area has been around 0.1 hectares since the 1980s (declining from 0.13 in 1970 to 0.08 in 1995). Despite limiting natural resources, agricultural production has grown at a remarkable rate over the past decades. The growth of agricultural production in China since the 1950s, particularly during the 1979 to 1995 reform period, has been one of the main accomplishments of the national development policy. Except during the famine years of the late 1950s and early 1960s, China has enjoyed production growth rates that have outpaced the increase in population.

Grain production reached a record level of 466.6 million tons in 1995, an increase of 146 million tons or 46 percent over 1980 (Table 14.2.3). A new record of 488 million tons was reached in 1996. Among grains, cereal production grew at an even higher rate. Cotton production hit a historical high of 4.77 million tons in 1995, an increase of 76 percent over the last 15 years. The production of oil-bearing crops reached 22.44 million tons in 1995, an increase of 14.75 million tons or about three times the production in 1980. Sugar-bearing crops, vegetables, and fruit production all grew at significant rates of more than 10 percent annually.

Table 14.2.3.   Selected major crop and livestock production, 1980 to 1995

	1980	1985	1990	1995
	(million tons)			
Total grains	320.6	379.1	446.2	466.6
Paddy rice	139.9	168.6	189.3	185.2
Wheat	552.1	85.8	98.2	102.2
Maize	62.6	63.8	96.8	112.0
Cotton	2.7	4.2	4.5	4.8
Oilseeds	7.7	15.8	16.1	22.4
Sugar crops	29.1	60.5	72.2	79.4
Fruit	6.8	11.6	18.7	42.1
Red meat	12.1	17.6	25.1	42.5
Pork	11.3	16.6	22.8	36.5
Poultry	1.1	1.6	3.2	9.4
Aquatic products	4.5	7.1	12.4	25.3

SOURCES: SSB, *Statistical Yearbook* 1995; *Statistical Survey* 1996.

The livestock sector grew at an even faster rate than the crop sector as incomes increased and demand for meat expanded. The total production of red meat (pork, beef, and mutton) was at a historical high of 42.54 million tons in 1995, with an annual growth rate of about 9 percent over the last 15 years, or 250 percent higher than the level in 1980. Fisheries production experienced more dramatic growth during the reform period. Total production of aquatic products increased to 25.25 million tons in 1995, nearly six times the production in 1980.

Agricultural trade has also grown substantially during the reform period, even though at a lower rate than for nonagricultural sectors. Total exports of food, beverages and tobacco, inedible materials, and oil and fats increased more than three times, from $4.9 billion in 1980 to $16.3 billion in 1995. Total imports grew at a similar rate, rising from $6.7 billion to $19.2 billion during the last 15 years. The pattern and structure of agricultural trade continues to change as the economic reforms progress.

Agricultural Development Strategy and Policies

Government Targets for Agriculture in 2000 and 2010

Food self-sufficiency has been and will continue to be the central goal of China's agricultural policy. The Ninth Five-Year Plan for 1996 to 2000 and the National Long-Term Economic Plan for 2010 envision continued growth in agricultural production and farmer income at 4 percent annually, maintaining food self-sufficiency levels and eliminating absolute poverty (Li 1996). These two plans strive to achieve the following targets by 2000 and 2010:

- increase grain sown area and yield, raising grain production to 490 to 500 million metric tons (mmt) (unprocessed grain; measured in traded grain this is about 430 to 440 mmt) with an annual increase of 10 mmt by 2000;

- develop sustainable growth in grain production through increased public investment in agriculture and in science and technology, with a target of 560 mmt of grain production by 2010;

- strengthen the cotton production base to increase cotton production to 5.25 mmt by 2000;

- increase agricultural output value (in 1990 prices) by 4 percent per year to reach 1,210 billion yuan by 2000, and maintain a similar rate of growth throughout the first decade of the twenty-first century;

- promote rural enterprise development and increase the township and village enterprise (TVE) output value to 5,290 billion yuan (1990 prices) with annual growth rates of 18 to 19 percent for the remainder of the 1990s; and

- increase poverty alleviation funds and eliminate poverty among 70 million absolutely poor by 2000.

The strategy for achieving these targets and goals includes measures to deepen rural economic and institutional reforms; improve incentives for farmers and the local government to invest in agriculture; protect natural resources, specifically land and water, and use them more efficiently; provide incentives through input and output prices to increase multiple cropping; take new measures to strengthen the application of scientific methods to the sector; undertake structural adjustment in the rural economy; optimize agricultural production linkages; strengthen anti-poverty programs; and further open China's agriculture sector to foreign investment and improve the efficiency of foreign capital use in agriculture.

Agricultural Policies

Improve incentives to invest in land. One recent significant change in the institutional arrangements for agricultural production is the renewal of the land contract system. The new land contract introduced in 1994/95 extended the term by 30 (or 50) years from the expiration of the original contract. While the collective land ownership regime remains unchanged, land-use rights can be transferred with payment. This policy change was designed to overcome the problem of farmers being increasingly unwilling to invest because of unclear land titles and the small scale of their holdings. It also has implications for land consolidation and the commercialization of agricultural production. It is expected that the development of a more efficient land contract system, together with the evolving institutional and legal framework, will have a significant impact on incentives for agricultural production and land investment.

Public financial and investment policies. China's progress in the development of rural infrastructure has been a major source of productivity gain. Among various components of the rural infrastructure, water control and capital construction are the primary areas of government intervention. While much of the labor for China's irrigation development was contributed by local residents, public irrigation expenditures financed a big part of the construction of the national water control network. Irrigated area increased from less than 18 percent of cultivated land in 1952 to about 50 percent in the early 1990s.

On the other hand, the ratio of public expenditure in agriculture to total public expenditure declined during the early reform period (Huang, Rozelle, and Rosegrant 1996). The decline in government expenditure for agriculture during the 1980s has focused great attention on the sustainability of future agricultural production growth. The investment policy was reviewed and agricultural investment was increased in the early 1990s. Both the Ninth Five-Year Plan (1996 to 2000) and China's National

Long-Term Economic Plan for 2010 envision the government increasing investment in agriculture, including rural infrastructure, and loans and credits for agricultural production. Irrigation and water control are the first priorities for the government investment within "base industries."

Research and technology development. Developed from almost nothing in the 1950s, China's research system grew rapidly after 1960. It has been successful in producing a steady flow of new varieties and other technologies since the 1950s. Farmers used semidwarf varieties developed in China several years before the start of the Green Revolution technology elsewhere. China was the first country to develop and distribute varieties of hybrid rice. Chinese-bred corn, wheat, and sweet potato technologies were comparable to the best in the world during the pre-reform period.

Since the 1980s, China has gradually implemented a series of science and technology reforms by shifting funding from institutional support to competitive grants. These grants support mainly scientific research useful for economic development, and encourage applied research institutes to support themselves financially by selling the technology they produce. Although competitive grant programs have increased the effectiveness of China's agricultural research system, the reliance on commercialization revenue to subsidize research and make up for a falling budgetary commitment has weakened the system. Realizing the recent weakening of the research system and the importance of science and technology in agricultural development, the Chinese government has set up several programs to stimulate agricultural technology development and farmers' adoption of new technologies through increasing public investment in agricultural research and extension. A new strategy of economic development was introduced in the national conference for science held in 1995; it was called "Building Up the Country through Development of Science and Education." Both the Ninth Five-Year Plan and the National Long-Term Economic Plan for 2010 foresee that China will rely largely on introducing new technologies to raise agricultural production, particularly new varieties of grain, oil-bearing crops, and cotton.

Price and marketing policies. Price and market reforms have been key components of China's development policy and the gradual shift from a socialist to a market-oriented economy. Although the process has been characterized by cycles of deregulation and reinstatement of controls, there has been a definite trend toward liberalization. Moreover, nominal protection rates, while in general high, have declined rather than increased with rapid economic development, as predicted by the economic theory of agricultural protection.

Other important developments in the agricultural marketing system have been the "two-line operation system" for marketed grain and the provincial governor's "Rice Bag" responsibility system. The former is designed to separate policy operation of state grain procurement from market operations. The system was tested in 1994 and the government decided in July 1995 to implement the reform countrywide in 1996. The provincial governor's "Rice Bag" responsibility system seeks to raise investment in agriculture by various local governments, particularly in grain-deficit provinces. More reforms and adjustments in this area will be made in the coming years to encourage grain market integration and efficiency of resource allocation.

Agricultural trade policy. There have been significant changes in the nature and extent of government trade interventions during the reform process. Trade policy reform has involved the introduction of greater competition in international trading and the gradual development of instruments for indirect control. Among agricultural products, there is also a significant degree of redundancy in tariff protection. At the Asia-Pacific Economic Cooperation (APEC) summit meeting in November 1995, China's President Jiang Zeming made a commitment to reduce tariffs by at least 30 percent in 1996. China also eliminated quotas, licensing, and other import controls on 176 tariff lines, or more than 30 percent of products subject to these restrictions.

Antipoverty policy. Both central and local governments have sustained their strong commitment to poverty alleviation. Emphasis has been placed on economic development programs through subsidized bank loans and on infrastructure improvement through the Food-for-Work program in poorer areas.

Prospects for China's Food Economy

One of the most closely watched debates by both administrators and researchers of China's agricultural economy, particularly the food economy, whether inside or outside the country, is the question: Will China be able to produce most of what it needs to feed itself in the twenty-first century? The preponderance of evidence produced by those who have seriously addressed this issue favors the viewpoint that China will be able to feed itself even though grain imports will likely rise over the next several decades (FAO 1993; IFPRI 1995; Rozelle, Huang, and Rosegrant 1996; Mei 1996). On the other hand, more pessimistic scholars (i.e., Brown 1995) forecast that China will face more than 100 or even several hundred million metric tons of grain shortage during the first three decades of the twenty-first century.

This section presents our assessment of China's food economy for the early twenty-first century. Our projections, which incorporate the most recent government policies, are based on a comprehensive projection framework developed by a research group in the Chinese Academy of Agricultural Sciences (Huang and Rozelle 1996). In the projection, beside population, income, and prices, a series of important structural factors and policy variables is accounted for explicitly. The latter include urbanization and market development on the demand side, technological change, agricultural investment, environmental trends, and institutional innovation on the supply side. After presenting these baseline projections, alternative scenarios are examined under different growth rates for income, population, and government investment. Key assumptions underlying these demand and supply projections are summarized in Table 14.2.4.

Table 14.2.4.   Assumptions for the growth of factors affecting grain suppy and demand in China, 2000 to 2020

	Low	Baseline	High
		(millions)	
Population			
2000	1,271	1,278	1,292
2010	1,359	1,376	1,411
2020	1,440	1,468	1,525
		(growth rate in percent)	
Per capita real expenditure			
Rural	2.00	3.00	4.00
Urban	2.50	3.50	4.50
Agricultural research and water control investment	2.50	3.50	4.50
		(percent)	
Share of urban population			
2000	34.00	34.00	34.00
2010	42.00	42.00	42.00
2020	50.00	50.00	50.00
		(index number)	
Rural market development index			
2000	0.60	0.60	0.60
2010	0.70	0.70	0.70
2020	0.80	0.80	0.80

NOTE: Assumptions were derived from the Ninth Five-Year Plan and China's National Long-Term Economic Plan to 2010 (Li 1996). Price trends were assumed to follow world market trends.

Baseline Projections

According to this analysis, per capita food grain consumption in China hit its zenith in the late 1980s and early 1990s, and will fall over the forecast period of 2000 to 2020 (Table 14.2.5). In contrast, per capita demand for red meat is forecast to rise sharply throughout the projection period. China's consumers will more than double their meat consumption by 2020, from nearly 20 kilograms per capita in 1994 to 43 kilograms. Rural demand will grow more slowly than overall demand, but urbanization trends will shift more people into the higher-consuming urban areas. Although starting from lower levels, per capita demands for poultry and fish rise proportionally more.

Table 14.2.5.   Projected annual per capita food grain consumption under two income growth scenarios for China, 2000 to 2020

	2000	2010	2020
	(kilograms)		
Baseline			
Food grain	223	214	203
Red meat	23	32	43
Poultry	3	5	8
Fish	10	17	28
Low-income growth			
Food grain	220	211	202
Red meat	22	27	34
Poultry	3	4	6
Fish	9	14	20
High-income growth			
Food grain	225	215	203
Red meat	25	36	53
Poultry	4	6	10
Fish	11	21	40

NOTE: Authors' estimates.

The projected rise in the demand for meat, poultry, fish, and other animal products will put pressure on aggregate feed grain demand. Feed grain as a proportion of total grain utilization will grow from less than 25 percent in 1995 to more than 30 percent in 2010 and about 40 percent in 2020. This process of moving from an agricultural economy that produces grain primarily for food to one that is becoming increasingly animal feed–oriented is typical of rapidly developing economies elsewhere in the world.

When considered along with the projected population rates, the projected per capita demands for food and feed grain imply that aggregate grain demand in China will reach 445 mmt (in trade grain form or trade weight) by 2000 (Table 14.2.6), an increase of about 15 percent over the

Table 14.2.6.   Projections of grain production, demand, and net imports under selected scenarios with respect to population, income, and technology, 2000 to 2020

Alternative scenarios	2000			2010			2020		
	Demand	Production	Net imports	Demand	Production	Net imports	Demand	Production	Net imports
				(million metric tons)					
Baseline	445	426 (485)[a]	19	507	486 (550)	21	588	570 (640)	18
High income and high investment	454	428 (487)	26	532	517 (538)	15	644	643 (717)	2
High income, high investment, and low population growth	452	428 (487)	24	525	517 (583)	8	626	643 (717)	-16

[a] Figures in parentheses are for grain production in unprocessed form.
NOTE: Authors' estimates.

level of the early 1990s. Although per capita food demand falls in the later projection period, total grain demand continues to increase through 2020, mainly because of population growth and the increasing importance of meat, poultry, and fish in the average diet. By the end of the forecast period, aggregate grain demand will reach 588 mmt, about 50 percent higher than the initial baseline demand (Table 14.2.6).

Aggregate grain supply is predicted to reach 426 mmt in trade weight (equivalent to 485 mmt in unprocessed grain form) by 2000. This projection implies an increase in grain output of about 11 percent over the early 1990s or 5 percent over 1995, a figure close to the government target in the Ninth Five-Year Plan, which is 430 to 440 mmt by 2000 (or 490 to 500 mmt in unprocessed grain weight). Aggregate domestic grain production is expected to reach 486 mmt in 2010 and 570 mmt by 2020.

Under the projected baseline scenario, the gap between the forecast annual growth rate of production and demand implies a slight deficit. Imports are projected not to greatly exceed the 1995 level, and are estimated at about 19 mmt in 2000 (Table 14.2.6). After peaking in 2010 at 21 mmt, grain imports remain at the 20 mmt level through 2020.

Alternative Projections

To test the sensitivity of the results to changes in the underlying forces driving the supply and demand balances, two alternative policy scenarios were produced. Baseline growth rates of the key variables were altered, including income, agricultural investment, and population. The results, shown in Table 14.2.6, indicate that with high income growth, projected total grain demand will increase by 9, 25, and 56 mmt in 2000, 2010, and 2020, respectively, compared with the baseline. Because high income growth may offer a high investment in agricultural technology and infrastructure, high investment growth raises production to 643 mmt in 2020, 73 mmt more than the baseline projection.

Simulation results for this scenario indicate that grain imports could increase by 7 mmt (from 19 to 26 mmt) by 2000, but they would be reduced by 16 mmt (from 18 to 2 mmt) in 2020, if the high income growth scenario is accompanied by the high investment assumption. The import level falls farther to only 8 mmt by 2010 and shows slight exports of grain after 2015, if the high income and high investment are combined with a low population growth assumption (Table 14.2.6).

Summary and Conclusions

Over the past several decades agriculture has made an important contribution to the development of the national economy in China. This is a broadly acclaimed achievement. China has been able to feed 22 percent of the world's population on only about 7 percent of the arable land. Food self-sufficiency has been the central goal of China's agricultural policy, and will continue to be so in the future. The Ninth Five-Year Plan for 1996 to 2000 and the National Long-Term Economic Plan for 2010 envision continued growth in agricultural production and farmer income at about 4 percent annually, maintaining food self-sufficiency levels and eliminating absolute poverty.

Projections from a model incorporating likely agricultural policies and economic trends show that under the most plausible expected growth rates in the important conditioning factors, China's imports will rise slightly throughout the next decade. By 2000, imports are expected to reach 19 mmt, similar to the 1995 level. The increasing imports arise mainly from the accelerating demand for meat and feed grains. However, after 2000, grain imports are shown to stabilize. Supply growth is sustained with an ongoing investment in agricultural research and irrigation.

Alternative projections also show that China will neither empty the world grain markets nor become a major grain exporter. It does seem likely, however, that China will become a more important player in world grain markets as an importer in the coming decades. Net grain import of 20 mmt is a likely annual level for China's grain, and is also a figure likely to be acceptable to the Chinese government. This level represents 4 to 5 percent of total domestic grain demand.

References

Brown, L. 1995. *Who Will Feed China? Wake-up Call for a Small Planet.* The Worldwatch Environmental Alert Series. New York: W. W. Norton.

Food and Agriculture Organization of the United Nations (FAO). 1993. *World Agriculture: Toward 2010.* Rome: FAO.

Huang, J., and S. Rozelle. 1996. Technological Change: Rediscovering the Engine of Productivity Growth in China's Agricultural Economy. *Journal of Development Economics* 49(2):337–69.

Huang, J., S. Rozelle, and M. Rosegrant. 1996. *China's Food Economy to the 21st Century: Supply, Demand, and Trade.* IFPRI 2020 Discussion Paper. Washington, D.C.: International Food Policy Research Institute.

International Food Policy Research Institute (IFPRI). 1995. *Population and Food in the Early Twenty-first Century: Meeting Future Food Demand of an Increasing Population,* edited by Nurul Islam. Washington, D.C.: IFPRI.

Li, Peng. 1996. *Guanyu Guominjinji He Shehui Fazhan Jiuwu Jihu He 2010 Nian Yuanjing Mubiao Gangyao De Baogao. (A Report of the National Economy and Social Development for the Ninth Five-Year Plan and 2010 Long-Term Goals).* Beijing: People's Press.

Mei, F. 1996. Food Development in China in the Early 21st Century. *Food and Nutrition in China* 2(1996):27–9.

Ministry of Agriculture (MOA). 1995. *China's Agricultural Development Report.* Beijing: MOA.

Ministry of Foreign Economic Relations and Trade (MOFERT). Various issues. *Almanac of China's Foreign Economic Relations and Trade, and China Custom Statistics.* Beijing: China's Social Publishing House.

Rozelle, S., J. Huang, and M. Rosegrant. 1996. Why China Will NOT Starve the World. *Choices* 6(1):18–24.

State Statistical Bureau (SSB). 1996. *A Statistical Survey of China.* Beijing: China Statistical Press.

_____. 1980–1995. Various issues. *Zhongguo Tongji Nianjian (Statistical Yearbook of China).* Beijing: China Statistical Press.

World Bank. 1992. *China: Strategies for Reducing Poverty in the 1990s.* A World Bank Country Study. Washington, D.C.: The World Bank.

Discussant ♦ *Hans R. Herren*

Although there are certainly differences among continents and regions in the reasons for the deficits in food production, there are also many commonalities. Different rates of population growth and agricultural production growth lead to acute food deficits, and these rates may vary as much within as among continents and regions. The important point, however, is that wherever there is a food and/or nutrition problem, there is initially a poverty problem. There are certainly also other problems such as the lack of education, poor health care, unavailability of credit, limited extension services, and lack of access to improved seeds and production technologies. The emphasis, therefore, must be on the need to address the food problem in a holistic way and to look at integrated solutions.

We are somehow in a vicious circle, a sort of a treadmill. But, as it has been shown by Dr. Yunus's project and similar national lending projects, there are fairly simple and promising ways of getting off of this treadmill.

Although there may be differences among countries and regions on how such schemes may perform, there is no doubt that this holistic approach is the way forward and up for the millions of poor people. The overexploitation of natural resources and the mining of capital are not geographically limited phenomena. They occur wherever people do not have access to the knowledge of or the means to find alternatives to the exploitative agricultural practices. The need to develop sustainable agricultural practices is global. The way to do it will vary according to regions and, in particular, to ecological zones. This was recognized some time ago and is reflected in the ecoregional approach to research and development in the International Agricultural Research Centers (IARC).

Much remains to be done, in particular in the area of collaboration among relevant institutions in the selected ecological zones. In Africa, for example, a major obstacle to progress in the development of new solutions to agricultural production remains the almost total lack of involvement by the universities. The whole burden of research remains with the National Agricultural Research Systems (NARS), whereas training and education is the domain of the universities or technical colleges. Compared with Latin America, for example, Africa certainly is lacking in farmer training. To assimilate new technologies and develop a strong partnership with scientists and extension workers, farmers must have a certain level of education. In addition, better education will raise the status of farmers and encourage them to stay in rural areas. This outcome is not only because of the education but also because of the increased income generated through the application of improved farming practices.

One major factor in achieving food security in different regions is each region's endowment with, or lack of, natural production resources. Africa is well known for its mostly poor soils, mainly lacking phosphorus. It is the responsibility of the government to invest in providing the farmers with this resource. Water availability is also a major limiting factor, depending on the region or ecological zone. The basic production infrastructure, in addition to the above-mentioned fertilizer and water, is essential for productive agriculture and needs to be made available to the farmers by the governments. Given these essentials, the farmers, with proper support, will be able to produce the food they need for their families and for the market.

Here we are touching on another area of investment by governments— storage of major staples to stabilize prices and ensure reserves for lean years. Such reserves will also allow countries or regions to help each other, as disasters do not usually strike on a continent-wide scale. A word of caution when it comes to regional food security, however, is needed with respect to the tendency of promoting, over large areas, single crop varieties that could easily fall under the pressure of diseases or insect pests. The use of seed material with narrow, or identical, genetic makeup is a great danger for food security and needs to be addressed. The use of local varieties in crop improvement programs is, therefore, very important to ensure the level of diversity that keeps major losses at bay.

From my experience and my own bias, the following are key elements to ensure food security, anywhere in the world:

- government policies that encourage production;

- research and extension systems that are responsive to farmers' needs and adapted to the local ecological and social requirements;

- efficient agricultural products marketing and transformation systems;

- income-generating activities in rural areas to ensure cash availability for input purchase;

- appropriate land allocation/tenure systems that promote the use of sustainable production practices; and

- access to credit and farm inputs.

Discussant ♦ *George H. Beaton*

Regional Perspectives on Problems and Solutions

Although there were many differences between the two background papers, both implied the need for regional/national solutions to regional/ national problems. The obverse of this, which many might contest, could be an inference that there is little to be learned from one region that is applicable to another.

Much has been said about Africa as a region of crisis. What is interesting is that no matter what the background of the speaker—agriculturist, economist, nutritionist, infectious disease and health care specialist, educator, or other—the discourse is very similar, and almost all speakers imply, if not openly state, sentiments akin to the opening sentence of Dr. Rukuni's background paper (but refer it to their own sector): "The failure to give priority to public sector investment in agriculture by African governments has been arguably their most serious error in political judgment."

To that error Dr. Rukuni attributes a predicted increase in the numbers of undernourished people from 100 million in 1996 to 300 million in 2020. In point of fact, he is talking about the expected shortfalls in food supply in comparison with projected population size and estimated energy requirements, which does not necessarily imply a comparable change in the numbers of malnourished people. I do not wish to take away from the importance of his message about the need to address the food production situation. I do want to be sure, however, that his message is interpreted in a broader perspective. Agricultural reforms must be seen as important, but not sufficient, steps towards achievement of our real goal—the betterment of the human condition.

The paper on China by Drs. Yansheng and Huang speaks to another "region," the most populous country in the world. China offers a situation where, for the agricultural sector, national solutions to national problems have had amazing impact. The statistics from China suggest that although "wants" may not yet be satisfied, there has been startling avoidance of a situation that many had predicted—needs grossly outstripping supplies. In part, the amazing difference has to be credited to the operation of central planning and control of supply and demand (macroeconomic management) as well as control of many aspects of human life and work. Africa, by contrast, presents a picture of strongly independent governments still trying to see if there are indeed areas (e.g., in agricultural research and in market development) where regional cooperation might be of benefit without compromise of independence and self-determination. Both regions exhibit major heterogeneity of agricultural and other resources, and of cultures and traditions, but in the African setting these differences are often expressed in political boundaries and all too often in civil strife.

The differences between the regions are seemingly much greater than the similarities, and it is very unlikely that lessons learned from the Chinese experience will find much application in the African setting.

It is essential that we be very clear about goals and not get lost in our own rhetoric. It is now evident that "regional" perspectives carry some advantages but they also carry some serious dangers. If we have learned nothing else in the past 50 years, we have learned that it is misguided and dangerous to assume that national problems find their origins and solutions within a single sector of government endeavor. Difficult as it may be we *must* think in multisectoral terms, and any actions we promote must be viewed as having multidimensional effects.

Surely, our real goal and the reason for symposia such as this one are not merely to look at agricultural production or the new buzz words "regional / national / household food security." They are simply concepts. Surely, our real goal is to relieve the condition of deprivation that plagues the great majority of people on the face of our planet—to better the human condition for future generations. Understand that by deprivation I do not mean food and nutrient inadequacy. I include that dimension but go much further to include economic and social deprivations and inadequacies of access to education and health services. I also argue that a population proxy for this syndrome of multifaceted deprivation is "stunting" (Beaton et al. 1990). Others see stunting as an indicator of "malnutrition." I see it as an indicator of the presence of multifaceted deprivation. The distinction is not trivial. If one sees this indicator as malnutrition, too many policy planners assume that the solution to the problem marked by stunting lies in the improvement of the food supply and of access to food. If one sees it as an indicator of multifaceted deprivation then policy planners are forced to think of multisectoral, multifaceted solutions.

Recently, Pelletier and Ethiopian colleagues published an interpretation of findings of the 1992 National Rural Nutrition Survey in Ethiopia (Pelletier et al. 1995). Their analyses showed that the rates of malnutrition did not correlate well with major conventional indices of food security. The "food-surplus" regions of the country did not have lower rates of stunting than did the "food-deficit" regions. Rates of change in the prevalence of stunting did not characteristically differ between these regions of the country. Within regions, the area of land owned had low association, sometimes negative, with stunting and wasting. Factoring in household size and altitude (a recognized variable of land productivity in Ethiopia) did not alter the picture. Something else was important in the etiology of stunting and that something had not been captured. The authors of that paper drew upon similar observations in other countries and geographic regions to present two rather critical conclusions:

> The existence of a food-first bias in the perceptions of malnutrition would be a serious hindrance to the formulation of effective nutrition policy. I would substitute "development policy" for nutrition policy;

and even more relevant to this symposium,

> Under this [more comprehensive] view of malnutrition, household food security is a necessary but not sufficient condition for [improving and] maintaining adequate nutritional status (Pelletier et al. 1995, 280, 294).

The goal is not just ensuring the supply of food to all geographic regions of the world and addressing inequities of distribution between and within the geographic regions. We must promote solutions that go well beyond that goal.

Many years ago, a group at Cornell University conducted an experiment in Peru. They acquired large land holdings and introduced cooperative ownership, farming, and marketing by the former finca workers. The goal was to demonstrate that cooperatives and agriculture extension approaches that had worked in North America could be exported to a developing country if land tenure systems were reformed. The experiment was generally successful in terms of agricultural production and income. However, a side observation was very telling. The improvement of target conditions had virtually no impact on sanitation, health, and health care—at least not in the short run. Again, one might draw the inference that the agricultural reforms were an advantageous, but not sufficient, action to better human conditions.

The paper by Pelletier and colleagues illustrates another important point. It is clear from their analyses that when one looks closely and objectively at a country, virtually any country, one finds major heterogeneity—intracountry heterogeneity. From trend analyses conducted and published by the United Nations Administrative Committee on Coordination/Subcommittee on Nutrition (ACC/SCN 1992, 1993, 1994; Gillespie, Mason, and Martorell 1996) it is apparent that while there are major differences among geographic regions of the world, there are also major differences among the individual countries within regions. Africa is no exception. In the period from 1975 to 1990, for sub-Saharan Africa as a whole, per capita total energy supply was relatively stable, the infant mortality rate was declining, immunization rates were increasing, and access to health services and potable water was improving. Conversely, the economic picture was worsening. Secondary education participation rates, particularly for females, remained very low and the prevalence of underweight children was falling very slowly (ACC/SCN 1992). These regional statistics masked the important intraregional differences among countries. Using the prevalence of underweight children as an indicator of general deprivation, it appeared that in areas such as Nigeria, Togo, and Zambia the situation was worsening while in Kenya, Tanzania, and Zimbabwe there appeared to be definite improvement (Gillespie, Mason, and Martorell 1996).

Although parallel statistics have not been widely available for China until recent years, one phenomenon is strikingly apparent. There is a major "secular trend" in very early child growth now under way (Beaton 1990,

1993). This trend has been under way for at least 10 to 15 years. I do not attempt to speculate on the changes in China that effected this apparent improvement in the situation of early growth and development, but I do note it as another important dimension of the apparent "success story" of recent China. Later in this paper, I will point out that it may be a success that carries a delayed cost—a cost that must not be neglected.

Although there are considerable dangers in focusing upon the large geographic regions, there are also advantages because countries in close geographic proximity often share important cultural traits and practices, have important trade links, and may face similar challenges of climate and infestation that affect food production (and income generation). In sub-Saharan Africa there are two commonalities that have not been emphasized today during this symposium: many of the countries share a history of colonialism, and major attempts have been made to reform the perceived injustices of that system. These issues contribute to the obvious political unrest that has plagued many parts of the region in recent decades and that has been noted in Rukuni's background paper. The colonial history and subsequent reform continue to impact national policy formulation and will for decades to come. It is one of the many African realities that we may fail to appreciate.

Unfortunately, another reality is that many African countries face an extremely high incidence of acquired immune deficiency syndrome (AIDS). It has been estimated that in some areas, up to 30 percent of mothers are HIV-positive and that for these women, the rates of transmission to their infants (in utero or postpartum) may be as high as 30 to 40 percent. This is a devastating picture. We have yet to see the full medium- and longer-term impacts of this disease. From accumulating information, one can predict that important infrastructures will be decimated, impacting the government's ability to implement programs in *any* sector. We already hear that some rural lands and their potential for agricultural production are being abandoned. The impact of this abandonment is yet to be realized. Demands on the health care and social support systems *must* be increasing. The AIDS virus certainly impacts the future of Africa: it cannot be ignored. It now appears that AIDS may be increasing very rapidly in some other regions and might have similar impacts there.

Earlier I mentioned the very strong secular trend in early child growth rates observed in China in recent years. If this trend really represents a release from environmental constraints affecting early child growth as I suspect (Beaton et al. 1990), then we can expect to see parallel increases in adult size beginning in the mid-1990s, 20 years after the secular trends in early growth became noticeable. The favorable implication of this size increase is that the change in early growth probably indexes a general improvement in health and development for coming generations. The potentially negative implication is that there could be a very significant increase in biomass (larger adults) and hence, a substantial increase in food demand to support these large bodies. Using the Food and Agriculture Organization/World Health Organization/United Nations University (FAO/WHO/UNU 1985) report as a base, a 5- to 10-kilogram increase in average adult body weight might be expected to result in a 5 to 10 percent increase in energy needs. I do not know whether the projections of future food supply and demand in China take this likelihood into account.

Conclusions

There can be no question that there is urgent need to address the problems of agricultural production. Addressing the problems must *not* be done at the cost of forgetting what we have learned in the 20 years since the 1974 World Food Conference. Improved agricultural production and regional, national, and household food security are necessary but not sufficient parts of the struggle to achieve betterment of human conditions. We must not forget the critical roles of health, education, and economic development sectors, and the impact of increasing demands on women's time and energy (e.g., as the source of labor for the household farm in many African settings) on the maintenance of household functions. The United Nations Children's Fund (UNICEF) has designated this cluster of related activities and functions under the rubric of "care" and has given wide publicity to this often forgotten dimension. Most discussions, such as this symposium's, focus upon the development of resources. Although resources are obviously necessary, I put to you the proposition that it is what people do with available resources, not the resources themselves, that is most important.

Returning to the two papers, although I appreciate the analysis and encourage intraregional examinations, I must stress that ultimately most of the regions are too large to be considered homogeneous planning units. Rather than "African solutions for African problems," we need "national solutions for national problems." In China, ultimately there may be subnational solutions for subnational problems reflecting the very real heterogeneity of that single nation. I freely admit that these national solutions must be set within regional and world perspectives. I suggest also that we should be looking towards regional systems of mutual support—perhaps a networking of national resources and experiences. There are signs that such networking is developing in the African setting. Where we find it, we must take care to foster and not displace it.

References

Beaton, G. H. 1990. Nutrition Research in Human Biology: Changing Perspectives and Interpretations. *American Journal of Human Biology* 4:159–77.

_____. 1993. Which Age Groups Should Be Targeted for Supplementary Feeding. In *Nutritional Issues in Food Aid.* ACC/SCN Symposium Report Nutrition Policy Discussion Paper No. 12, 37–54. Geneva: ACC/SCN.

Beaton, G. H., A. Kelly, J. Kevany, R. Martorell, and J. Mason. 1990. *Appropriate Uses of Anthropometric Indices in Children: A Report Based on an ACC/SCN Workshop.* ACC/SCN State-of-the-Art Series Nutrition Policy Discussion Paper No. 7.

Food and Agriculture Organization/World Health Organization/United Nations University (FAO/WHO/UNU). 1985. *Energy and Protein Requirements: Report of a Joint FAO/WHO/UNU Expert Consultation.* WHO Technical Report Series No. 724. Geneva: WHO.

Gillespie, S., J. Mason, and R. Martorell. 1996. *How Nutrition Improves.* A report based on an ACC/SCN Workshop held on 25–27 September 1993 at the 15th International Union of Nutritional Sciences (IUNS) meeting, Adelaide, Australia. ACC/SCN State-of-the-Art Series Nutrition Policy Discussion Paper No. 15.

Pelletier, D. L., K. Debneke, Y. Kidane, B. Halle, and F. Negissie. 1995. The Food-first Bias and Nutrition Policy: Lessons from Ethiopia. *Food Policy* 20(4):279–98.

United Nations Administrative Committee on Coordination/Subcommittee on Nutrition (ACC/SCN) 1992. *Second Report on the World Nutrition Situation, Vol. I: Global and Regional Results.* Geneva: ACC/SCN.

_____. 1993. *Second Report on the World Nutrition Situation, Vol. II: Country Trends, Methods and Statistics.* Geneva: ACC/SCN.

_____. 1994. *Update on the Nutrition Situation, 1994.* Geneva: ACC/SCN.

Discussant ♦ *Uma Lele*

I am going to be provocative and make two sets of comments that I hope lead to some debate. I want to contrast Asia and Africa by saying what several people have already said: Political will is absolutely fundamental in achieving food security. It was present abundantly in Asia and has often been lacking in Africa. Also, the quality of external assistance that Asian countries received was often measurably superior to that which Africa received. I want to elaborate on both these points.

Obviously, Asia is not more self-sufficient in food today than it was in the 1960s. Indeed, Asia's dependence on imports is quite high. However, its capacity to import food has increased tremendously because of its ability to increase both its food production and its exports. Food imports, as a percentage of exports, have declined from 26.7 percent in the early 1970s to 10.7 percent in the early 1990s. Africa's import dependence, when measured as the proportion of food consumed, is comparable to that of Asia. But when measured as the proportion of imports, it has gone up tremendously because export volumes have not increased. Yet, the terms-of-trade effects in both regions have been more or less the same. As Dr. Rukuni pointed out, many of the commodities that Africa exports are also exported by Asia. Asians have basically taken over the markets from Africa. The question is *why*.

First, there is the fundamental issue of political will to solve the problem of food security. Large countries in Asia, such as China, India, and Indonesia, have always recognized that their ability to command food at the national level is fundamental for their overall sociopolitical and economic development. Their perception of food security as being equivalent to "national security" has greatly influenced the steps they have taken to invest in their agricultural sector.

Second, there are no simple solutions to agricultural development, no silver bullets! It is not a soils problem alone, nor a problem of extension, although each must form a part of the larger strategy towards African agricultural development.

Third, agricultural development is a complex phenomenon requiring a combination of policies, institutions, infrastructure, human capital, and technology, all of which must be cleverly combined and continuously fine-tuned.

Investment needs of rural infrastructure alone are tremendous in Africa. Nigeria in the 1980s had only one-fifth of the level of rural infrastructure of India in the 1950s, and local institutions in Nigeria simply do not work to maintain roads. The question is, where are the policies, institutions, and resources going to come from to develop markets and to bring in technological packages and inputs that are badly needed in the African continent?

Fourth, the sustainability of policies is critical. The big difference between what is occurring in the Sasakawa program in Africa and the accomplishments of Dr. Borlaug in South Asia is the ability of Asian policymakers to sustain a set of policies and institutions over a long period of time to achieve a Green Revolution. There are exceptions in sub-Saharan Africa, such as Kenya, but these exceptions are few and far between. Good policies often are not formulated and implemented over a sustained period. The question is, can sustained policies occur in Africa given little political will?

The final issue I want to discuss briefly is the role of donors. After having done a great deal of comparative work and being engaged in lending from the World Bank in Asia and Africa over the last 25 years, I have concluded that not only was Asia fortunate in the quality of political leadership and the technical and human capital it brought to bear (e.g., in the forms of the Swaminathans, Kuriens, and Yunuses of the world), but also Asia received excellent technical advice and assistance in agriculture.

I compare the quality of technical advice that countries such as India received from the U.S. Agency for International Development (USAID) and the World Bank in the mid-1960s and thereafter with the assistance and advice African countries have been receiving over the past 20 years. As a professional, I feel sad that Africans simply have not received good advice. In many cases, the kinds of standard solutions that we have brought to bear—for example, the tremendous changes in macroeconomic policies and the tremendous shifts away from donor support for public sector distribution systems to the extremes of private sector reliance—have been very destabilizing in Africa.

To summarize, it is the lack of political support and poor policies domestically combined with the poor quality of external assistance in large volumes that Africa has been getting that has been its major problem. Unless something is done in a very honest and soul-searching way to ask these questions both *for* Africa and *in* Africa, I am afraid that, much as I would like to, I cannot be optimistic about Africa. The time has come for us to ask some very honest and searching questions.

The paper on China, which I found very interesting, comes to conclusions that are not too dissimilar from the ones that our colleagues in the World Bank, the Food and Agriculture Organization of the United Nations (FAO), and the International Food Policy Research Institute (IFPRI) have reached: namely, that China might be quite capable of feeding itself, and the growth in its imports may not be significant enough to make a major difference in world food prices.

There is one element of the paper, however, that I want to focus on—namely, the issue of reduced investments in agricultural research in China. The authors point out that there have been a number of "reforms" in agricultural research in China. Recent studies on China by colleagues at Stanford University reinforce the point made in Drs. Yansheng and Huang's paper. It is also a phenomenon that can be noted in various parts of the world, including Brazil, where I happen to be working currently. An important element of the reform process is that public sector research institutions are increasingly being asked to commercialize their research results and to use these resources to finance research while public expenditures on research slacken.

One of the consequences of the reduction of public support for research is that researchers are devoting most of their time to commercial activities and little time to the conduct of research. This consequence is true even in China, a country that has given substantial emphasis to agricultural development and agricultural research. I come back to the issue of political will. As I have argued before, the policies in China over the past 30 years have been driven by the perception of national security being related to food security as it is in India and Indonesia, and driven by considerations of command over food and the ability to import food on one's "own" terms.

PART IV

Harvesting Ideas and Recommendations

CHAPTER FIFTEEN

New Opportunities for Development Assistance Programs

Peter McPherson

Vast problems remain in the developing world, and unfortunately there is shrinking public support for foreign assistance. Accordingly, we have an obligation to review the effectiveness of and the manner in which donors are delivering that assistance. In this chapter I focus on *delivery* rather than on the sectors of an economy in which donors might work best. I look at which delivery approaches have not worked and which have worked. I argue that forces are coming together that provide leapfrog opportunities for donors; in other words, opportunities for jumping over some of the incremental steps that have been required in the past. I suggest that the U.S. Agency for International Development (AID), the World Bank, and other donors might alter their delivery approach somewhat to take greater advantage of these new forces.

What Has Not Worked

What has not worked, i.e., what has not generally provided an adequate return on investments, includes the following practices.

1. Assistance to countries lacking sound policies or political stability. Bad policies in a sector or country often make progress almost impossible, but we frequently provide assistance anyway. There were reasons for Cold War assistance, and it often worked well for the Cold War period. However, there are fewer reasons today. I think conditionality can work but usually only to support indigenous reformers.

2. Resource transfers, especially without conditionality. Resource transfers of goods and services have not provided a high return, unless they were part of a focused and time-limited transfer of technology. Disaster relief systems are, of course, a special case. We talk about teaching people how to fish but we have often

found it easier to give people fish, food aid, fertilizer, and so on. I do not debate that some people are aided by resource transfers, but generally it does not optimize the number of lives that are helped.

3. Government delivery of goods and services. In addition, we have relied too much on government organizations to deliver goods and services. Governments have an indispensable role, but they are usually very weak institutions in developing countries.

4. Large and complicated integrated development efforts. Multisector projects have not produced good returns. They have been too complicated and usually lacked focus. When we have worked in the whole range of sectors on a project, we have been tempted to put resources in sectors with poor policies or where we did not have any special technology. Overall, we have had poor rates of return on our resources when there was too much money to spend too quickly in a country or region.

What Has Worked

There have been problems, but we must keep in mind that much has worked. Assistance has contributed to increased food production, longer lives, and reduced child mortality. These outcomes suggest that there have been approaches to delivery of the donor assistance that have, in fact, been valuable. I would list as examples the following categories.

1. The creation of technology. Good examples are the Green Revolution and the development of other important agricultural technology, and the creation of oral rehydration therapy (ORT) by the International Diarrheal Diseases Research Center in Bangladesh. There are substantial data suggesting that technology creation does have a good return for the dollars spent.

2. Focused efforts to disperse technology. Examples here are the dissemination of ORT and immunization efforts, many population programs such as those in Mexico and Brazil, the smallpox efforts a generation ago, and polio work in Latin America by Rotary International, among other initiatives.

3. Training and institutional building. There is great dispute about just what went into the economic success in Korea and Taiwan, but there is substantial evidence that the early and major efforts for training in those countries were an important foundation for their successes. The research of Dr. Ted Schultz a generation ago established a recognition of the high rate of return on investment for training.

4. Infrastructure. We have all seen the great impact of many road and electricity investments.
5. Policy work. Policy analysis completed with relatively little funding has often been a key in building indigenous support. Policy advocacy in areas such as economic policy, population, environment, and women's issues has often been very effective.

The Future

Keeping in mind what has worked and what has not worked in the past, let us now look to the future. We have a new world of opportunities with the end of the Cold War. The Cold War was an important consideration in resource allocation decisions that often overrode development criteria. More recently, forces have come together that enable us to deliver assistance more effectively. We have tools to make even faster progress. I would list these as including at least the following forces.

1. Biotechnology. Some of the remarkable, practical progress in biotechnology made by the Monsanto Corporation has been well documented. Monsanto stock was up over 70 percent in 1996 in large part because of their biotechnology work. I believe that the political and the patent issues in biotechnology can and will be overcome. Biotechnology is important not only in agriculture but also in medicine, the environment, and many other areas.
2. Information technology. Information technology gives us important tools for development. They will help us greatly in both the creation and the transfer of technology. The old issue has been how we can link science located in developing countries to a world scientific community. We are certainly not far off in having libraries and other sources of knowledge globally tied together with scientists and students. As for training, the day is already here. For example, in January 1997 at Michigan State University (MSU) we offered several courses through the World Wide Web and e-mail from the students back to the faculty member. This kind of linkage will not only provide degrees, but just as importantly, it will keep people who are already trained updated on important scientific issues.

    Institution building in this age of information technology will be easier. When MSU was building universities in India and Nigeria and Brazil in the 1950s, we sent large numbers of professors to those countries to stay for long periods of time. This was effective but expensive. Today, we might perform the function by engaging one senior person and a technician to keep equipment fully operational. Institutions need to be built or strengthened in a range of areas, such as agriculture, research, education, the environment, medical delivery, policy analysis, etc.

3.   The unprecedented spread of mass media. It was noted earlier that the mass transfer of technology has sometimes been very effective. The increased level of mass media, literacy, and entrepreneurship should make mass transfer an even stronger delivery mechanism. With mass media we have some examples of development successes. For example, AID population work was greatly facilitated by conveying the message through soap operas in Mexico. In Egypt, where the use of ORT was initially about 10 percent, AID used television and other efforts to increase usage to 85 percent in one year. So we see that in Egypt at the time—the mid-1980s—there was substantial penetration of television to the whole population. There is now television over much of China and many other parts of the world. Of course, radio is almost pervasive.

4.   Higher rates of literacy. Mass mobilization is greatly facilitated by literacy. Literacy rates are up in most parts of the world, and there is an increased literacy rate among women. Jim Grant and the United Nations Children's Fund (UNICEF) showed how mass mobilization can be used in medicine. It is clear that mass mobilization can be a greater tool with family planning, agriculture extension, possibly even microcredit. Faster progress may be made than we have had in the past. I would note here that many mass mobilization decisions should be made with involvement of the people who are to be affected.

5.   The global economy and the spread of entrepreneurship. The global market has augmented entrepreneurship. There was probably always more entrepreneurship than was recognized, but the pressures have been such that it is clearly a greater factor than ever. Entrepreneurship can be very important in distributing family planning, medicine, agricultural inputs, and other goods and services.

Development is essentially about creating and transferring technology, and these forces greatly help in that effort. Moreover, our most important asset is the drive of the individual for a better life. These forces can mobilize the aspirations of individuals.

These new technologies and forces fit very nicely into what, historically, has been successful. The technology creation successes of the past can be augmented with the biotechnology information technologies. The training and institutional building is significantly enhanced by information technology. The mass transfer of technology in focused and time-limited programs can be greatly facilitated by mass media, literacy, and entrepreneurship.

Accordingly, I believe that these new forces should have a real impact upon our allocations of the assistance dollars. For example, for biotechnology and for agriculture we need a sizable fund, perhaps $50 million or more per year. It could be provided by the World Bank or regional development banks or some combination. This absolutely must not take resources away from the Consultative Group on International Agricultural Research (CGIAR) system. A similar fund should be available for biotechnology work in health and population. The two funds should provide grants only on a peer-reviewed basis. In addition, the information technology should augment training and institutional building efforts, and money should be allocated accordingly.

The mass mobilization possibilities should somewhat change how we provide agricultural extension and health care delivery. Private volunteer organizations (PVOs) and the private sector should both have a major role in this process. Mass mobilization needs to have some time limits so that the efforts do not become a permanent transfer of resources.

Implications for the World Bank and the U. S. Agency for International Development

As the premier development organization in the world, how the World Bank organizes and delivers assistance is of the greatest importance. The Bank programs need to take into greater consideration the forces I have mentioned. In addition, the Bank can probably be more effective if it continues to move in the direction of having more people in the field. It will be difficult, but the Bank needs to continue to work at how it can reduce the role of government guarantees, since that inevitably reinforces the government's role in the delivery of goods and services. The coordinating role of the Bank for development assistance is extremely important, since no one else is in the position to do it.

As for AID, I feel strongly that we should keep the agency. The United States needs an assistance organization to focus on this work. To eliminate AID and then try to replace it with a new organization would be slow and dangerous to the development mission given the politics of Washington. Also, a new agency as a practical matter is unlikely to produce any more than what a concentrated reform of AID can produce. On the other hand, the Administration has tools to make substantial changes through the presidential executive order and the delegation authority of the secretary of state. For example, the name of AID could be changed to include the word "partnership" or "cooperation." In addition, AID could be given authority to change its procurement and contract procedures—for example, baseline procurement and contract procedures to rework the total picture. It may well be that some of the concepts that I have set forth here concerning the new forces could be presented in an address by the secretary of state or the president. In any case, the nature and the policies of AID could be greatly changed both in substance and in form.

In addition, at least some part of AID funding should be used for foundation-like work. There are fundamental things that we can do with AID to make it the great organization it once was, and the World Bank needs a strong U.S. bilateral foreign aid program to complement it and to help with the important development job.

I have tried to lay out what I think we have not done so well and also to speak about the things that we have done well. I believe deeply that events have come together in this world to give us an opportunity to mobilize and to make greater progress than we have made during the last decade or two. I have suggested how those opportunities need to be translated into dollars and programs, and have presented my ideas about the World Bank and AID.

CHAPTER SIXTEEN

Summaries of the Panel Discussions

- ◆ MODERATOR
- ◆ SUMMARIES

Moderator ◆ *A.S. Clausi*

The tenth anniversary of The World Food Prize was a milestone of several dimensions. It celebrated the coming of age of this most prestigious recognition of humankind's efforts toward alleviating food problems in the world. It also expressed hope and vision for the future and our ability to solve many of the remaining food security issues of the globe. The laureates recognized by The World Food Prize to date have all been great visionaries and inspirations for us to emulate. To a person, they are not only dedicated and committed to their work but they all are sensitive, feeling people who have persisted with dogged and often creative determination to solve an important food problem of our times. They are, every one of them, tough acts to follow! Yet I am convinced that the best is yet to come, and their examples will inspire many more to take up the challenge of further advancing the quality, quantity, and availability of the world's food supply.

The next century, which we are about to enter, promises to be a more peaceful one—at least from the standpoint of no major conflagrations—and as such could provide a venue for progress in food security never experienced before. We know how to grow the food—we certainly have shown that with the Green Revolution, and more advances will be made with Green Revolutions yet to come. We have the technology and the infrastructure know-how to store, process, and distribute the food to the tables of the world. We have the ingenuity to make further advances given the time and resources, and in this age of information highways we have the means to transfer all this knowledge globally and locally without difficulty.

In the past, however, we have been our own worst enemy; politics, wars, religious differences, selfishness, and greed all have managed to get in the way, one way or another. Let us hope as we move into the twenty-first century we can set some of these human frailties aside long enough to be resolute and even more effective in feeding the world's population.

The Symposium on Food Security: New Solutions for the Twenty-first Century dwelt heavily on the issues of population control and, to be sure, this is an issue we must face and manage. This is clearly an overriding problem, but an equally key issue is one of food accessibility. The peoples of the earth must have the financial capacity to be able to afford the basic necessities of food and shelter. Poverty is the greatest enemy of food security.

Furthermore, we must improve the preservation and distribution parts of the food chain to match up with the production part more effectively around the world. We must control pre- and postharvest losses and get the food from the farm to the cities.

The delegates to the seventh general assembly of the International Union of Food Science and Technology (IUFOST), referencing the joint Food and Agriculture Organization of the United Nations (FAO)/World Health Organization (WHO) international conference on health and nutrition held in Rome in 1992, issued a Budapest Declaration, which recognized that access to nutritionally adequate and safe food is the right of every individual. They declared their commitment to work with all other organizations to ensure sustained well-being for all people in a peaceful, just, and environmentally safe world. They recognized the central role of food science and technology in ensuring the year-round availability of the quantity, quality, and variety of safe and wholesome foods necessary to meet the nutritional needs of the world's growing population. The delegates went further to say there is an urgent need to strengthen the food science and technology base, particularly in low-income, food-deficient countries, in order to expand and diversify food supplies, create income-earning opportunities, and generate local resources for development in those countries.

To me, the keys are *the need to create income opportunities to expand and diversify food supplies and to generate local resources for development.* To achieve this we will need to employ all of the observations and recommendations coming out of the symposium.

- We must think globally but act locally;
- we must think consumer and processor as well as farmer;
- we must make decisions based on the poor, women and children, and the environment;

- while we need more Green Revolution, we must manage population growth;
- we must find new institutional approaches, infrastructures, and policies; and
- we need dollars, technologies, and patience.

Now that we are achieving a "global village" perhaps we will see more clearly that feeding that village is not as complicated or as perplexing as it has appeared in the past, if we give it a chance to happen! There appears to be more reason for optimism than pessimism.

Summary ♦ *Kenneth J. Frey*

Our group's topic was "Agricultural Science and the Food Balance." Four persons provided insights on this subject. They were Per Pinstrup-Andersen, who gave the resource paper, and John Niederhauser, Vernon Ruttan, and Luc Maene, who served on the panel.

Dr. Pinstrup-Andersen opened by suggesting that food security could be judged from two viewpoints: (1) from the adequacy of production to provide some mean number of calories per day, per person, or per capita, and (2) from the availability of food to individuals. These are two quite different ways of looking at food security. Number 1 takes into account only agricultural productivity, and Pinstrup-Andersen was optimistic that enough food can be produced for meeting the needs of the human population in 2015–2020. Number 2 takes agricultural productivity into account but it also takes into account all of the forces involved in food distribution, including poverty. He pointed out that food has been produced at falling cost per unit and this helps the consumer, but attention needs to be given to even cheaper food. Furthermore, production needs to be looked at as a source of employment and, by all means, agricultural production needs to be done in a more sustainable way.

Pinstrup-Andersen touched on several other points that are salient to the question of food security.

1. He proposed that most new resources be spent on research for less productive land areas. Currently, agricultural production on these lands leads to rapid land degradation; research should address this problem.

2. Most of the technology needed for poor farmers must be provided by public sector research because it will not come from private industry.
3. We need to speed up the pace of the use of biotechnology. Many forces keep biotechnology from rapid acceptance—return on investment, activities of a negative nature by nongovernment organizations (NGOs), and so forth.
4. He closed by saying that we need to use all of the tools available to sustain food security.

Dr. Niederhauser, who is an activist by nature, challenged us to project needs and resources for the year 2100 instead of 2020. But then, he advised, we should become pragmatists and plan for the year 2020 as the first step towards 2100. He made a plea for more extension work to get technology to farmers, and for government policy that is friendlier to agricultural development.

Dr. Ruttan presented a case for pessimism about food security. He thinks that (a) we are running into practical yield ceilings for major grain crops, and (b) governments are unwilling to fund agricultural research. In fact, he averred that a world calamity on food security, such as occurred in the 1970s, may be necessary to jar governments into a mode to fund agricultural research adequately.

Dr. Maene represented private industry via the fertilizer industry. He indicated that industry research has done a good job of reducing the negative environmental effects of fertilizer products, but that much research still needs to be done on application procedures.

All four speakers were unanimous on the need for more investment in agricultural research. Recent economic studies have shown that one dollar of investment by a developed country for agricultural research in a developing country will lead to four dollars in new export trade for the developed world. Audience member Les Swindale called our attention to the phosphate deficiency in soils of Africa. Several audience members made a plea for a larger proportion of the research grant dollar to filter to the researchers.

Summary ◆ *Samuel Enoch Stumpf*

Looking back upon the wide-ranging discussions of our panel, several leading ideas can be harvested as follows.

1. Appearing in all discussions was the obvious fact that our population exists under circumstances of limited resources. This

situation also raised the question of the need for stabilizing populations by various means, including voluntary family planning.

2. Along with overpopulation there is the problem of poverty, which in turn prevents access by the poor to adequate food. To break the cycle of poverty and lack of access to food raises the critical need for the creation of jobs to insure incomes that can guarantee access to nutritional resources.

3. In order to make rational judgments and decisions while attempting to solve the problems of the relationship between populations and food resources, it is necessary to have solid, accurate information. This is particularly the case when predicting the size of populations and even more important when predicting the quantity of resources that will be available in the future. This exercise in prediction is hazardous because the impact of technology on the future can rarely be known in advance of its development.

4. Another notable point was the absolute significance of the nutritional component in food. Not only is it important that a sufficient quantity of food be made available, but it is equally important that the food supply contain adequate nutrients. For this reason, many speakers reported on research activities in various centers where our scientists pursue studies on the most nutritious varieties of food.

5. And finally, a subtle theme throughout the discussions was the element of the moral responsibility of affluent nations to share their wealth in appropriate ways with those who suffer from the worst ravages of poverty and from the privation of adequate quantities and quality of food.

Summary ♦ *Donald Winkelmann*

The session on "Food Security and Poverty" offered a rich and varied discussion. We started with the observation that one of life's major ironies is that most of those with food security problems are in agriculture, where their role is precisely to produce food!

Irma Adelman, based on data and analysis from Korea and her reading of events in other places, affirmed the Kuznets hypothesis that with income growth from low levels, income distribution becomes first less, then more, equal. She pointed out that effective policy can shorten the period during which distributions become less equal. She observed that the findings of research have been influenced strongly by the choice of countries and the data sets employed. Her findings favor agriculturally led industrial growth and export-led industrial growth.

Verghese Kurien asked if poverty is intractable and went on to give examples that it need not be. He expressed concern about the possibility that strategies resting on comparative advantage might collide with food security, and wondered if theories of economics have replaced "common sense and decency." He opined that the agriculture of developed countries should not be shackled, but rather should support progress for the sector in developing countries, where control should be "in the hands of producers."

Muhammad Yunus, noting that poverty is not a creation of the poor, was optimistic about arresting its impacts, especially through the development of institutions where individuals play a strong role. Such concepts will, he asserted, require a new vision about institutions. A key part of that vision would include the encouragement of individuals to take control of their own destinies. The world, he remarked, will be very different 25 years hence.

Robert Herdt noted the progress already made in food production and asked what would be required to assure its continuation. He described the role of three dimensions: technologies that are well oriented and well managed; policies that level the playing field for agriculture; and institutions that provide new avenues to frame solutions to important problems. National and social will is also needed, he said, if progress is to continue. He suggested that each situation would require its own balance among the three dimensions.

Discussion from the floor brought forth these observations: that one is best advised to see comparative advantage in a dynamic context and that some of the multilateral lenders are shaping loans in that context; that dual-price systems (which attenuate the role of comparative advantage) have brought little advantage to the poor; that new data and analysis suggest that with growth, in approximately 50 percent of cases, income distributions actually become more equal, not less, but that in most cases, the absolute income of the poor increases; that more attention might be focused on postharvest themes; and that the world community must search through its experience for "generalizable" lessons about the creation of wealth and the formation of useful institutions.

Several questions for Dr. Kurien and Dr. Yunus focused on particular experiences and their possible application in other settings. Both panelists were optimistic about the potential for further extrapolation of the innovations with which they have been associated.

For the most part, optimism prevailed about the possibilities for reducing poverty and improving food security, as well as for the pursuit of solutions that were touched upon. Perhaps the two things that sat least easily with the group were the role of comparative advantage, even dynamic comparative advantage, and the extent of the correlation between income growth and poverty reduction.

Summary ◆ *Robert L. Thompson*

Several important distinctions were made during our group's discussion of "Problems of Food Shortage." I will summarize the distinctions first and then mention some of the key points that were made. These distinctions included: the importance of distinguishing between food security and nutritional security; the importance of distinguishing among individual, national, and global food security; and the importance of distinguishing between the availability of food and access to food.

Our discussion started with the observations that 20 percent of the world's population, or about 800 million people, have insufficient food available to them and that there are problems of increasing malnutrition in sub-Saharan Africa and in South Asia. Micronutrient malnutrition and protein malnutrition are greater problems than a deficit of calories. It was observed that the education of girls plays an extremely important role in improving the quality of the diet and ensuring adequate nutrition to children.

Individual food insecurity reflects principally a lack of purchasing power. Poor people cannot access available food supplies. The rich in no country go hungry except in times of war, natural disaster, or politically imposed famine. But to reduce poverty in low-income countries we need broad-based economic development that raises the income of the largest possible number of the poor people in a society. During this transition to more rapid and more broadly based economic development, however, there needs to be a nutritional safety net.

Much of the poverty in the world is in rural areas. No country in the world has solved the problem of rural poverty on the farm or in agriculture. The only countries that have effectively reduced rural poverty have not only raised productivity in agriculture, but also have created nonagricultural income-earning opportunities by means of the development of cottage industries and/or off-farm employment opportunities. However, rural development is lagging in most developing countries. There is often a strong antirural bias in public infrastructure investment policies, with a resulting inadequate investment in rural roads, communications, schools, clinics, and wells. There is an urgent need to correct this imbalance between cities and rural areas in national infrastructure investment in order to raise agricultural productivity and generate greater earning opportunities off the farm in rural areas. Otherwise, poor people will continue to migrate to the cities, and the rural areas, where the greatest concentration of poverty exists, will continue to lag behind the more urban and affluent parts of countries.

The observation was made that national food security can be ensured by domestic production of food, by food aid, or by commercial imports. Each country should use its own agricultural resources to the fullest extent it can efficiently, but population and agricultural resources are not distributed uniformly across the planet. Therefore, countries that have a larger percentage of the world's population than of the world's arable land are likely to be net importers of food.

Food aid availability is declining due to changes in national agricultural policies in the United States and the European Union, which are likely to result in these governments not accumulating inventories of commodities. This is happening at the very time when the world is likely to need more food aid. Our group was confident that, if we make the necessary investments in agricultural research and development, the world's farmers can produce enough food through increasing productivity using environmentally benign technologies such as biological controls, integrated pest management, and biotechnology to ensure global food security. However, the availability of improved technologies is not enough. The governments of many developing countries must remove the antifarm bias that exists in their pricing policies that keeps the prices of commodities so low and the prices of inputs so high that it does not pay farmers to adopt the improved technologies that are available.

So the panel concluded that, because the natural resources of agriculture are distributed across the planet in different proportions to the population distribution, a larger fraction of the world's food production is likely to move through international trade in the future. However, the poorest developing countries have an analogous situation to the poorest members of society. Those countries have to have international purchasing power to access the food that is available in the world market. Those countries have to be able to export some products to earn the necessary foreign exchange to import part of their food needs commercially. However, third world countries frequently encounter significant barriers to exporting the products in which they do have a comparative advantage. These barriers directly impede the countries' ability to access global supplies of food to ensure their national food security as well as individual food security.

In conclusion, the panel was confident that individual, national, and global food security can be achieved if we make the necessary investments now in agricultural research and in rural infrastructure. During the transition a nutritional safety net is needed.

It should be noted, however, that there was a wide range of views in our group over whether we, in both the high-income and low-income countries, have the will to make these necessary investments in both agricultural research and rural development.

Summary ◆ *Richard L. Hall*

Our session was about what happens and what must happen to food between harvest and consumption, and it turned out to be an unexpectedly lively session. Only a small portion of our food supply is produced at the time and the place where it is consumed. Therefore, it must be stored and transported to consumers as well as stabilized and protected for storage and transport. Except for the cereal grains, an important exception, most raw foods are not naturally stable. They deteriorate rapidly after harvest or slaughter. That instability becomes even more important when, as Nevin Scrimshaw pointed out, we must move our attention beyond simply caloric sources toward improved diets emphasizing fruits and vegetables. These foods are better sources of the micronutrients we all require, but their losses in developing countries customarily run 20 percent or more and, in my own experience, can run as high as 50 percent. That is a situation we should not tolerate and cannot afford. In addition, processing also makes many foods safe, nutritious, easy to consume and digest, and more attractive, whereas the unprocessed foods would lack one or more of those important qualities.

In our panel's discussion, we looked ahead to what the technologies available in the more advanced countries will provide. Certainly, information technology will let us model better, and let us digest more readily the information we have. Sensor technology will provide us with much more immediate feedback, by less invasive methods, for better control of process streams. Material technology will, in some instances at least, permit smart materials. Packages now inform the purchaser about the general nature of the contents. In the future, they often will tell us about the stability and quality of the food before we purchase or open the package.

There are important differences between the developed and the developing countries in terms of their perspectives. This is inevitable, and we need to bear these differences in mind. The developed countries are concerned largely about the safety and quality of their food supplies. Those countries that are still developing are far more concerned about access and availability. Indeed, some of us in the developed countries can afford to worry about nonexistent hazards while the rest of the world worries about getting enough to eat.

Postharvest technology, unfortunately, often seems to be dropped between the ship and the wharf. It rarely is transferred successfully. It needs to be moved, yes, but redeveloped and adapted in situ for each new environment, each location, and each need. In that connection, there is a real need for more extension services directed to rural processing industries. We have such services in the United States. Virtually every land grant university has an extension service directed to the food processing industry. How much more are they needed in those areas where the technology does not yet exist? There is also the matter of the safety and quality of water; this is important for food processing, as well as for direct consumption.

A key in all of this, particularly in the access of developing countries to industrial technology, is to give local producers and processors a large and continuing stake in their success. An aspect urgently and cogently pointed out by Verghese Kurien was a result of his experience with Operation Flood. We need to be aware of, and oppose, the universal, pervasive, and perverse ingenuity we have shown in erecting nontariff barriers to trade. There is now a move within the Food and Agriculture Organization of the United Nations (FAO) to try to redefine milk so that it must come only from a cow. How we will define milk that comes from mothers and from goats is not entirely clear. But it would suggest that food processing gets international attention only when it succeeds and becomes a competitive threat.

I conclude with these key thoughts: we must have this postprocessing technology, first, to avoid the waste that would otherwise occur and, second, to improve the utility of the agricultural commodities we do produce. Beyond that, as development proceeds, the food processing industry becomes more of an economic engine, even, than agriculture itself.

Summary ◆ *Charles E. Hess*

Two resource papers were presented at the in-depth session on "Geographic Food Security." The first, "Food Crisis: The Need for an African Solution to an African Problem," was presented by Mandivamba Rukuni. The second, "Agricultural Development and Food Policy in China: Past Performance and Future Prospects," was written by Yang Yansheng, Vice President of the Chinese Academy of Agricultural Science, and Jikun Huang. The paper was presented by Zhou Wei. The panelists who discussed the papers from their own perspectives were 1995 World Food Prize laureate Hans Herren, and George Beaton and Uma Lele.

It was said in many presentations during this session that the African continent poses a great challenge in terms of food security. However, an African solution to the food crisis is needed and is feasible according to Dr. Rukuni. The solution must be approached from a holistic approach involving politics and governance as well as economic issues, hunger, and malnutrition. The last two are long-term problems requiring long-term solutions including a major investment in getting agriculture going for the next 20 years.

The gains in agricultural productivity in China over the past several decades have been a major contribution to the growth of the national economy. The national policy continues to maintain food self-sufficiency and eliminate absolute poverty with a 4 percent annual growth in agricultural production and farmers' income. These goals will be achieved by rural economic and institutional reforms, including renewal of land contracts, incentives to invest in agriculture, increased use of science, and the opening of China's agricultural sector to foreign investment. The paper by Drs. Yansheng and Huang suggests that although China's imports will rise slightly through the next decade due to increased demand for meat and feed grains, China will "neither empty the world grain markets, nor become a major grain exporter" (p. 370).

Some key ideas presented by the panelists and by the audience include the following.

- Food security is a multidimensional problem and it requires multidimensional solutions which are integrated. Although this theme was repeated a number of times, one member of the audience reminded the group that a lot of progress has been made in the past by a simpler approach, such as the development of a new, more productive, and/or disease-resistant crop. The whole agricultural sector was not taken on, but with the introduction of a new variety with a two-fold increase in yield, other things fell into place.

- Political will is an essential component of the solution of food security problems. A panel member suggested that the goal should be broader than food security; rather, it should be the enhancement of the quality of life.

- Research is essential to help achieve food security but by itself is not sufficient.

- Research and extension must be demand driven, not just supply driven. However, to form true partnerships among researchers, extensionists, and farmers, the level of education of the farmer must be raised as part of the multidimensional approach to achieving food security and enhancing the quality of life.

- Infrastructure and the welfare of rural people have to be considered. A subsidy is not recommended, but there must be a level playing field. In many developing countries, rural people pay a premium for inputs and receive less for their products because of a lack of infrastructure such as roads to bring in supplies and take out products. Liberalization of markets must continue.

- Given the size and variation within the continent of Africa, regional approaches must be used with care because of differences in culture, policies, climate, and similar factors.

- Panelists suggested a contrast between developments in Asia and those in Africa. In Asia, political will was fundamentally important to the success that has been achieved. First, a goal of food self-sufficiency was established and once this goal was realized, then other sectors of the economy developed. The key was that the agricultural sector of the economy provided the foundation on which other sectors could be built. In Africa, many leaders attempted to skip over agricultural development and move directly to industrialization, and this approach was cited as part of the reason that the challenges of food security and quality of life in the rural sector are in difficulty.

- A question was raised if there was a bias against science and technology in Africa. Panelists answered with a strong "no." However, as mentioned previously, although research and technology are essential, they are not sufficient alone. It was then suggested that some nongovernment organizations (NGOs) have concerns about the use of new technologies, such as the release of beneficial insects as part of an integrated pest management program. Although there have been cases of such opposition, the NGOs overall were regarded as an important partner, particularly in bringing new technologies to the farmers and rural people.

- The statement that China's agricultural policy will continue the goal of food self-sufficiency was welcomed, but there were concerns whether or not there would be sufficient land and water to permit the proposed increases in productivity of 4 percent per year. The response was that China is developing a new and strong land-use policy and that new lands and irrigation are being developed in Northwest and Northeast China. Another concern was raised in regard to the current policy of commercializing agricultural research so that it, too, could be self-sufficient, or nearly so. The panel viewed this policy as jeopardizing longer-term research, and felt that it also could jeopardize the ability of China to maintain a policy of food self-sufficiency.

CHAPTER SEVENTEEN

Observations and Recommendations

Stanley R. Johnson, Helen H. Jensen,
Amani E. El Obeid, and Lisa C. Smith

The symposium has provided a rich set of perspectives for the issues of food security in the twenty-first century. These perspectives were informed by data and projections assembled from available sources. In general, these data and projections address issues of food availability, food access, and nutritional status. The projections for the future are driven largely by technology, population, and demographics and by assumptions on the success of the policy and institutional changes designed to improve the performance of the national and global economies and to support more balanced participation in the fruits of economic growth.

While these data and projections are of value in assessing the food security issues of the twenty-first century, they are not in themselves the unique contribution of the book. The latter is provided by the insights and analyses of the experienced and distinguished contributors to the symposium. These, of course, include the current and past Food Prize Laureates, those who by their intellect and deeds have contributed significantly to progress in improving food security during the latter part of the twentieth century. It is the perspectives of these laureates and the other distinguished participants in the symposium that we have assembled in this final chapter.

Following the introduction, we discuss what we term the "moral dilemma" that is suggested by the prevalence of food and nutritional insecurity in a global community that produces sufficient food and generates adequate income to assure that the challenge of food security could, at the current date, largely be successfully met. How can a global community that has benefited from significant economic growth, successful policy and institutional reform, and an almost startling rate of technical progress in food production as well as a clearer understanding of nutrition, accept the fact that a large share of the population are and will be food and nutritionally insecure well into the twenty-first century. The costs of addressing food and nutritional security problems in both the developed and developing nations would appear small compared with the costs of military armament, space exploration, and even segments of the entertainment sector. Our distinguished contributors ask, Why is food security not a higher national and international priority?

In the third section, we highlight the conclusions drawn from the data and projections that have been assembled. Essentially, these are status quo projections, based on a continuation of trends in agricultural and related technologies, policy and institutional reforms, and investments in health, infrastructure, and the other determinants of population dynamics, economic growth, poverty, and food production and distribution. The fourth section includes a collection of the central recommendations or conclusions that emerged from the symposium. Here the idea is not to be encyclopedic. In fact, this would not be possible in the space allotted. As well, in many cases the recommendations or conclusions require the context of the plenary or special sessions. These must be harvested by a careful reading of the material. Still, it is possible to highlight themes that emerge and that have major implications for the issues of food and nutritional security in the twenty-first century.

Finally, we take the license to make a few of our own observations, drawn from the opportunity to help organize and participate in this symposium celebrating the tenth anniversary of The World Food Prize and attended by the current Laureates Gurdev Singh Khush and Henry M. Beachell, as well as by past Laureates Robert F. Chandler, Jr., Verghese Kurien, John S. Niederhauser, Nevin S. Scrimshaw, Edward F. Knipling, Muhammad Yunus, and Hans R. Herren. These observations in large measure reflect on the dialogue that was experienced by those who attended and participated in the Tenth Anniversary Symposium.

The Moral Dilemma

"I feel disgraced—and I hope all of you do as well—by our failure to address the most fundamental of needs for nearly 1 billion human beings across the globe. It is, indeed, morally outrageous." Perhaps this quote from the plenary address of Robert S. McNamara (chapter 1) most vividly characterizes a strong undercurrent of the symposium, the moral dilemma of the prevalence of food insecurity in a period of unprecedented global abundance—abundance reflected by the level and growth of agricultural production and productivity compared with the population increase, and abundance of incomes for the "haves" compared with the "have nots" both among nations and among populations within nations. This same theme was evident in the plenary addresses of Ismail Serageldin and Jacques Diouf.

What is added to the dilemma by these latter two plenary addresses is the contrasting trend toward decreased public investments in research and development for agricultural and food systems and declining commitments by the developed nations to provide assistance aimed at improving economic growth and directly assisting the food needs of the at-risk populations. Unfortunately, it seems that the problem of food and nutritional insecurity remains significant, while at the same time the nations that are in position to lead the global effort to improve food security are reducing their financial commitments to the bilateral and multilateral agencies that have the capacity to successfully address the issue.

This dilemma is further emphasized by the recent record of success in reducing food insecurity. The share of the global population that is food insecure has decreased significantly during the past few decades. We know that with appropriate investments in research and development, agricultural production and productivity can be increased. This was exemplified by Norman Borlaug in his work and in his comments introducing this symposium, as well as in the resource paper by Pinstrup-Andersen and Pandya-Lorch and the other contributions to the session on "Agricultural Science and the Food Balance," chapter 9.

Additionally, there is a significant opportunity for postharvest technology in reducing losses of food and improving food availability and quality (see the background paper by Dennis and the commentary in the session summarized in chapter 13). Also, policies and institutions to foster the participation of the private sector are better understood and can add to success in reducing food insecurity. Perhaps the success story that stands out most prominently over the past three decades is that of India—see Swaminathan, chapter 6, and Kurien, chapter 11. Finally, there are innovations that have resulted in success in addressing poverty, the consensus and primary cause of food insecurity—see Yunus, chapter 7, and the resource piece by Adelman and the associated commentary in chapter 11. In short, approaches already tried and proven successful in reducing food insecurity are available for use in the twenty-first century.

An added sense of urgency for more rapidly addressing food and nutritional insecurity is developed in the plenary address of Scrimshaw, chapter 8, and in the resource paper by Kennedy and related commentary from the session summarized in chapter 12. The downstream costs of nutritional insecurity, the "hidden hunger," although not fully documented are clearly of major proportions, especially in the case of malnutrition of children. Ample evidence of impacts that involve both physical and mental aspects of life is available. Stunting, susceptibility to chronic diseases, poor performance, and retarded development of cognitive and other indicators of human potential are among the results of malnutrition for children.

What this evidence of impacts of malnutrition means in the most stark terms is that the costs of poor nutrition to society and to the families and individuals involved are felt directly, and those costs imply enduring limitations for both the individuals and for national development. The hidden costs are likely higher than is recognized, even by those who work with and study food insecurity. Malnutrition can have the cumulative effect of trapping societies and nations in a circle of poverty and food insecurity.

To summarize, the moral dilemma stems from the continuing prevalence of food and nutritional insecurity; the accumulating evidence that these problems can be solved with adequate resources and attention to modern developments in institutions and policy; the decreases in resources allocated to public agricultural research and development (and to development assistance); and the compelling evidence of the downstream and irreversible damage to those (particularly children) who experience serious malnutrition. There is a twenty-first century food security agenda and at least a latent sense of moral commitment to addressing it.

The Food Security Assessment

The symposium was designed to include presentations on data and projections to be used in assessing food security and the attendant condition of the global population. This evidence involved population and other demographics, food availability, food access, malnutrition, and environment. The general questions addressed were the current situation and projections for the future, conditioned on the assumption of a continuation of current trends. Results of the assessments are both encouraging and discouraging. Progress has been made especially during the past three decades (see the Winkelmann and Herdt contributions to chapter 11), but the available projections indicate substantial food security problems for the twenty-first century and particularly acute conditions in two broad geographic areas.

Food availability has been running ahead of population growth during the past three decades of the twentieth century (Diouf, chapter 3 and El Obeid et al., chapter 4). In large measure this has been due to the Green Revolution, significant gains in yields for the staple crops grown in temperate regions, more open trading regimes that permit fuller expression of comparative advantage, and expansion and improved technology for irrigated agriculture. There is concern about the slowing of the growth of the global food supply during the 1990s (Swaminathan, chapter 6). But, the consensus from the projections is that at least in the intermediate run, there will be capacity to meet the food availability requirements of the global population.

Threats to this capability that emerged during the symposium and in the plenary papers were the environment, changed demands for food brought about by urbanization and increased incomes, and inadequate investments in research and development for the food system. These themes were evident throughout the symposium. For the papers and presentations that most explicitly make the case for these threats, see the following chapters: for the environment, Swaminathan, chapter 6 and Myers, chapter 10; for changed demands due to income and population dynamics, Myers and the other contributors to chapter 10; and for research and development, Serageldin, chapter 2 as well as Pinstrup-Andersen and Pandya-Lorch and the other contributors to chapter 9. An alternative assessment of food availability, based less on projections and more on a simulation, was reviewed by Guthrie in her contribution to chapter 10. This analysis comes to the same general conclusion, but indicates that with appropriate research and development, investments, and national policy priorities, the global food availability capacity is well beyond the demands indicated by current population and related projections.

The most serious food availability problems are in the developing nations, especially those in sub-Saharan Africa and South Asia. These are the areas where food insecurity is most prevalent and/or where the largest populations are food insecure. The nation-by-nation and regional assessments of food balances for these two regions are, moreover, a source of major concern. For sub-Saharan Africa, domestic production as a share of total food consumption is decreasing (El Obeid et al., chapter 4). These are mainly agrarian nations. And, they have per capita incomes that are low and have been decreasing. Meeting food requirements from imports will be increasingly difficult, unless the development course is reversed (which may require very different approaches than in the past according to Rukuni, chapter 14). Projections for these African nations suggest continuing and increased risk of deteriorating food availability.

For South Asia, the food security problems are less related to national food balances than to food access (Smith et al., chapter 5). In these nations there are large numbers of individuals in poverty, and thus facing current and future food insecurity. At the same time there is evidence from China that in these developing nations appropriate national priorities can yield increases in availability and food access (Yansheng and Huang, chapter

14). Food security in both of these regions is further complicated by inadequate infrastructure, health care, and other factors that make for limited access and nutritional problems.

The consensus for the major cause for food insecurity coming from the symposium was poverty. Poverty characterizes the food insecure populations in whatever region—and is the critical factor in food access (Diouf, chapter 3 and Smith et al., chapter 5, as examples). This implies that the broader development agenda will have to be successfully addressed if significant progress is to be made in reducing the food insecurity of the 840 million people (about 20 percent of the global population) who currently suffer chronic undernourishment. Including the 840 million who are at serious food security risk, approximately 2 billion people globally live on less than $2 per day (approximately 40 percent of the population). Many of this larger share of the global population are likely at risk for food accessibility and related malnutrition of various types.

The implications of malnutrition are summarized in the papers by Scrimshaw, chapter 8 and Smith et al., chapter 5. These implications suggest even more urgent attendance to the issues of food security in the twenty-first century. The growing body of scientific evidence on the implications of malnutrition, especially for children, is devastating. And, the indicators of this condition presented at the symposium suggest that well over the 840 million of the global population that are food insecure may be at significant nutritional risk, due, for example, to problems of intrahousehold distribution of food, seasonal access to food, poor sanitation, and other factors that go hand in hand with poverty.

The in-depth presentation on population growth projections and other demographic issues was by Myers and the other contributors to chapter 10. The implications of these projections further underscore the importance of the priority for more actively addressing food security in the twenty-first century. First, the most rapid rates of population increase are in the developing and low income nations. Second, the rates of increase are especially high in sub-Saharan Africa and in other nations where food insecurity is most prevalent. In addition, there are reasons to believe that the environmental constraints that may emerge for food production will most likely be felt first in these nations. Given their incomes, these are the very nations with the most limited means for dealing with environmental constraints to food production. Last, the presentation on population dynamics and food security focused on increased urbanization. On net, the population increase during the next 25 years will be urban. Moreover, this marked increase in urbanization will come in the developing nations. Here, too, there are significant implications for food processing and distribution of the postharvest food systems, as discussed by Dennis and the other contributors to chapter 13.

The final set of major issues contributing to the assessment of the food security situation was highlighted in the plenary paper by Swaminathan, chapter 6 and in the paper by Myers and the other contributions to chapter 10. Swaminathan focuses on the United Nations Conferences of 1972 in Stockholm and 1992 in Rio de Janeiro. The issues emerging from these conferences and from related lines of research suggest significant environmental problems with a continuation of current food production technologies, especially in the geographic areas of high population density. The implication is for research and development strategies that are much more sensitive than in the past to the integrity of the local and global ecological systems. The specter of an ecological crash is gaining increased scientific support. The evidence of accumulations of greenhouse gases and climate change is symptomatic of the possibly severe consequences that pursuing the current course will have on increasing food availability. In chapter 10, these same issues are raised, but with an addition of the potential for changed and more environmentally demanding diets that may emerge with the increasing incomes that are so important to the elimination of poverty and improved food security. The implications for the developed nations are far reaching as well. These nations are presently, and by a wide margin, the major contributors to environmental contamination.

In summary, the assessment yields evidence of major problems for food security (both availability and access) in the twenty-first century. At the heart of the food security problem is poverty. Poverty is also closely associated with malnutrition, or hidden hunger, a less visible but perhaps more significant threat to the fuller achievement of human potential for a large share of the global population. Initiatives to improve food security and reduce poverty in the twenty-first century will have to deal more directly with environmental and ecological factors. Finally, and of particular significance for near-term policy and other interventions to address food security, there are clear hot spots—sub-Saharan Africa, South Asia, and the rapidly urbanizing populations in the developing nations. In the latter case, improvements in agricultural productivity and agricultural production can have a double impact. They both reduce the push to urbanization and increase the availability of food.

Recommendations and Conclusions

The recommendations and conclusions that emerged are presented here consistent with the structure of the symposium and this book: Pillars of Food Security—Part Two; Special Food Security Issues—Part Three; and Harvesting the Ideas—Part Four. In each case, the recommendations and conclusions were informed by the presentations and papers from Part One of this book—Food Security Problems and Solutions. For purposes of convenience and brevity, we present these conclusions and recommendations in outline form.

Pillars of Food Security

Agricultural science and the food balance. To date, the agricultural sector has been very responsive to private incentives and technological change. The major conclusions in this area were:

- With continuing attention to agriculture and food science, and to policy reform, food production can grow at rates consistent with world population growth.
- Food production per capita in many of the poorer nations, however, is declining, suggesting that the status quo for agricultural research policies and policy reform will not adequately address problems of regional food insecurity.
- A large share of the funding for agricultural research and technology traditionally has come from multinational corporations and governments of richer nations, raising questions on how poorer nations will benefit.
- Biotechnology holds great potential for increases in global food production.

The major recommendations that arose from these conclusions were:

- Investments in science and technology, which have decreased following the Green Revolution, must be maintained and, perhaps, accelerated.
- Technology development and adoption must target less productive land areas and poorer nations.
- The increasing role of the private sector in agricultural technology implies a requirement for adjusting public sector agricultural research and development policy.
- New research and technology transfer institutions should be established for low-income nations.
- Sustainability is a critical issue; the environment must be protected as food production increases throughout the world.

Population and the environment. Projections of population growth have important implications for global and regional food security. Major conclusions of the symposium were:

- Population growth is increasing most rapidly in the poorer and food-insecure nations and regions.
- Population growth in the richer nations is occurring more rapidly among the lower-income, at-risk segments.
- As poverty, which is a source of food insecurity, is alleviated, food demand will likely exceed the growth in population.
- Nutritional insecurity is a major issue for the expanding populations in low-income countries and may have serious impacts on the future performance of these populations.
- Technologies for family planning and birth control are available and have been implemented in many countries, for example in China. However, these measures are likely to be less successful in the longer term than more indirect measures such as increased income, education, increased opportunity for women, and policies targeting poverty alleviation.
- In select countries, reductions in population growth are encouraging. For example, in Bangladesh economic empowerment has been important in reducing fertility rates from 7 births per woman in 1975 to 3.6 births per woman in 1994.
- Population data are of questionable reliability, especially in the poorer nations where growth projections are highest, suggesting improved monitoring.

The major recommendations were:

- The best approaches to population management are indirect: increasing income, generating greater economic opportunity for women, compulsory education, and institutional changes that increase civil, political, and economic rights.
- Special measures, such as education, the development of technologies for producing diversity in the diet, and food fortification are necessary for nutritional security.

Food security and poverty. The general conclusion related to income and poverty is that economic growth strategies involving broader participation and less inequality have led to more rapid growth and development. The major conclusions were:

- Policies for increasing income that have a positive effect on food security emphasize agricultural-led industrialization and export-led growth.
- The growth of urban populations in poorer nations may put added pressure on food distribution policies and mechanisms.

- Pricing policies and subsidies used together have been effective in increasing food supplies.
- Income subsidies transferred by use of artificially low prices and other market mechanisms often create political constituencies that make the policies difficult to reverse.
- The administration of assistance programs in less-developed nations is often frustrated by local institutions and bureaucratic delays.

The major recommendations were:

- Poverty is a major cause of systemic food insecurity; therefore, food security policies must target sources of income.
- Balanced economic growth can be achieved through technical assistance, improving education, and reducing dualism between rural and urban areas.
- Although technical assistance and income transfers are helpful, the lower-income nations must ultimately generate economic growth from within.

Special Food Security Issues

Problems of food shortage. The problems of food shortage in times of civil disorder and natural disaster are likely to continue. The major conclusions were:

- Transition economies, particularly those in Eastern Europe and the former Soviet Union, have left substantial populations at risk, especially pensioners, children, and displaced persons.
- Refugees and displaced populations are the groups most vulnerable to acute food shortage. Those refugees affected by civil strife face especially high mortality rates due to the unavailability of food.
- Food aid is and will continue to be an essential requirement in natural disaster situations.
- Policies to increase incomes of the poor and increase agricultural production are necessary to improve chronic food security problems.
- Micronutrient and protein deficiencies are larger problems today than insufficient energy intake.
- Catastrophic food shortages resulting from pest damage have less impact today than in earlier years.

The major recommendations were:

- Donor agencies cannot be fully responsible for refugee and food shortage problems; new institutions and approaches are necessary.

- Early warning systems and advanced preparedness should be used to prevent food shortage during and following national disasters, and in cases of civil disorders and political instability.
- Virtual food banks are a possibility as a developed nation cooperative approach to food shortage.

Food processing and distribution. In an increasingly urban world, and especially for the poorer nations, processing and distribution assumes greater importance in assuring adequate food supplies for the households at risk. Postharvest technology also has a more critical role for solving food security problems. The major conclusions in this area were:

- Government ministries of agriculture, often seen as having responsibility for national food supplies and food security, have focused primarily on production.
- Technical assistance programs for food processing and postharvest technology can have substantial benefit for improving food security, but these programs are often not given appropriate consideration in national and multinational assistance efforts.
- Institutions that foster efficient markets and broad participation in processing and distribution will gain importance as populations become more urbanized and agriculture more specialized.

The major recommendations were:

- Existing preservation methods and distribution technologies must be increasingly used and widely disseminated.
- New developments in food preservation such as ultrasound, irradiation, non-thermal processing, rapid heating, and new packaging should be encouraged to reduce losses and thereby increase the food supply.
- Attention to the entire integrated "food chain" is essential for improving global food systems, especially in poorer nations.
- Increased job opportunities in processing and distribution can be a significant source of income for segments of rural populations living in poverty.

Geographic food security. Africa and China both have current and projected problems with food supplies. China, and other densely populated nations in Southeast Asia, also will have an important influence on international food commodity markets. The African countries and China (and its neighbors) are in the midst of significant political and economic reforms. This transition places them at higher risk for food insecurity. At the same time, food security is considered to be essential for successful economic transition, sustained national development, and a stable political environment.

Major conclusions for Africa and China were:

- In Africa, urban centers control political power and often exploit agriculture, leading to low productivity and high levels of migration from rural to urban areas. Structural adjustment programs aimed at improving economic growth in Africa have failed to concentrate on agriculture.
- Many African countries have become dependent on food donations from the developed nations. Food donations may wane as changes take place in agricultural policies in higher-income nations.
- China has food security problems largely related to economic transition, limited resources, and high population density.
- China's agricultural and food policies have emphasized food self-sufficiency and agricultural development. Grain production increased by 46 percent between 1980 and 1995, despite several natural disasters. With increased income resulting from economic reforms, however, China may find it difficult to continue its current self-sufficiency policy.
- A change in the policy for agriculture, and continued economic growth in China, could have major implications for international food commodity markets.
- Environmental threats to food security both in Africa and China are real.
- There are significant interregional food security differences within China. These are becoming more aggravated by the concentration of income growth in provinces with access to international markets.

Recommendations that emerged related to geographic food security were:

- Seldom, if ever, can food security problems be assigned to one sector or one government ministry. Similarly, not all of the solutions lie within one source. Participation by a majority of citizens and institutions of these nations is essential to the realization of political stability, economic growth, and food security.
- Sustainable policies are required, especially in Africa, which is resource rich but with many frail ecosystems, and in China, which is resource poor. Environmental protection, albeit for different reasons, will be essential to long-term food security.
- In Africa, major institutional changes will be required, involving market reforms and democratic political institutions. Rural institutions are weak and need strengthening; prerequisites include major expenditures in infrastructure, increased education, and improved governance.

Harvesting the Ideas

A number of solutions and recommendations generated from the plenary and panel sessions as well as informal exchanges among the participants were "harvested" during the two closing sessions. There are many important lessons to be learned from the failed policies and initiatives for addressing food security, as summarized by McPherson, chapter 15.

Approaches that have not functioned effectively include:

- Assistance provided in countries without sound policies and political stability.
- Transferring food and other resources to local governments without placing conditions on how these resources are to be used.
- Implementing inappropriate policies using large, complicated, integrated development programs.

Policies that have worked and are working include:

- Technology generation and technology transfer.
- Incentives to encourage the adoption of technologies, training, and institution building.
- Education and the protection of women's rights.
- Support for inclusive indigenous policy reforms leading to broad participation of citizens in political decisions and economic activity.

These ideas led to summary solutions and recommendations:

- More comprehensive approaches to policy and institutional change sensitive to the history and culture of the nations will be necessary. Blunt policies, those that simply attempt to "buy" food security, are highly unlikely to be effective in the twenty-first century.
- Institutional change deserves high priority for addressing food security globally and among the nations in which the systemic problem is most prevalent. Institutional change is evident in these past successes: The Green Revolution, the network of international research institutions (CGIAR), the Operation Flood project in India, and the Grameen Bank in Bangladesh.
- A proposal has been suggested for major institutional change in food assistance programs in the developed nations, for example PL-480 in the United States. Instead of simply making food grants to nations with food security problems, major grants could be used to establish "food grant universities" similar to U.S. land grant universities. These food grant universities would be sensitive to local cultures and conditions and would emphasize broad access and integrated teaching, research, and extension programs.

- New approaches to providing aid for nations experiencing food shortage must be developed, including approaches that would essentially privatize the delivery of short-term and longer-term development assistance. Public sector delivery systems must change or be rejected. An approach consistent with privatization of development assistance would be to establish a fund (a bottom-up rather than a top-down approach) that would be available on a competitive basis. Agencies, private and public, would propose packages of institutional changes, policy reforms, technical assistance, and other measures. This approach would be quite different from that of the World Bank, the U.S. Agency for International Development, and other donor institutions in the developed nations.

- Added attention should be given to nutritional status by the multinational organizations and national governments concerned with improving food security. Tools for addressing nutritional status are different from those used to solve food security problems.

- Food access is a persistent and common problem for both developed and developing nations. Policies to deal with access are deserving of added attention, especially in nations that have adequate resources to solve the problem.

A Final Comment

The recommendations and conclusions from the symposium have been detailed in the previous section. By design, they are related to the specific topics addressed in the symposium and in the corresponding chapters of this book. In this concluding section, we offer some more general observations that were developed from the symposium, and we take some latitude ourselves in suggesting priorities for addressing food security in the twenty-first century.

One of the recommendations that was put forth by Robert McNamara in chapter 1 and reflected in the comments on the FAO World Food Summit included in the plenary address of Jacques Diouf, chapter 3, involved the idea of better tracking of food security problems and the progress being made in addressing them. Such a comprehensive data base and set of indicators could be managed by the multilateral donor organizations, cooperating with FAO, and could provide a clear and easily communicable basis for reporting on the status of global food security. In addition to being useful to national and multinational organization policymakers, this set of indicators, regularly reported, could help to energize the political support necessary for the levels of national and multinational investment to more successfully address food insecurity.

A second overriding issue emerged most directly in the paper by M. S. Swaminathan, chapter 6. In retrospect, the symposium could even have had a special session on ecology and food security. The upshot of this chapter and the related developing science is that food systems will increasingly have to be viewed in greater environmental or ecological contexts. What we have come to call sustainability in the latter part of the twentieth century will likely take on greater scope in the twenty-first century as the relations among agriculture and local and global ecological systems become better understood. The immediate conclusion from this observation is clearly not that agriculture will become somehow more constrained. Instead, many of the environmental problems of agriculture relate to waste. This would imply greater attention to external impacts of these wasted inputs and, in general, better management of agriculture in the context of systems that are of larger scope than have concerned those studying and advocating sustainability in the latter part of the twentieth century.

Poverty as the critical factor in food and nutritional security received a great deal of emphasis in the symposium. The focus on poverty in a sense expands the dimensions of the food security problem. No longer, for example, is food security simply equated with improving agriculture. New institutions and policies to address poverty need to be an integral part of the attack on food insecurity. These issues were perhaps most elaborated in the session summarized in chapter 11. To overly summarize, it is not at all evident that the way to address poverty and sustainable growth of economies is with policies that are aimed at the most rapid short-term national growth rates. There is real question as to whether the benefits of this growth trickle down to the disadvantaged populations. Other policies that focus more directly on the portions of the populations that are in poverty may be the better avenue for addressing sustainable growth and reductions in food insecurity (see the plenary address by Yunus, chapter 7).

Malnutrition is more difficult to measure than food availability or access. Still, malnutrition, especially for children, has enormous consequences for the disadvantaged populations and the perpetuation of food insecurity through poverty and limited human capacity. The downstream effects of child malnutrition are only starting to be fully understood. But the evidence available indicates that malnutrition should become a target of much more active policy intervention in the twenty-first century. This is underscored by the fact that the projections are for the most rapid population growth among the households and nations that are the most poverty prevalent—a large share of the new entrants to the population of the twenty-first century are likely to have to deal with the maladies of malnutrition.

Geography is important for setting the priorities for addressing food security. The consensus target regions are sub-Saharan Africa and South Asia. Interestingly, these are also regions in which basic institutions of civil society appear to be the weakest. This was emphasized in the paper by Mandivamba Rukuni, chapter 14. What this observation and Rukuni's paper suggest again is that the focus should be poverty. Moreover, the implication is for greater attention to the institutions that can enable the at-risk populations. Ruttan, chapter 9, raises the potential constraint of health problems, including new infectious diseases, control of important parasitic diseases, and the environmental health effects of agricultural intensification. The broad development of societies will be a prerequisite for food security in these nations, if the real limitations to economic opportunity lie in the basic institutions.

Finally, the familiar theme of the importance of research and development for agriculture was crosscutting in the symposium. But even here, there were open issues for the twenty-first century. These relate to the appropriate roles for the public and private sectors, the growing concentration of the agricultural technology engine for the global community within a few multinational corporations, and the idea of focusing increasingly on yield improvements for indigenous staple foods in contrast to temperate climate commodities. The latter of these is driven by the geography of food security and the concentrations of prevalence of food insecurity in areas that are not within the temperate climatic zone. How will the public and private sectors address the improvement of the technologies for these food staples that are largely not internationally traded and for which markets are typically local?

We look, as did the symposium participants, to the twenty-first century with optimism. There are food security problems and they are of major magnitude. But, we know more about how to solve these problems than before. And, symposia such as this one will continue to add to the moral commitment and resolve of nations and other participants in the international community to successfully address food security in the twenty-first century.

ISBN 0-8138-2910-0